Current Topics in Microbiology

200 # and Immunology

Editors

A. Capron, Lille · R.W. Compans, Atlanta/Georgia
M. Cooper, Birmingham/Alabama · H. Koprowski,
Philadelphia · I. McConnell, Edinburgh · F. Melchers, Basel
M. Oldstone, La Jolla/California · S. Olsnes, Oslo
M. Potter, Bethesda/Maryland · H. Saedler, Cologne
P.K. Vogt, La Jolla/California · H. Wagner, Munich
I. Wilson, La Jolla/California

Apoptosis
in Immunology

Edited by G. Kroemer and C. Martinez-A.

With 14 Figures and 9 Tables

Springer-Verlag

Berlin Heidelberg New York
London Paris Tokyo
Hong Kong Barcelona
Budapest

GUIDO KROEMER

CNRS-UPR 420
Génétique Moléculaire et
Biologie du Développment
19, rue Guy Môquet – BP8
94801 Villejuif Cedex, France

CARLOS MARTINEZ–A.

Centro Nacional de Biotecnologia CSIC
Campus de Cantoblanco
Universidad Autonoma
28049 Madrid, Spain

Cover Illustration: Detection of apoptosis and associated changes. At the upper left, nuclei from thymocytes cultured in the presence of glucocorticoids have been stained with propidium iodine and subjected to cytofluorometric analysis, allowing for the detection of hypoploid cells that are undergoing apoptosis (A). At the upper right, oligonucleosomal DNA fragmentation detectable by horizontal agarose gel electrophoresis and staining with ethidium bromide are demonstrated. The lower part shows an electron transmission micrograph of thymocytes exhibiting the typical morphology of apoptosis.

Cover design: Harald Lopka, Ilvesheim

ISSN 0070-217X
ISBN 3-540-58756-X Springer-Verlag Berlin Heidelberg New York

Production: PRODUserv Springer Produktions-Gesellschaft, Berlin
Typesetting: Thomson Press (India) Ltd, New Delhi, India
SPIN: 10479950   27/3020–5 4 3 2 1 0 – Printed on acid-free paper.

Il n'y a qu'un problème philosophique vraiment sérieux: c'est le suicide. Juger que la vie vaut ou ne vaut pas la peine d'être vécue, c'est répondre à la question fondamentale de la philosophie. Le reste, si le monde a trois dimensions, si l'esprit a neuf ou douze catégories, vient ensuite.

Albert Camus, *Le Mythe de Sisyphe*

There is one really serious problem in philosophy: suicide. To judge whether life is or not worth to be lived implies to respond to the fundamental question of philosophy. The rest, whether life has three dimensions, whether the spirit has nine or twelve categories, comes later.

Albert Camus, the Mythe of Sisyphus

Preface

In any movement of their life , immune cells, especially T and B lymphocytes, are confronted with an essential choice: to continue their existence or to commit a sort of metabolic suicide that is referred to as apoptosis or programmed cell death. In contrast to most philosophers, lymphocytes and their precursors are constantly susceptible to suicide, and it even appears that the usual cause of T or B cell elimination is suicide rather than death from natural causes, accidents or murder. This book provides a vast overview of lymphocytes suicide: external triggers and internal motives leading to suicidal impulses, accomplices in self-destruction, weapons implicated in self-execution, removal of dead bodies and pharmacological prevention of suicide.

Most of the chapters in this book are devoted to the physiology of apoptosis. The goal is to unmask the external triggers of apoptosis, unravel the signal transduction processes involved therein and describe the role of oncogenes, "death genes" and effector molecules in the apoptotic cascade. The remaining chapters deal with the pathophysiological aspects of lymphocyte apoptosis, namely, as a host contribution to HIV-induced lymphopenia, and therapeutic strategies for the avoidance of lymphocyte death.

We are confident that this compendium will contribute to the exploration of cellular suicide, not only from a basic scientist's viewpoint but also with regard to the possible clinical implications of apoptosis (dys)regulation. Far from having a depressing effect on the reader, cellular suicide may thus provide a source of both intellectual excitement and therapeutic inspiration.

GUIDO KROEMER and CARLOS MARTÍNEZ-A.

List of Contents

List of Contributors

(Their addresses can be found at the beginning of their respective chapters.)

T Cell Apoptosis Triggered via the CD3/T Cell Receptor Complex and Alternative Activation Pathways

D. KABELITZ, T. POHL, and K. PECHHOLD

1 Introduction

Programmed cell death (apoptosis) can be triggered in immature thymocytes and (under certain conditions) in mature T lymphocytes by several distinct signals including: (1) γ-irradiation (SELLINS and COHEN 1987); (2) glucocorticoids (WYLLIE 1980; NIETO and LÓPEZ-RIVAS 1989); (3) cell surface signaling via the T cell receptor (TCR)/CD3 Complex (SMITH et al. 1989; TAKAHASHI et al. 1989; McCONKEY et al. 1989; SHI et al. 1991; RUSSELL et al. 1991, 1992), via the CD2 antigen (BIERER et al. 1991; WESSELBORG et al. 1993a), or the Fas antigen (OWEN-SCHAUB et al. 1992; KLAS et al. 1993), or (4) by removal of essential survival and/or growth factors (DUKE and COHEN 1986; TSUDA et al. 1993; PERANDONES et al. 1993). In most instances, programmed cell death is associated with the fragmentation of genomic DNA into oligonucleosomal fragments of approximately 180 bp in length, a hallmark of apoptosis. Even though the various death signals may all lead to the same final stage of the cell (i.e., apoptosis associated with the characteristic morphological features and DNA fragmentation), they are likely to be differentially regulated. It is conceivable that intracellular signaling pathways are differentially activated if a T cell is stimulated, e.g., by anti-CD3/TCR antibodies, glucocorticoids, or lack of interleukin-2 (IL-2) i.e., failure to perceive a signal via a functional IL-2 receptor. In

Department of Immunology, Paul-Ehrlich Institute, Paul-Ehrlich-Strasse 51–59, 63225 Langen, Germany addition, differences in the level of apoptosis sensitivity exist between immature thymocytes and mature peripheral T lymphocytes. Whereas thymocytes rapidly undergo apoptosis in response to glucocorticoid or anti-CD3 antibody treatment (WYLLIE 1980; SMITH et al. 1989), mature peripheral T cells need to be activated before they acquire sensitivity to apoptosis signals (RUSSELL et al. 1991; WESSELBORG et al. 1993b; RADVANYI et al. 1993).

Substantial evidence supports the hypothesis that apoptosis is the mechanism by which potentially harmful developing T lymphocytes are deleted intrathymically (negative selection). Staphylococcal enterotoxin (SE) superantigens delete in thymic organ culture those thymocytes that express the corresponding SE-reactive TCR Vβ family (JENKINSON et al. 1989). Moreover, endogenous superantigens such as the mouse mammary tumor virus (MMTV) also induce, intrathymic deletion (by apoptosis) of developing-thymocytes expressing the reactive TCR Vβ elements (ACHA-ORBEA et al. 1991). Clonal deletion of specific thymocytes can also be achieved by intraperitoneal injection of the relevant antigen, as shown in an immunoglobulin idiotype-specific TCR transgenic mouse model (BOGEN et al. 1993). It appears that the capacity to induce clonal deletion in the thymus is not restricted to bone marrow-derived cells but can also be attributed to thymic epithelial cells (HUGO et al. 1994). Taken together, there is little doubt that apoptosis is an important physiological process contributing to the ordered development of the immune system and the shaping of the TCR repertoire. However, recent results from many laboratories indicate that the susceptibility to programmed cell death is not restricted to immature thymocytes or transformed T cells. It is quite clear now that mature peripheral T cells similarly undergo apoptosis under certain conditions (KABELITZ et al. 1993). These results have raised a recent burst of interest in apoptosis, due to its possible role in the regulation of cellular immune responses including the establishment of peripheral tolerance. In the following sections, we will discuss some of the issues related to the induction of apoptosis in mature T lymphocytes via the CD3/TCR complex and alternative activation pathways. Special emphasis is given to the question under which conditions can antigen induce death of responding T lymphocytes.

2 Apoptosis Induced by Anti-CD3/T Cell Receptor Antibodies

BREITMEYER et al. (1987) were the first to note growth inhibitory effects of anti-CD3 antibodies on human T lymphocytes. However, in this study, growth inhibition was reportedly not associated with significant cell death. More recently, anti-CD3/TCR antibodies were reported to trigger apoptosis associated with DNA fragmentation in murine and human thymocytes (SMITH et al. 1989; SHI et al. 1991; McCONKEY et al. 1989; MERKENSCHLAGER and FISHER 1991), transformed T cells (UCKER et al. 1989; MERCEP et al. 1989; ODAKA et al. 1990; TAKAHISHI et al. 1989), and activated mature T lymphocytes (RUSSELL et al. 1991, 1992; NEWELL et al. 1990; LENARDO 1991; JANSSEN et al. 1992; WESSELBORG and KABELITZ 1993; WESSELBORG et al. 1993a; RADVANYI et al. 1993; DAMLE et al. 1993a; BOEHME and LENARDO 1993; GROUX et al. 1993). The general consensus is that *resting* mature T cells do not undergo apoptosis in response to anti-CD3/TCR antibodies per se; they need to be primed in order to respond to anti-CD3/TCR signaling with programmed cell death. Surprisingly, one of the signals that can program mature T cells for apoptosis is IL-2 (LENARDO 1991). It appears that the level of susceptibility to apoptosis correlates with the level of cell cycling induced by the T cell growth factors IL-2 and IL-4; interestingly, cells blocked in the GI phase of the cell cycle were resistant to TCR-induced apoptosis, whereas cells blocked in S phase were susceptible (BOEHME and LENARDO 1993). Reports from several laboratories highlight the important observation that mature T lymphocytes need to be activated (by antigen or anti-CD3/TCR antibodies) and perhaps proliferate through several rounds of cell division before they acquire sensitivity to apoptosis inducing signals (RUSSELL et al. 1991; KLAS et al. 1993; RADVANYI et al. 1993; WESSELBORG et al. 1993a). It is not well understood what exactly happens during the time lag which is required before activated peripheral T cells can undergo apoptosis in response to anti-CD3/TCR antibodies. However, accumulating data suggest a relationship between Bcl-2 expression, Fas expression, CD45RO expression, and susceptibility to apoptosis: Bcl-2 expression is known to protect from apoptosis in several distinct experimental systems (ITOH et al. 1993; VEIS et al. 1993; HOCKENBERY et al. 1993; SCHWARTZ and OSBORNE 1993). Interestingly, the primed T lymphocytes which express CD45RO on their surface are characterized by low Bcl-2 expression (AKBAR et al. 1993). Moreover, the progressive in vitro differentiation of human CD4 T cells is associated with an increased expression of CD45RO and Fas but a reduced expression of Bcl-2 (SALMON et al. 1994). This fits in nicely with the observations in patients with infectious mononucleosis in which ex vivo isolated lymphocytes are characterized by expression of CD45RO, lack of Bcl-2 expression, and rapid apoptosis upon culture in vitro (TAMARU et al. 1993).

While cross-linking of cell surface CD3/TCR molecules by (immobilized) antibodies is sufficient to trigger apoptosis in preactivated T lymphocytes, additional signals may accelerate the acquisition of sensitivity to apoptosis of resting T cells. In this context, cross-linking of CD4 has been shown to facilitate subsequent apoptosis of mature T cells in response to anti-CD3/TCR antibodies (NEWELL et al. 1990). Importantly, cross-linking of CD4 molecules by HIV gp120 similarly primes T cells for apoptosis triggered through the CD3/TCR molecular complex (BANDA et al. 1992) thus pointing to the possible involvement of apoptosis in the progressive depletion of CD4 T lymphocytes in HIV-infected individuals (GROUX et al. 1992; OYAIZU et al. 1993; GOUGEON et al. 1993; GOUGEON and MONTAGNIER 1993). Apoptosis resulting from CD4 cross-linking also seems to be involved in the depletion of CD4 T cells in vivo following the administration of certain anti-CD4 antibodies (HOWIE et al. 1994).

Apoptosis of mature T lymphocytes triggered by anti-CD3/TCR antibodies can be influenced by multiple signals. The interaction of cells of the immune system is governed by cell adhesion molecules. Antibodies against such molecules can modulate T cell activation in a positive or negative manner. DAMLE et al. (1993a) reported that coligation of the TCR with ICAM-1 or VCAM-1 enhanced the activation-induced death of allospecific human CD4 T cells. An additional important interaction between T cells and antigen-presenting cells (APCs) is mediated via the CD28/CTLA4 molecules (on T cells) and the B7-1/B7-2 molecules on APCs (LINSLEY and LEDBETTER 1993). T cells receive via the CD28/B7 interaction the necessary costimulatory signal required for successful T cell activation. If T cells are triggered via the CD3/TCR molecular complex in the absence of this costimulatory signal, they become anergic. So far, the potential role of the interaction between CD28/CTLA4 and B7-1/B7-2 for the induction or prevention of T cell apoptosis has not yet been intensively investigated. Whereas DAMLE et al. (1993a) reported that anti-CD28 antibody did not affect TCR-dependent, integrin-facilitated T cell death, GROUX and coworkers (1992) observed that anti-CD28 antibody prevented apoptosis of lymphocytes taken from HIV-infected individuals. The recent discovery that there are at least two different ligands (B7-1, B7-2) for CD28/CTLA4 (AZUMA et al. 1993; BOUSSIOTIS et al. 1993) suggests that a combination of antibodies directed against the various ligands should be used in order to reveal the impact of a blockade of costimulatory signals on CD3/TCR-dependent T cell apoptosis.

The role of cytokines in the regulation of CD3/TCR-dependent T cell apoptosis remains controversial. LENARDO (1991) reported on a priming effect of IL-2 for induction of apoptosis in murine splenic T cells, whereas the addition of IL-2 prevented apoptosis in other experimental systems (GROUX et al. 1993). Similarly, interferon-γ (IFN-γ) played a decisive role in the anti-CD3/TCR antibody-triggered apoptosis of a particular murine Th1 clone (LIU and JANEWAY 1990) and in some studies using human thymocytes or activated T cells (GROUX et al. 1993), whereas no such role of IFN-γ could be revealed in other studies (DAMLE et al. 1993a). In line with the latter observations, we could not see any effect of neutralizing anti-IFN-γ antibodies on the CD3/TCR-dependent apoptosis of activated peripheral blood T lymphocytes, nor did the addition of exogenous IFN-γ modulate activation-induced T cell death (unpublished observations). In contrast to this, exogenous IFN-γ seemed to counteract antigen-induced T cell death in other studies (COHEN et al. 1993). The triggering of activated T cells by anti-CD3/TCR antibodies under conditions in which apoptosis ensues may well be associated with the induction of a wide range of cytokine genes. It is conceivable that the respective cytokines somehow contribute to the induction or prevention of apoptosis. Nevertheless, it appears that the role of known cytokines in this process is not a priori clear. Thus, the effect of cytokines in his context may depend on the type of T cell under investigation (CD8, CD4/Th1, CD4/Th2) and on the timing of CD3/TCR and cytokine signaling (LENARDO 1991).

3 Apoptosis Induced by Antigen

Although monoclonal antibodies directed against the CD3/TCR complex are useful reagents for the analysis of T cell activation, it is quite clear that they do not accurately mimic the stimulation of T cells by nominal antigen presented by APCs. The discovery of the potent T cell-stimulating activity of superantigens has been followed by the observation that such superantigens are also potent inducers of apoptosis (KAWABE and OCHI 1991; GONZALO et al. 1992; LUSSOW et al. 1993; MACDONALD et al. 1991; KABELITZ and WESSELBORG 1992; DAMLE et al. 1993b). This topic is adequately addressed in other chapters of this volume. We have used the staphylococcal enterotoxin E (SEE) reactivity of a human CD4 T cell clone as a model system to investigate the role of APCs in the induction or prevention of superantigen-induced T cell apoptosis. The results revealed that the addition of APCs (EBV-transformed lymphoblastoid cells) did not prevent the SE superantigen-triggered death of a fraction of T clone cells; rather, it helped to rescue the surviving cells and to initiate a vigorous proliferative response in them (KABELITZ and WESSELBORG 1992). In this and other clonal models of T cell apoptosis, it is apparent that not all cells of a given T cell clone undergo apoptosis following signaling via CD3/TCR. It is not yet clear how the susceptibility to apoptosis is controlled, but the cell cycle seemingly is important (BOEHME and LENARDO 1993). In this context, the activation of the serine-threorine kinase p34^{cdc2} might be a critical checkpoint (SHI et al. 1994).

Superantigens activate T cells by directly cross-linking the MHC class II molecule on APCs with the TCR Vβ region, thereby bypassing the need for antigen processing (MARRACK and KAPPLER 1990). The observation that super-antigens are potent inducers of apoptosis in activated T lymphocytes stimulated great interest in the question of whether conventional nominal antigen can similarly trigger death in reactive mature CD4 and/or CD8 T cells. An accumulating body of evidence suggests that this is indeed the case. It has been known for some time that peptide epitopes can induce self-destruction of murine and human cytotoxic T cells (WALDEN and EISEN 1990; SUHRBIER et al. 1993). Importantly, the injection of antigenic peptides also induces death of responsive peripheral T cells in vivo as has been elegantly shown in TCR transgenic mouse models by KYBURZ et al. (1993) and MAMALAKI et al. (1993). More generally, activation of T cells during the process of viral infection seems to predispose T cells to subsequent apoptosis triggered by signaling through the CD3/TCR complex (TAMARU et al. 1993; RAZVI and WELSH 1993).

Apoptosis of mature T cells triggered by nominal antigen is not restricted to viral antigens but can also be revealed in response to MHC antigens. Immobilized MHC class I antigens triggered the death of responding murine CD8 T cells (UCKER et al. 1992), and exposure of HLA-DR6-reactive human CD4 T cells to APCs expressing the relevant DR6 molecules induced the death of responding T cells (DAMLE et al. 1993a).

In order to investigate the mechanism(s) of antigen-induced death of mature T lymphocytes, we have established a new flow cytometry method, Standard cell dilution assay (SCDA), which allows for rapid determination of the absolute number of viable cells of any given phenotype within heterogeneous cell populations (PECHHOLD et al. 1994). Using SCDA, we have followed the fate of alloantigen-stimulated, polyclonal, CD8, short-term T cell lines upon restimulation with specific or third party stimulator cells. As shown in Fig. 1, restimulation with the specific alloantigen reduced the number of viable responder cells after 24 h by 25%–30%; no such reduction of responder cells was observed upon restimulation with third party stimulator cells. Typically, we found a reduction of responder cells upon specific restimulation in the range of 15%–40% (Pohl et al., manuscript in preparation). A reduction of the proliferative response upon restimulation, which is often taken as a parameter to estimate the extent of cell death or anergy induction by mitogens, proved inadequate for the investigation of cell death induction in our (allo)antigen-specific system (Fig. 1). Further analysis indicates that, in general, the extent of the proliferative response (in relation to the number of viable cells) and the induction of cell death are closely correlated. Thus, a strong stimulation of CD8 T cells, most likely based on the high antigen dosage and high affinity of the cell:cell interaction, results in an increased induction of cell death (up to 40%) within the first 24 h and in an increased subsequent proliferative response of the surviving cells, as long as optimal culture conditions (e.g., growth factors) are supplemented. This has been observed by other investigators to prevent apoptosis in preactivated CD8 T cells (KIRBERG et al. 1993). However, helper factors, such as IL-2, had no obvious influence on the activation-induced cell death. Therefore, mechanisms that are involved in the regulation of the proliferative response of antigen-specific T cell lines (e.g., the induction of growth factor responsiveness) may also govern the induction of the deletional process itself. This view is supported by the inhibitory effect of antibodies known to

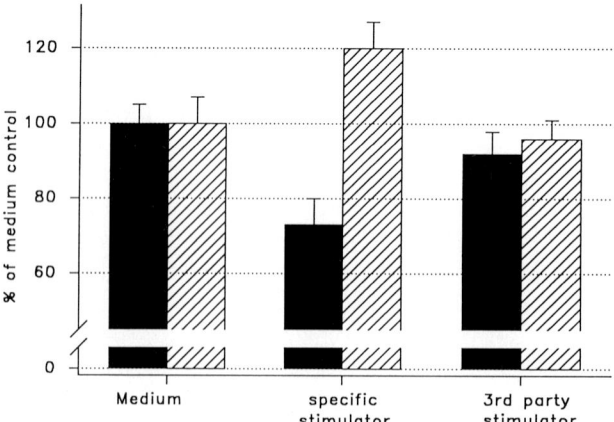

Fig. 1. Effect of restimulation on cell viability and proliferation of alloreactive T cell lines after 24 h. A polyclonal alloreactive T cell line was restimulated with the specific or HLA-mismatched third party stimulator cells. After 24 h, the viability of responding T cells was determined by standard cell dilution assay (SCDA), and the proliferative activity was measured by uptake of [³H]TdR. *Solid bars*, cell viability; *hatched bars*, proliferation

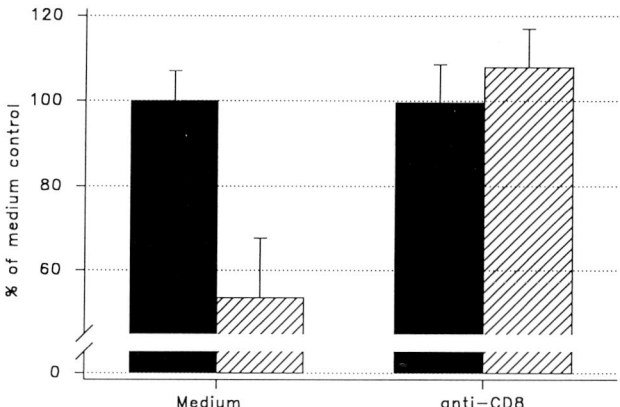

Fig. 2. Inhibition of cell death upon restimulation in the presence of anti-CD8 monoclonal antibodies: A polyclonal alloreactive T cell line was restimulated with the specific stimulator cells in the absence or presence of immobilized anti-CD8 antibody. The viability of responding T cells was determined by standard cell dilution assay (SCDA) after 24 h. *solid bars*, medium; *hatched bars*, specific stimulator

negatively interfere with antigen-specific T cell activation. Figure 2 shows that anti-CD8 antibodies almost completely blocked the induction of cell death of CD8 T cells by alloantigen. The sensitivity to alloantigen-induced cell death does not appear to be limited to CD8 T cells. A shift towards the accumulation of CD8 T cells is noted as a general feature on repeated stimulation of polyclonal alloreactive T cell lines in vitro indicating that cell death and/or anergy induction may be even more critically regulated in CD4 T cells. In fact, the results obtained using purified polyclonal CD4 T cell lines indicate that a fraction of antigen-reactive CD4 T cells is deleted upon reexposure to alloantigen or tetanus toxoid (Oberg et al., unpublished observation).

Does apoptosis also contribute to the deletion of mature CD4 T cells in vivo? Perhaps the best studied model is experimental allergic encephalitis (EAE), an autoimmune disease mediated by myelin basic protein (MBP)-specific CD4 T cells. Ultrastructural analysis of T cells infiltrating the parenchyma of the spinal cord suggested that a fraction of T cells was undergoing apoptosis (PENDER et al. 1992). Combining ultrastructural analysis with in situ nick translation to reveal DNA fragmentation, SCHMIED et al. (1993) confirmed the appearance of apoptotic T cells in EAE lesions. The assumption that antigen can kill antigen reactive activated T cells formed the basis for the impressive study of CRITCHFIELD et al. (1994), showing that the injection of high concentrations of MBP can delete (by apoptosis) MBP-reactive CD4 T cells in vivo, thereby abrogating the clinical and pathological signs of autoimmune encephalitis. This raises the promising prospect that antigen-induced apoptosis of T cells involved in pathological processes may eventually turn out to be a feasible goal.

4 Apoptosis Induced by Anti-CD2 Antibodies

In addition to the CD3/TCR-dependent activation pathway, alternative T cell activation pathways have been described. In this context, the activation of human T cells via the 50 kDa CD2 molecule is of particular interest. CD2 is the surface receptor for the LFA-3 (CD58) molecule. Combinations of two antibodies directed against different epitopes of the CD2 molecule in concert stimulate IL-2 production and T cell proliferation (MEUER et al. 1984). Certain anti-CD2 antibodies trigger apoptosis in murine T cell hybridomas transfected with human CD2 cDNA (BIERER et al. 1991) and in human thymocytes (LI et al. 1992). Using polyclonal IL-2-dependent human T cell lines and established T cell clones, we were also able to induce apoptosis by a combination of anti-CD2 antibodies (WESSELBORG et al. 1993b). The extent of cell death did not differ from apoptosis triggered by anti-CD3/TCR antibodies. Although these experiments suggested that susceptibility to anti-CD2-mediated apoptosis is a general feature of activated T lymphocytes, more recent data from ROULEAU et al. (1993) would suggest that this may not be the case. By separating human CD8 T cells into CD57$^+$ and CD57$^-$ subsets ROULEAU et al. (1993) observed that anti-CD2 antibodies triggered apoptosis in CD8$^+$CD57$^+$ but not in CD8$^+$CD57$^-$ subsets. Whether CD57 expression is also relevant for antigen-induced apoptosis of CD8 T cells remains to be investigated.

5 Relevance of Fas Expression for CD3/T Cell Receptor-Dependent Apoptosis

The Fas antigen, a member of the tumor necrosis factor (TNF) receptor family, is the target molecule for an efficient apoptosis pathway. Anti-Fas antibodies in particular trigger cell death in activated, but not resting, human T lymphocytes (OWEN-SCHAUB et al. 1992; KLAS et al. 1993). Mouse strains carrying a defect in the expression of Fas (*lpr*) or Fas ligand (*gld*) suffer from a generalized lymophoproliferative disease (WATANABE-FUKUNAGA et al. 1992; TAKAHASHI et al. 1994). *Lpr* and *gld* mice have a defect in the CD3/TCR-dependent cell death pathway, suggesting that functional Fas/Fas-ligand expression is a prerequisite for apoptosis mediated via the CD3/TCR complex (RUSSELL et al. 1993; RUSSELL and WANG 1993; BOSSU et al. 1993).

We have attempted to investigate the interdependence of Fas-dependent and CD3/TCR-dependent apoptosis in a clonal T cell system using variants of the CD3$^+$ human T cell line JM. Stimulation of Vβ8$^+$ JM cells with SEE in the presence but not absence of MHC class II-positive APCs induces growth arrest and death of a fraction of JM cells. Continuous exposure to SE superantigens plus APCs selects for JM variants lacking cell surface expression of CD3/TCR. A second set of JM variants was established by continuous treatment with anti-Fas antibody. The resulting JM variants still expressed Fas antigen but had completely lost the functional responsiveness to anti-Fas treatment (i.e., induction of apoptosis). The analysis of JM variants indicated that the sensitivity to anti-Fas-mediated apoptosis was not impaired in CD3/TCR-negative JM variants. Moreover, the anti-Fas-resistant JM variants were still susceptible to SE superantigen-triggered cell death, suggesting that the CD3/TCR-dependent and the Fas-dependent cell death pathways are not functionally linked to each other in the JM T cell line (unpublished observations). However, since anti-Fas-resistant JM variants still express Fas antigen on their surface, it is possible that the death signaling cascade triggered by SEE via the CD3/TCR complex is intracellularly connected to the Fas signal transduction pathway. Further analysis is required to address this issue.

6 Relevance of Antigen-Induced T Cell Death for the Regulation of Cellular Immune Responses and the Establishment of Peripheral Tolerance

As outlined above, signaling via the CD3/TCR molecular complex as achieved by anti-CD3/TCR antibodies, superantigen or nominal antigen can induce death by apoptosis in activated mature T lymphocytes. This implies that antigen-induced death of reactive T cells is an important parameter in regulating cellular immune responses. Several examples in favor of such a hypothesis have been discussed above. If it is true that contact with antigen can initiate seemingly opposite patterns of reactivity in T cells (i.e., activation associated with cytokine production in resting T cells vs deactivation associated with cell death in activated T cells), then one must postulate stringent control of both pathways in order to ensure a well balanced T cell reactivity. Some of the molecular mechanisms controlling the death pathways in lymphocytes are now being unraveled (SCHWARTZ and OSBORNE 1993). It appears that antigen dosage is of additional critical importance. In the studies of UCKER et al. (1992), quantitative differences alone determined the alternative cellular responses of cell death and cell proliferation in nontransformed murine T cells, with cell death being initiated by higher concentrations of antigen. Similar conclusions were reached by AUPHAN et al. (1992) using an anti-H-2K^b-reactive transgenic TCR mouse model. In their experiments, the density of the H-2K^b antigen expression determined the degree of deletion of T cells expressing the anti-H-2K^b-reactive transgenic TCR. Together with the recent observation that a high dosage of MBP deletes MBP-reactive T cells in vivo (CRITCHFIELD et al. 1994), it appears that death of activated mature T cells is more readily triggered by higher concentrations of antigen. Taken together, apoptosis of antigen-reactive T cells may help terminate an ongoing cellular immune response, thus preventing the continuous expansion of specific T cells. Figure 3 illustrates that a number of different parameters determine the outcome of restimulation of activated T cells by antigen (proliferation, anergy, memory, apoptosis).

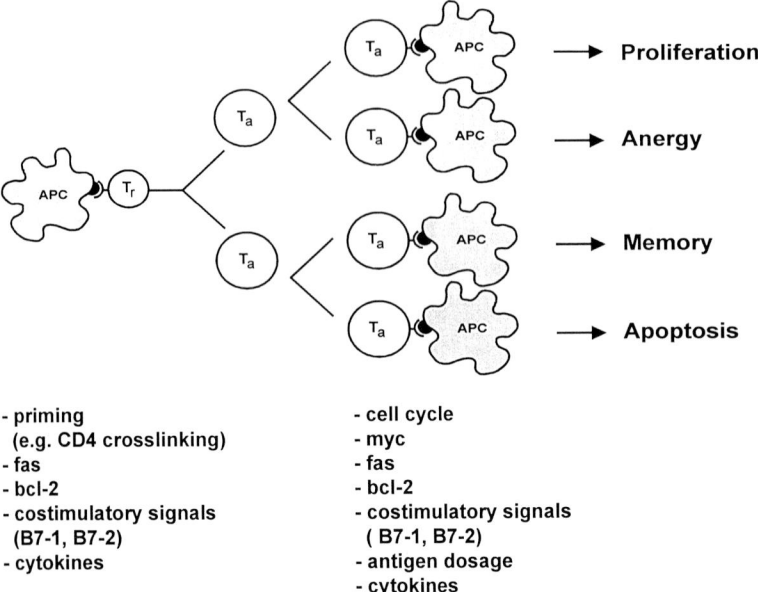

- priming
  (e.g. CD4 crosslinking)
- fas
- bcl-2
- costimulatory signals
  (B7-1, B7-2)
- cytokines

- cell cycle
- myc
- fas
- bcl-2
- costimulatory signals
  (B7-1, B7-2)
- antigen dosage
- cytokines

Fig. 3. Factors controlling the fate of activated T cells upon encounter of antigen. CD4 cross-linking, costimulating signals, and cytokines influence the sensitivity to apoptosis of resting T cells (T_r), in addition to the level of bcl-2 and *Fas* expression (*left*). Although the same parameters also affect activation-induced death of activated T cells (T_a), additional parameters such as cell cycle, *myc* expression, and antigen dosage are important (*right*)

Finally, the question arises whether apoptosis of antigen-reactive T cells contributes to the establishment of peripheral tolerance. Induction of tolerance in the mature immune system is a multifactorial process involving modulation and down-regulation of TCR molecules, induction of anergy, and deletion of specific T cells (Schönrich et al. 1991; Arnold et al. 1993). The beneficial effect of donor-specific pretransplant blood transfusion is associated with a reduction of the frequency of circulating donor-specific cytotoxic T cell precursors (Hadley et al. 1992), presumably resulting from alloantigen-induced apoptosis of a fraction of donor reactive T cells. The priming effect of CD4 cross-linking for subsequent CD3/TCR-dependent T cell apoptosis has been successfully explored by Pearson et al. (1992) in an organ transplantation model. This group demonstrated that brief treatment of C3H/He mice with anti-CD4 antibody together with C57BL10 donor cells induced specific tolerance of subsequent C57BL/10 cardiac allografts in C3H/He recipients. Although not formally proven, death of donor-reactive T cells may well have been triggered by the combination of anti-CD4 antibody plus donor antigens in C3H/He mice. More recently, several groups have sucessfully exploited the strategy of intrathymic inoculation of donor cells as a means of deleting donor-reactive T lymphocytes in experimental organ transplantation (Markmann et al. 1993; Campos et al. 1993; Nakafusa et al. 1993). Taken as a whole, the preliminary results suggest that intrathymic inoculation of donor cells (perhaps in combination with other forms of treatment) is a powerful strategy of inducing specific tolerance of at least certain MHC-disparate organs. Again, it is likely that the deletion of donor-reactive T cells in the periphery following intrathymic inoculation of donor cells is at least in part due to alloantigen-induced apoptosis of reactive T cells (MARKMANN et al. 1993).

A precise understanding of the regulation of T cell apoptosis at the molecular level is required before programmed cell death of mature T cells can be successfully manipulated by pharmacological agents (KROEMER and MARTÍNEZ-A 1994). Nevertheless, it can be anticipated that modulation of T cell apoptosis in a positive or negative manner will eventually provide a powerful immunotherapeutic strategy in the field of autoimmune diseases and transplantation medicine.

Acknowledgements. The work from our laboratory was supported by the Alfried Krupp Award for young professors (to D. K.). We thank Constanze Taylor for expert assistance in preparing the manuscript.

References

Acha-Orbea H, Shakhov AN, Scarpellino L, Kolb E, Müller V, Vessaz-Shaw A, Fuchs R, Bölchlinger K, Rollini P, Billotte J, Sarafidou M, MacDonald HR (1991) Clonal deletion of Vβ14-bearing T cells in mice transgenic for mammary tumour virus. Nature 350: 207–211

Akbar AN, Borthwick N, Salmon M, Gombert W, Bofill M, Shamsadeen N, Pilling D, Pett S, Grundy JE, Janossy G (1993) The significance of low bcl-2 expression by CD45RO T cells in normal individuals and patients with acute viral infecfions. The role of apoptosis in T cell memory. J Exp Med 178: 427–438

Arnold B, Schönrich G, Hämmerling GJ (1993) Multiple levels of peripheral toterance. Immunol Today 14: 12–14

Auphan N, Schoenrich G, Malissen M, Barad M, Haemmerling G, Arnold B, Malissen B, Schmitt Verhulst AM (1992) Influence of antigen density on degree of clonal deletion in T cell receptor transgenic mice. Int Immunol 4: 541–547

Azuma M, Ito D, Yagita H, Okumura K, Phillips JH, Lanier LL, Somoza C (1993) B70 antigen is a second ligand for CTLA-4 and CD28. Nature 366: 76–79

Banda NK, Bernier J, Kurahara DK, Kurrle R, Haigwood N, Sekaly R-P, Finkel TH (1992) Crosslinking CD4 by human immunodeficiency virus gp120 primes T cells for activation-induced apoptosis. J Exp Med 176: 1099–1106

Bierer BE, Schreiber SL, Burakoff SJ (1991) The effect of the immunosuppressant FK-506 on alternate pathways of T cell activation. Eur J Immunol 21: 439–445

Boehme SA, Lenardo MJ (1993) Propriocidal apoptosis of mature T lymphocytes occurs at S phase of the cell cycle. Eur J Immunol 23: 1552–1560

Bogen B, Dembric Z, Weiss S (1993) Clonal deletion of specific thymocytes by an immunoglobulin idiotype. EMBO J 12: 357–363

Bossu P, Singer GG, Andres P, Ettinger R, Marshak-Rothstein A, Abbas AK (1993) Mature CD4+ T lymphocytes from MRL/lpr mice are resistant to receptor-mediated tolerance and apoptosis. J Immunol 151: 7233–7239

Boussiotis V, Freeman GJ, Gribben JG, Daley J, Gray G, Nadler LM (1993) Activated human B lymphocytes express three CTLA-4 counterreceptors that costimulate T-cell activation. Proc Natl Acad Sci USA 90: 11059–11063

BreitmeyerJB, Oppenheim SO, Daley JF, Levine HB, Schlossman SF (1987) Growth inhibition of human T cells by antibodies recognizing the T cell antigen receptor complex. J Immunol 138: 726–731

Campos L, Alfrey EJ, Posselt AM, Odorico JS, Barker CF, Naji A (1993) Prolonged survival of rat orthotopic liver allografts after intrathymic inoculation of donor-strain cells. Transplantation 55: 866–870

Cohen PA, Kim H, Fowler DH, Gress RE, Jakobsen MK, Alexander RB, Mule JJ, Carter C, Rosenberg SA (1993) Use of interleukin-7, interleukin-2, and interferon-gamma to propagate CD4+ T cells in culture with maintained antigen specificity. J Immunother 14: 242–252

Critchfield JM, Racke MK, Zúñiga-Pflücker JC, Cannella B, Raine CS, Goverman J, Lenardo MJ (1994) T cell deletion in high antigen dose therapy of autoimmune encephalomyelitis. Science 263: 1139–1142

Damle NK, Klussman K, Leytze G, Aruffo A, Linsley PS, Ledbetter JA (1993a) Costimulation with integrin ligands intercellular adhesion molecule-1 or vascular cell adhesion molecule-1 augments activation-induced death of antigen-specific CD4+ T lymphocytes. J Immunol 151: 2368–2379

Damle NK, Leytze G, Klussman K, Ledbetter JA (1993b) Activation with superantigens induces programmed death in antigen-primed CD4+ class II+ major histocompatibility complex T lymphocytes via a CD11a/CD18-dependent mechanism. Eur J Immunol 23: 1513–1522

Duke RC, Cohen JJ (1986) IL-2 addiction: withdrawal of growth factor activates a suicide program in dependent T cells. Lymphokine Res 5: 289–299

Gonzalo JA, de Alborán IM, Alés-Martínez JE, Martínez-A C, Kroemer G (1992) Expansion and clonal deletion of peripheral T cells induced by bacterial superantigen is independent of the interleukin-2 pathway. Eur J Immunol 22: 1007–1011

Gougeon M-L, Montagnier L (1993) Apoptosis in AIDS. Science 260: 1269–1272

Gougeon M-L, Garcia S, Heeney J, Tschopp R, Lecoer H, Guetard D, Rame V, Dauguet C, Montagnier L (1993) Programmed cell death in AIDS-related HIV and SIV infections. AIDS Res Hum Retrovir 9: 553–563

Groux H, Torpier G, Monté D, Mouton Y, Capron A, Ameisen JC (1992) Activation-induced death by apoptosis in CD4+ T cells from human immunodeficiency virus-infected asymptomatic individuals. J Exp Med 175: 331–340

Groux H, Monte D, Plouvier B, Capron A, Ameisen J-C (1993) CD3-mediated apoptosis of human medullary thymocytes and activated peripheral T cells: respective roles of interleukin-1, interleukin-2, interferon-γ and accessory cells. Eur J Immunol 23: 1623–1629

Hadley GA, Anderson CB, Mohanakumar T (1992) Selective loss of functional antidonor cytolytic T cell precursors following donor-specific blood transfusions in long-term renal allograft recipients. Transplantation 54: 333–337

Hockenbery DM, Oltvai ZN, Yin X-M, Milliman CL, Korsmeyer SJ (1993) Bcl-2 functions in an antioxidant pathway to prevent apoptosis. Cell 75: 241–251

Howie SEM, Sommerfield AJ, Gray E, Harrison DJ (1994) Peripheral T lymphocyte depletion by apoptosis after CD4 ligation in vivo: selective loss of CD44− and 'activating' memory T cells. Clin Exp Immunol 95: 195–200

Hugo P, Kappler JW, Godfrey DI, Marrack PC (1994) Thymic epithelial cell lines that mediate positive selection can also induce thymocyte clonal deletion. J Immunol 1022–1031

Itoh N, Tsujimoto Y, Nagata S (1993) Effect of bcl-2 on fas antigen-mediated cell death. J Immunol 151: 621–627

Janssen O, Wesselborg S, Kabelitz D (1992) The immunosuppressive action of OKT3. OKT3 induces programmed cell death (apoptosis) in activated human T cells. Transplantation 53: 233–234

Jenkinson EJ, Kingston R, Smith CA, Williams GT, Owen JJT (1989) Antigen-induced apoptosis in developing T cells: a mechanism for negative selection of the T cell receptor repertoire. Eur J Immunol 19: 2175–2177

Kabelitz D, Wesselborg S (1992) Life and death of a superantigen-reactive human CD4+ T cell clone: staphylococcal enterotoxins induce death by apoptosis but simultaneously trigger a proliferative response in the presence of HLA-DR+ antigen-presenting cells. Int Immunol 4: 1381–1388

Kabelitz D, Pohl T, Pechhold K (1993) Activation-induced cell death (apoptosis) of mature peripheral T lymphocytes. Immunol Today 14: 338–339

Kawabe Y, Ochi A (1991) Programmed cell death and extrathymic reduction of Vβ8+ CD4+ T cells in mice tolerant to Staphylococcus aureus enterotoxin B. Nature 349: 245–248

Kirberg J, Bruno L, von Boehmer H (1993) CD4+8− help prevents rapid deletion of CD8+ cells after a transient response to antigen. Eur J Immunol 23: 1963–1967

Klas C, Debatin KM, Jonker RR, Krammer PH (1993) Activation interferes with the APO-1 pathway in mature human T cells. Int Immunol 5: 625–630

Kroemer G, Martínez-A C (1994) Pharmacological inhibition of programmed lymphocyte death. Immunol Today 15: 235–242

Kyburz D, Aichele P, Speiser DE, Hengartner H, Zinkernagel RM, Pircher H (1993) T cell immunity after a viral infection versus T cell tolerance induced by soluble viral peptides. Eur J Immunol 23: 1956–1962

Lenardo MJ (1991) Interleukin-2 programs mouse αβ T lymphocytes for apoptosis. Nature 353: 858–861

Li J, Campbell D, Hayward AR (1992) Differential response of human thymus cells to CD2 antibodies: fragmentation ot DNA of CD45RO+ and proliferation of CD45RO− subsets. Immunology 75: 305–310

Linsley PS, Ledbetter JA (1993) The role of the CD28 receptor during T cell responses to antigen. Annu Rev Immunol 11: 191–212

Liu Y, Janeway CA Jr (1990) Interferon γ plays a critical role in induced cell death of effector T cell: A possible third mechanism of self-tolerance. J Exp Med 172: 1735–1739

Lussow AR, Crompton T, Karapetian O, MacDonald HR (1993) Peripheral clonal deletion of superantigen-reactive T cells is enhanced by cortisone. Eur J Immunol 23: 578–581

MacDonald HR, Baschieri S, Lees RK (1991) Clonal expansion precedes anergy and death of Vβ8$^+$ peripheral T cells responding to staphylococcal enterotoxin B in vivo. Eur J Immunol 21: 1963–1966

Mamalaki C, Tanaka Y, Corbella P, Chandler P, Simpson E, Kioussis D (1993) T cell deletion follows chronic antigen specific T cell activation in vivo. Int Immunol 5: 1285–1292

Markmann JF, Odorico JS, Bassiri H, Desai N, Kim JI, Barker CF (1993) Deletion of donor-reactive T lymphocytes in adult mice after intrathymic inoculation with lymphoid cells. Transplantation 55: 871–877

Marrack P, Kappler J (1990) The staphylococcal enterotoxins and their relatives. Science 248: 705–711

McConkey DJ, Hartzell P, Amador-Pérez JF, Orrenius S, Jondal M (1989) Calcium-dependent killing of immature thymocytes by stimulation via the CD3/T cell receptor complex. J Immunol 143: 1801–1806

Mercep M, Noguchi PD, Ashwell JD (1989) The cell cycle block and lysis of an activated T cell hybridoma are distinct processes with different Ca^{2+} requirements and sensitivity to cyclosporine A. J Immunol 142: 4085–4092

Merkenschlager M, Fisher AG (1991) CD45 isoform switching precedes the activation-driven death of human thymocytes by apoptosis. Int Immunol 3: 1–7

Meuer SC, Hussey RE, Fabbi M, Fox D, Acuto O, Fitzgerald KA, Hodgdon JC, Protentis JP, Schlossman SF, Reinherz EL (1984) An alternative pathway of T-cell activation: a functional role for the 50 kd T 11 sheep erythrocyte receptor protein. Cell 36: 897–906

Nakafusa Y, Goss JA, Mohanakumar T, Flye WM (1993) Induction of donor specific tolerance to cardiac but not skin or renal allografts by intrathymic injection of splenocyte alloantigen. Transplantation 55: 877–882

Newell MK, Haughn LJ, Maroun CR, Julius MH (1990) Death of mature T cells by separate ligation of CD4 and the T-cell receptor for antigen. Nature 347: 286–288

Nieto MA, López-Rivas A (1989) IL-2 protects T lymphocytes from glucocorticoid-induced DNA fragmentation and cell death. J Immunol 143: 4166–4170

Odaka C, Kizaki H, Tadakuma T (1990) T cell receptor-mediated DNA fragmentation and cell death in T cell hybridomas. J Immunol 144: 2096–2101

Owen-Schaub LB, Yonehara S, Crump III WL, Grimm E (1992) DNA fragmentation and cell death is selectively triggered in activated human lymphocytes by fas antigen engagement. Cell Immunol 140: 197–205

Oyaizu N, McCloskey TW, Coronesi M, Chirmule N, Kalyanaraman VS, Pahwa S (1993) Accelerated apoptosis in peripheral blood mononuclear cells (PBMSs) from human immunodeficiency virus type-1 infected patients and in CD4 cross-linked PBMCs from normal individuals. Blood 82: 3392–3400

Pearson TC, Madsen JC, Larsen CP, Morris PJ, Wood KJ (1992) Induction of transplantation tolerance in adults using donor antigen and anti-CD4 monoclonal antibody. Transplantation 54: 475–483

Pechhold K, Pohl T, Kabelitz D (1994) Rapid quantification of lymphocyte subsets in heterogeneous cell populations by low cytometry. Cytometry 16: 152–159

Pender MP, McCombe PA, Yoong G, Nguyen KB (1992) Apoptosis of alpha beta T lymphocytes in the nervous system in experimental autoimmune encephalomyelitis: its possible implications for recovery and acquired tolerance. J Autoimmun 5: 401–410

Perandones CE, Illera VA, Peckham D, Stunz LL, Ashman RF (1993) Regulation of apoptosis in vitro in mature murine spleen T cells. J Immunol 151: 3521–3529

Radvanyi LG, Mills GB, Miller RG (1993) Religation of the T cell receptor after primary activation of mature T cells inhibits proliferation and induces apoptotic cell death. J Immunol 150: 5704–5715

Razvi ES, Welsh RM (1993) Programmed cell death of T lymphocytes during acute viral infection: a mechanism for virus-induced immune deficiency. J Virol 67: 5754–5765

Rouleau M, Bernard A, Lantz O, Vernant J-P, Charpentier B, Senik A (1993) Apoptosis of activated CD8$^+$/CD57$^+$ T cells is induced by some combinations of anti-CD23 mAb. J Immunol 151: 3547–3556

Russell JH, Wang R (1993) Autoimmune gld mutation uncouples suicide and cytokine/proliferation pathways in amivated mature T cells. Eur J Immunol 23: 2379–2382

Russell JH, White CL, Loh DY, Meleedy-Rey P (1991) Receptor-stimulated death pathway is opened by antigen in mature T cells. Proc Natl Acad Sci USA 88: 2151–2155

Russell JH, Rush BJ, Abrams SI, Wang R (1992) Sensitivity of T cells to anti-CD3-stimulated suicide is independent of functional phenotype. Eur J Immunol 22: 1655–1658

Russell JH, Rush B, Weaver C, Wang R (1993) Mature T cells of autoimmune lpr/lpr mice have a defect in antigen-stimulated suicide. Proc Natl Acad Sci USA 90: 4409–4413

Salmon M, Pilling D, Borthwick NJ, Viner N, Janossy G, Bacon PA, Akbar AN (1994) The progressive differentiation of primed T cells is associated with an increasing susceptibility to apoptosis. Eur J Immunol 24: 892–899

Schmied M, Breitschopf H, Gold R, Zischlor H, Rothe G, Wekerle H, Lassmann H (1993) Apoptosis of T lymophocytes in experimental autoimmune ancephalomyelitis. Evidence for programmed cell death as a mechanism to control inflammation in the brain. Am J Pathol 143: 446–452

Schönrich G, Kalinke U, Momburg F, Malissen M, Schmitt-Verhulst A-M, Malissen B, Hämmerling GJ, Arnold B (1991) Down-regulation of T cell receptors on selfreactive T cells as a novel mechanism for extrathymic tolerance induction. Cell 65: 293–304

Schwartz LM, Osborne BA (1993) Programmed cell death, apoptosis and killer genes. Immunol Today 14: 582–590

Sellins KS, Cohen JJ (1987) Gene induction by γ-irradiation leads to DNA fragmentation in lymphocytes. J Immunol 139: 3199–3206

Shi L, Nishioka WK, Th'ng J, Bradbury EM, Litchfield DW, Greenberg AH (1994) Premature p34^{cdc2} activation required for apoptosis. Science 263: 1143–1145

Shi Y, Bissonnette RP, Parfrey N, Szalay M, Kubo RT, Green DR (1991) In vivo administration of monoclonal antibodies to the CD3 T cell receptor complex induces cell death (apoptosis) in immature thymocytes. J Immunol 146: 3340–3346

Smith CA, Williams GT, Kingston R, Jenkinson EJ, Owen JJT (1989) Antibodies to CD3/T-cell receptor complex induce death by apoptosis in immature T cells in thymic cultures. Nature 337: 181—184

Suhrbier A, Burrows SR, Fernan A, Lavin MF, Baxter GD, Moss DJ (1993) Peptide epitope induced apoptosis of human cytotoxic T lymphocytes. Implications for peripheral T cell deletion and peptide vaccination. J Immunol 150: 2169–2178

Takahashi T, Tanaka M, Brannan CI, Jenkins NA, Copeland NG, Suda T, Nagata S (1994) Generalized lymphoproliferative disease in mice caused by a point mutation in the fas ligand. Cell 76: 969–976

Takahishi S, Maecker HT, Levy R (1989) DNA fragmentation and cell death mediated by T cell antigen receptor/CD3 complex on a leukemia T cell line. Eur J Immunol 19: 1911–1919

Tamaru Y, Miyawaki T, Iwai K, Tsuji T, Nibu R, Yachie A, Koizumi S, Taniguchi N (1993) Absence of bcl-3 expression by activated CD46RO$^+$ T lymphocytes in acute infectious mononucleosis supporting their susceptibility to programmed cell death. Blood 82: 521–527

Tsuda H, Huang RW, Takatsuki K (1993) Interleukin-2 prevents programmed cell death in adult T-cell leukemia cells. Jpn J Cancer Res 84: 431–437

Ucker DS, Ashwell JD, Nickas E (1989) Activation driven T cell death I. Requirements for de novo transcription and translation and association with genome fragmentation. J Immunol 143: 3461–3469

Ucker DS, Meyers J, Obermiller PS (1992) Activation-driven T cell death. II. Quantitative differences alone distinguish stimuli triggering nontransformed T cell proliferation or death. J Immunol 149: 1583–1592

Veis DJ, Sorenson CM, Shutter JR, Korsmeyer SJ (1993) Bcl-2 deficient mice demonstrate fulminant lymphoid apoptosis, polycystic kidneys, and hypopigmented hair. Cell 75: 229–240

Walden PR, Eisen HN (1990) Cognate peptides induce self-destruction of CD8$^+$ cytolytic lymphocytes. Proc Natl Acad Sci USA 87: 9015–9019

Watanabe-Fukunaga R, Brannan CI, Copeland NG, Jenkins NA, Nagata S (1992) Lymphoproliferation disorder in mice explained by defects in Fas antigen that mediates apoptosis. Nature 356: 314–317

Wesselborg S, Kabelitz D (1993) Activation-driven death of human Tcell clones: Time course kinetics of the induction of cell shrinkage, DNA fragmentation and cell death. Cell Immunol 148: 234–241

Wesselborg S, Janssen O, Kabelitz D (1993a) Induction of activation-driven death (apoptosis) in activated but not resting peripheral blood T cells. J Immunol 150: 4338–4345

Wesselborg S, Prüfer U, Wild M, Schraven B, Meuer SC, Kabelitz D (1993b) Triggering via the alternative CD2 pathway induces death by apoptosis in activated human T lymphocytes. Eur J Immunol 23: 2707–2711

Wyllie AH (1980) Glucocorticoid-induced thymocyte apoptosis is associated with endogenous endonuclease activation. Nature 284: 555–556

B Cell Activation and Apoptosis

P. Sarthou, L. Benhamou, and P.-A. Cazenave

1 Introduction

The antigen recognition potential of lymphocytes is theoretically so vast that, without appropriate regulation, the immune system should inevitably react to self-antigens. The development of B and T lymphocytes is therefore controlled at multiple levels in order to ensure that individuals will be able to respond to a large array of foreign antigens while being tolerant to self. Under some circumstances, the encounter between a foreign antigen and a naive lymphocyte will result in the emergence of a clonal population of effector cells able to eliminate this antigen. Such activation-induced positive selection not only involves the antigen-dependent step, but also depends on the differentiation state of the cell, and often requires additional stimuli from the environment. By contrast, activation-induced negative selection of lymphocytes refers essentially to the process whereby the immune system eliminates potentially harmful self-reactive lymphocytes. Again, the state of differentiation of lymphocytes and/or environmental signals play a critical role in this self-tolerance process. Viewed in this way, the distinction between activation-induced positive and negative selection is not straightforward. However, studies conducted recently had shed new light on this area and contribute to a better understanding of the underlying processes which keep the immune system under control.

Negative selection of B lymphocytes is now considered to occur by (at least) two distinct pathways (Nossal 1994). B cells can be silenced by antigens without

Unité d'Immunochimie Analytique, Institut Pasteur, 25 rue du Docteur Roux, 75724 Paris Cedex 15, France being deleted, or, alternatively, can be eliminated from the repertoire. The former situation is referred to as clonal anergy, a process which in some cases may be reversible upon removal of antigen and which allows some B lymphocytes to replace their self-reactive antigen receptors by non-self-reactive surface immunoglobulins, a phenomenon termed receptor editing (GAY et al. 1993; RADIC et al. 1993; TIEGS et al. 1993). In the latter situation, self-reactive lymphocytes are irreversibly silenced through activation-induced apoptosis. The manners in which lymphocytes are anergized or deleted may involve partially overlapping mechanisms. First, self-antigens may be weakly or strongly ligated by antigen receptors and may thereby induce anergy or deletion. Second, a single lymphocyte may differentially respond to an antigen according to its maturation state. The signal transducing capacity of the antigen receptor may vary during ontogeny and therefore may be critical in this respect. Third, environmental conditions may facilitate either anergy or deletion, according to the location where antigen encounter takes place, and/or the involvement of auxiliary cells and cytokines.

Although self-antigen-driven clonal deletion of autoreactive lymphocytes has been shown to involve apoptotic cell death, apoptosis is also believed to be responsible for purging the immune system of lymphocytes with nonfunctional antigen receptors and of lymphocytes which do not interact with antigens and/or other ligands at some stages of their developmental program. Thus, continuous elimination of abnormal or useless lymphocytes through apoptotic death is an essential part of shaping the immune repertoire and regulation of immune response.

Cumulative evidence for the essential role of apoptosis in the functional regulation of the lymphoid system has been provided over the recent years, with a special emphasis on the T cell lineage. However, it has now become clear that B cell tolerance is mediated by multiple mechanisms including activation-induced apoptosis. B cell development from early progenitors to peripheral B lymphocytes is mediated by interactions involving pre-B and B cell receptors, as well as accessory receptors, which receive signals from the environment. The specific functions of these molecules should not be examined independently of each others, since cross-talk between such receptors may influence the decision of the cell which has to integrate the information in a coherent way. This would result in the activation of genes which under some circumstances could induce either apoptotic cell death or resistance to apoptosis. Although the molecular bases of such critical events are yet largely unknown, a series of recent findings has provided some convincing evidence that is likely to increase our knowledge in this field.

B lymphocytes provide a suitable material to study the cellular and molecular events which make the decision between activation and tolerance. An increasing body of evidence strongly suggests that the extent of the B cell antigen receptor (BCR) cross-linking plays a pivotal role in the activation vs inhibition decision. If the tolerance thresholds of mature and immature B cells were to differ, the degree of ligation of surface immunoglobulins (sIgs) could markedly influence the fate of these cells. Thus, the question arises whether intracellular signals originating from the BCR may vary according to developmental and environmental conditions. Under most circumstances, BCR-dependent signaling provides a necessary but incomplete stimulus. Proliferation and further differentiation require additional signals. Coreceptors contribute either to amplify or to inhibit signals generated from the antigen receptor, and therefore cytokines and B cell-T cell contacts may dictate, or at least modulate, the choice between activation and tolerance. Recent evidence suggests that the CD40 receptor expressed on B cells plays a prominent role in this regard, but other B cell surface molecules such as CD45, which appear to interact with the BCR, are likely candidates for such a regulatory function in B cell responsiveness. Once delivered, primary and secondary signals have to be integrated by the cellular machinery as to result in gene expression in a way that may be beneficial or harmful to the cell. Apoptosis has been the subject of extensive investigations over the recent years. Within the lymphoid system, genes involved in the multiple forms of apoptosis encode the Bcl-2, APO-1/Fas, c-Myc, nur 77 and Pim-1 proteins (KRAMMER et al. 1994). It appears from most recent studies that *bcl*-2 and c-*myc* genes play an essential role in the control of apoptotic death, although full characterization of such regulatory pathways on a molecular basis awaits further investigation.

Some of the topics related to the role of apoptosis in the immune system have been reviewed in the last 2 years (COHEN et al. 1992; GREEN et al. 1992; KRAMMER et al. 1994; NEMAZEE 1993; NOSSAL 1994). The purpose of this review is therefore to highlight the latest developments which illustrate the contribution of apoptosis in the specific regulation of B cell responsiveness. As mentioned above, the strength of interactions between antigens and the BCR, the maturation status of the cells, the contribution of coreceptors and cytokines, and the role of genes implicated in apoptosis are essential pieces of a complex puzzle of metabolic pathways on which should be focussed the attention of investigators.

2 Antigen-Receptor Cross-Linking and B Cell Responsiveness

Maturation of the B cell antigen receptor proceeds throughout B cell differentiation from early B cell progenitors to mature B lymphocytes (JONGSTRA and MISENER 1993). Compelling evidence has accumulated which shows that pre-B cells require membrane expression of the μ heavy chain in association with other proteins in order to complete differentiation (MELCHERS et al. 1993). It seems likely that the μ chain-including complexes mediate interactions with the environment of the bone marrow through ligation of so far unidentified self-ligand(s). Selection and maturation of the developing B cell progenitors may therefore depend on the signal transducing ability of these complexes. Such signals may be necessary to rescue the cells from programmed cell death, and positive selection would ensure that only those cells which express functional μ chains on the surface are allowed to proceed to further differentiation. Although negative selection of B cells has been commonly viewed to involve B cells rather than pre-B cells (see below), the possibility that the μ chain complexes on pre-B cells may play a significant role in this respect has been suggested recently (SCHWARTZ and STOLLAR 1994). Partial skewing of the primary B cell repertoire could thus result from negative selection at the pre-B cell stage, provided that high-affinity interactions with self-antigens could occur within the bone marrow.

However, most of our knowledge in the field of B cell tolerance susceptibility comes from the large number of investigations which addressed this issue in sIg-positive B cell populations. From earlier studies (METCALF and KLINMAN 1977; NOSSAL and PIKE 1975), it appeared that sIg-positive immature B cells in the bone marrow were especially sensitive to inhibition after antigenic encounter, and further investigations revealed that these cells were rendered tolerant through clonal anergy (NOSSAL and PIKE 1980). Mice transgenic both for a soluble form of hen egg lysozyme (HEL) and anti-HEL-specific B lymphocytes were later used to demonstrate that self-reactive B cells were again silenced through clonal anergy (GOODNOW et al. 1988). However, when the soluble form of HEL was replaced by a membrane-anchored form, B cell tolerance was accomplished by arrested development and soon followed by cell death (HARTLEY et al. 1991, 1993). Interestingly, constitutive expression of the counter-apoptotic bcl-2 gene delayed cell death in chimeric transgenic mice, so that a large number of immature self-reactive B cells were shown to accumulate within the bone marrow and in the periphery (HARTLEY et al. 1993). It should be mentioned that clonal deletion of self-reactive transgenic B cells was first demonstrated when mice transgenic for an antibody specific for MHC class I H-2K^k molecules were crossed with H-2K^k mice (NEMAZEE and BÜRKI 1989a,b). However, recent work from the same group has now shown that such self-reactive B cells were not immediately deleted and that delayed elimination enabled some of them to escape death by means of receptor editing (TIEGS et al. 1993). Taken together, these experiments clearly show that anergy and deletion within the primary B cell repertoire may be variations around a theme and that arrested development is not a mandatory step towards cell death. In addition, and of special concern for this review, these and other findings support the view that while soluble antigens appear to promote B cell tolerance mainly through clonal anergy, membrane bound antigens are likely to readily induce B cell death. Since the latter form of antigenic stimulation most probably involves extensive cross-linking of ligand-receptor pairs, one may postulate that multimerization of sIgs generates stronger and/or more sustained signals than oligomerization does, with rapidly irreversible consequences in terms of cell death induction. However, this concept may appear hardly tenable if one considers that insolubilized anti-Ig antibodies or anti-Ig-dextran conjugates, which are believed to promote extensive cross-linking of sIg, are indeed in vitro powerful mature B cell stimulators, in contrast to their unconjugated counterpart (BRUNSWICK et al. 1988; MONGINI et al. 1992; PARKER 1975; PURE and VITETTA 1980). These seemingly paradoxical observations might reflect different signaling thresholds for activation vs inhibition in immature and mature B cells, due to intrinsic signaling properties and/or to the contribution of environmental factors. Indeed, it has been reported that the ligand binding requirements for growth inhibition of immature murine B cell lines were less stringent than those required for mature B cell activation (UDHAYAKUMAR et al. 1991a). More specifically, these studies demonstrated first that growth inhibition of immature B cells was achieved by much lower doses of soluble anti-Ig antibodies than those needed for stimulation of resting mature B cells. Second, while Sepharose-coupled anti-Ig antibodies were required for optimal mature B cell stimulation, both soluble and immobilized anti-Ig antibodies were equivalent in causing growth inhibition in immature B cells. In line with this study, it has recently been shown that malignant human B cell lines could either be inhibited or stimulated, depending on whether anti-Ig antibodies were used in a soluble or a Sepharose form respectively (VAN ENDERT and MOLDENHAUER 1992). It should be stressed however, that while growth arrest was clearly demonstrated in the latter two reports, none of them adressed the possibility that apoptosis was the final outcome of the inhibition process. Of interest, therefore, is the recent finding that extensive cross-linking of sIg receptors by plastic-immobilized anti-Ig antibodies, or biotinylated anti-Ig antibodies plus avidin, induced apoptosis in mature resting B cells (PARRY et al. 1994a, b). Thus, it appears that B cells respond to antigenic encounter in hardly predictable ways according to the read out systems employed. In this regard, the elaboration of a comprehensive model of tolerance which would take the influence of aggregation of sIg receptors into account is probably far from completion. Nonetheless, B lymphocytes are undoubtedly a valuable model for the study of the crucial effects of membrane molecular aggregation on the control of cellular responses (BACHMANN et al. 1993; METZGER 1992), though a more systematic investigation of this process is clearly required.

Apparent from the data discussed above, whether immature B cells are more sensitive to sIg ligation than mature cells remains an open question. Newly differentiated immature B cells are sIgM-positive and sIgD-negative, and it is commonly accepted that they mature further along with surface expression of IgD molecules (GOODNOW 1992, JONGSTRA and MISENER 1993; ROLINK and MELCHERS 1991). Surface IgM and IgD molecules are associated with some identical and other unique signaling components, which upon antigen binding may convey similar but not identical informations into the cytoplasm (CAMBIER et al. 1993). However, the physiological consequences of signaling through these two isotypes are unclear. For example, previous evidence suggested that these molecules played unique roles in the activation vs inhibition decision, when it was shown that antigen binding was stimulatory for mature IgM+/IgD+ B cells and inhibitory for immature IgM+/IgD− B cells (NOSSAL 1983). Phenotypically immature B lymphomas were also shown to be growth-inhibited by sIgM- but not sIgD-cross-linking (ALES-MARTINEZ et al. 1988; TISCH et al. 1988). These conclusions are further supported by recent studies conducted on the murine B lymphoma cell line WEHI-231. This lymphoma has been commonly used as a model to study anti-Ig-dependent apoptosis in immature B cells (BENHAMOU et al. 1990; HASBOLD and KLAUS 1990) and was initially characterized as sIgM+/sIgD−. However, it was recently shown that this cell line actually expresses IgD on its surface, but is not growth inhibited by anti-δ antibodies (GOTTSCHALK et al. 1994; HAGGERTY et al. 1993; and our unpublished observations). Interestingly, the δ-chain on WEHI-231 appears differentially glycosylated when compared to its counterpart in splenic B cells (HAGGERTY et al. 1993), and some features of the early biochemical signals generated upon sIgD ligation differ from those elicited by sIgM ligation (HAGGERTY et al. 1993; and our unpublished observations). Therefore, although these observations cast doubt on the so-called immature status of WEHI-231, they clearly demonstrate that sIgM molecules, but not sIgD, have the unique potential of inducing apoptotic death in this tolerance susceptible cell line. However, some equally convincing evidence has been provided which demonstrates that sIgD receptors can deliver inhibitory signals in immature B cells. Splenic B cells from neonatal mice display an immature phenotype on the basis of their susceptibility to the inhibitory effects of sIg ligation (BRINES and KLAUS 1991,1992,1993; CARSETTI et al. 1993; CHANG et al. 1991; NOSSAL 1983; YELLEN et al. 1991; YELLEN-SHAW and MONROE 1992). Although such cells are predominantly sIgM$^+$/sIgD$^-$, even the minority of sIgM$^+$/IgD$^+$ neonatal splenic B cells can be rendered unresponsive to lipopolysaccharide (LPS) after anti-μ or anti-δ treatment (BRINES and KLAUS 1992,1993). Surface IgM ligation in adult splenic B cells was recently shown to result in unresponsiveness to subsequent antigenic or LPS challenge, and whereas sIgD alone failed to induce tolerance in this system, this isotype could synergize with sIgM in the generation of negative signals (GAUR et al. 1993). While establishing that tolerance could be achieved through sIgM ligation in mature conventional B cells, the latter study did not address the issue of whether apoptosis resulted from such treatment. This possibility has now been tested in more recent studies which demonstrated that extensive cross-linking of both sIgM and sIgD receptors on mature B cells causes apoptosis (PARRY et al. 1994a,b). The latter findings strongly support the notion that polymerized antigens or membrane antigens which are likely to cause hyper-cross-linking of sIg receptors are powerful inducers of clonal deletion of mature B lymphocytes (see above). Moreover, they conclusively demonstrate that sIgD molecules have a killing potential, provided that they are extensively ligated by antigens. These observations are also consistent with the recent demonstration that B cells from HEL-specific Ig-transgenic mice were anergized or deleted, irrespective of the anti-HEL IgM- or IgD-isotypes expressed on the cell surface (BRINK et al. 1992).

B-1 (or Ly-1$^+$) lymphocytes may constitute a separate lineage from conventional (or B-2) lymphocytes and are found predominantly in the peritoneal cavity in mice. Interestingly, B-1 cells from mice transgenic for an anti-erythrocyte autoantibody underwent apoptotic cell death upon injection of the relevant antigen into the peritoneal cavity (MURAKAMI et al. 1992). Peritoneal B-1 cells from normal mice also underwent apoptosis when their sIg receptors were extensively cross-linked in vivo by injection of anti-Ig antibodies, but interestingly, B-1 cells from autoimmune disease prone NZB mice strains were resistant to such treatment (TSUBATA et al. 1994). Although these findings support the view that mature B-1 cells can be deleted in vivo by cross-linking antigens such as surface

erythrocyte molecules or anti-Ig antibodies, results from in vitro studies carried out by other groups have demonstrated that B-1 cells are resistant to sIg-mediated growth inhibition (LIOU et al. 1992; MORRIS and ROTHSTEIN 1993). Investigation of the influence of cytokines and/or cell-to-cell contacts in B-1 cell responsiveness to sIg ligation should help to resolve this paradox.

Altogether, the bulk of evidence discussed above provides some clues about how B cell antigen receptors may convey stimulatory or inhibitory signals. The picture emerges that both the strength of sIg-derived signals and the maturation status of B cells are of critical importance in the choice between clonal ignorance, clonal anergy, or clonal deletion. Simply put, immature B cells appear exquisitely sensitive to the inhibitory effect of antigenic encounter, with a tolerance threshold far lower than that of more mature B cells. However, this is not an all-or-nothing response, and according to the strength or duration of the stimulus, immature B cells may be irreversibly committed to apoptotic cell death, or alternatively may be given a chance to survive upon reception of external help. Once passed through this developmental stage, the cells would further mature and enter a second tolerance window. At this point, B cells should die by default unless rescued by antigen and/or accessory signals, or should be stimulated or killed by antigens according to the strength of the antigenic stimulus. If the signaling capacity of the BCR was to vary along with B cell ontogeny, this receptor could play a unique role in the decision that must be taken. Cosignals originating from the environment are also likely to markedly influence this decision, acting either in synergy with, or in opposition to, the initial signals.

Thus, these processes depend on a complex array of intracellular signals, which may interact with each others. Though the study of such metabolic pathways is still in its early stages, numerous reports have been recently published which provide valuable information in this field. The next section will focus on the regulation of apoptotic death by BCR- and accessory receptor-dependent signals.

3 Signals and Cosignals in the Control of B Cell Responsiveness

Upon antigen binding to sIg receptors, B cells undergo a signal transduction cascade originating from the BCR complex. This complex consists of an antigen binding subunit which is noncovalently but stably associated with a signal transducing subunit composed of disulfide-linked Igα and Igβ molecules. The cytoplasmic domains of the latter proteins carry a tyrosine-based activation motif which couples the BCR to protein tyrosine kinases (PTKs). After antigen binding, these PTKs phosphorylate a series of substrates including PTK, Igα and β molecules themselves, and phospholipase Cγ1 and γ2 (PLC), which in turn activate the phosphatidyl-inositol (PI) pathway, thereby initiating the calcium and protein kinase C (PKC)-dependent cascades. Other identified PTK substrates include guanine nucleotide exchange proteins such as Vav and the p21^{ras} GTPase-activating protein, PI-3 kinase, and MAP kinase (BAIXERAS et al. 1993; CAMBIER et al. 1993; DESIDERIO 1994; RETH 1994).

Whether BCR-mediated signaling properties vary during maturation of B cells has been the subject of numerous investigations and controversies. Since a comprehensive survey of this topic has been recently published (BAIXERAS et al. 1993), we will focus on some recent developments which may contribute to a better understanding of the mechanisms implicated in BCR-triggered growth inhibition and apoptosis.

B cell precursors express pseudo-Ig complexes on the cell surface (MELCHERS et al. 1993). Whether or not these complexes display unique signaling properties is still debatable, although recent evidence suggests that it might be the case. The λ5 surrogate light chain transduces early biochemical signals from surface μ chain-positive or -negative pre-B cell lines, suggesting that this protein may convey information inside the cell, even at the earlier stages of progenitor B cell differentiation (JONGSTRA and MISENER 1993; MISENER et al. 1991). The μ/pseudo-light chain complex in pre-B cells was found to be associated with the Igα and β molecules and to display the functional characteristics of a signal transduction unit (BOSSY et al. 1993; BROUNS et al. 1993; MATSUO et al. 1993). The interesting possibility that Igα and β molecules may transduce unique signals that vary along with the differentiation program of B lymphocytes has been suggested (NAKAMURA et al. 1993), and expression of structurally distinct Igα/Igβ-like heterodimers appears to change as a function of differentiation (ISHIHARA et al. 1993). Together with the reports that pre-B cell receptors may transduce incomplete signals when compared to mature B cells (BOSSY et al. 1993) and that surface μ chains fail to transduce growth inhibitory signals in pre-B cell lymphomas (TSUTSUMI et al. 1992a), in contrast to immature B lymphomas (see below), these observations strongly suggest that developmental maturation of BCR-like molecules may be specially relevant to the susceptibility of B cell precursors to antigenic encounter. It should be stressed, however, that activation-induced apoptotic death of precursor B cells has yet to be demonstrated, especially regarding the strength of the stimulus. This would further define the molecular requirements for positive vs negative selection, which may shape the early and late pre-B cell repertoires.

Most of the current knowledge of the signaling requirements which may condition activation or inhibition of B lymphocytes comes from studies of the unique susceptibility of immature, sIg-positive B cells or B cell lines to antigen-induced tolerance. Neonatal splenic B cells, which are known to be particularly susceptible to tolerance induction, have been shown to be deficient in phosphoinositide hydrolysis following sIg ligation (YELLEN et al. 1991). Immature B cells may therefore display alteration(s) in the BCR-dependent early signal transduction machinery. A BCR-related signaling defect has now been evidenced using the HEL/anti-HEL double transgenic approach (COOKE et al. 1994). When mice transgenic for anti-HEL sIg receptors were mated with transgenic mice expressing soluble HEL, self-reactive anti-lysozyme B cells developed but were tolerant to HEL. Biochemical analysis revealed that activation of part of the BCR-dependent PTK signaling cascade was prevented in these anergized cells. Interestingly, the signaling block could be overcome by extensive slg cross-linking by membrane bound HEL, in agreement with the notion that anergy can be reversed upon appropriate antigenic stimulation.

PTK and protein tyrosine phosphatase (PTPase) activities are critically involved in antigen-induced signal transduction and during development. Dysregulation of the subtle signaling balance which ensures appropriate positive B cell responses may therefore allow dominant negative signals to arise. Although the available evidence in this area is so far scarce, it has been recently shown that one of the PTKs involved in early signal transduction from the BCR, p55[blk], may play a significant role in this respect. Pretreatment of the immature CH31 B lymphoma with antisense oligonucleotides to blk effectively prevented anti-Ig-induced apoptosis (Yao and Scott 1993). Though these findings suggest that blk may convey a death signal in B cells, whether this kinase plays a similar role in other B lymphomas or normal B cells remains to be established, and blk gene targeting inactivation should contribute to test this hypothesis. Our recent observations suggest that PTK substrates such as p75[HS1] may play a similar role in this context. The HS1 protein binds to the SH2 domain of p53/56[lyn] PTK and is phosphorylated upon IgM ligation (Kitamura et al. 1989; Yamanashi et al. 1993). Mutants of WEHI-231 B cells resistant to anti-Ig-induced apoptotic death were shown to be deficient in HS1 expression (Benhamou et al. 1994), and complementation of a mutant cell line by the HS1 protein restored the anti-Ig-induced apoptotic phenotype (Fukuda et al. 1994). Moreover, B lymphocytes from HS1 knock-out mice were found to be resistant to anti-Ig-induced apoptosis (Taniushi et al., unpublished observations). Taken together, these observations raise the interesting possibility that PTK activities and PTK substrates may play a unique role in the tolerance susceptibility of B cells and that redundant mechanisms operate in this regard.

PTPases such as the CD45 molecule are expressed as multiple isoforms on the cell surface of B and T lymphocytes (Alexander et al. 1992; Fearon 1993). Tyrosine dephosphorylation by CD45 of proximal components of the signal transduction cascade is believed to be crucial for anti-Ig-induced B cell activation (Justement et al. 1991; Reth 1992). Mice defective in the expression of CD45 displayed a block in T cell development, but not in B cell development, although in this case B lymphocytes did not proliferate in response to slg cross-linking (Kishihara et al. 1993). Since the number of peripheral B cells was apparently normal in CD45 knock-out mice, the role of CD45 molecules in shaping the primary B cell repertoire remains questionable. However, CD45 negative variants from the WEHI-231 B cell line were recently found to be more susceptible to anti-Ig-induced apoptosis than the parental cells, suggesting that this molecule may nonetheless play a role in the regulation of tolerance in B cells (Ogimoto et al. 1994).

Downstream from PTK activation, calcium and PKC-dependent steps may also contribute to the regulation of apoptosis in B cells. Since apoptotic cell death is a Ca^{2+}-dependent process (COHEN et al. 1992; TRUMP and BEREZESKY 1992), and given that BCR ligation generates both early and sustained elevations of intracellular Ca^{2+} concentration, it is tempting to speculate that the amplitude and/ or duration of Ca^{2+} signaling may be an important parameter in the biological response of B cells. Although this proposal seems reasonable in view of the known susceptibility of lymphocytes to Ca^{2+} ionophore-induced apoptosis, the contribution of BCR-dependent Ca^{2+} signals to apoptosis is so far mostly speculative. However, an imbalance in sIg-dependent signaling pathways, which could favor sustained intracellular Ca^{2+} increase, has been proposed to be causal for the apoptotic death of murine B lymphoma (BENHAMOU et al. 1990,1994; SARTHOU et al. 1989), malignant human B-CLL cells (McCONKEY et al. 1991), and Burkitt's lymphoma cell lines (KNOX et al. 1992). Moreover, the immunosuppressive drug cyclosporin A (CsA), which was initially thought to most notably affect T cell functions, has now been shown to protect B cell lines from some forms of apoptotic death, including anti-Ig treatment (BONNEFOY-BERARD et al. 1994; KANAZASHI et al. 1994; MUTHUKKUMAR et al. 1993; UDHAYAKUMAR et al. 1991b). The Ca^{2+}-calmodulin-dependent protein phosphatase calcineurin is the target of the CsA/cyclophilin complex, and CsA inhibits several Ca^{2+}-dependent pathways (SCHREIBER and CRABTREE 1992). These findings therefore reinforce the notion that Ca^{2+} signals may play a unique role in sIg receptor-triggered apoptosis, perhaps from dysregulation of the early signaling cascade, with subsequent alterations in more distal Ca^{2+}-dependent events.

Although PKC activation is generally believed to be part of the initial signaling pathways involved in antigenic stimulation of B cells (CAMBIER et al. 1993), it has been suggested that PKC-independent signals may be generated from sIg receptor ligation (MOND et al. 1987). From previous examination of the proximal signaling pathways triggered by sIg receptor ligation in WEHI-231 B lymphoma cells, we suggested that insufficient activation of PKC could be related to sIg-induced growth inhibition and apoptotic death in these cells (BENHAMOU et al. 1990; SARTHOU et al. 1989). These findings were in agreement with earlier experiments which demonstrated that PKC activators such as phorbol diesters afforded protection from anti-Ig-mediated growth inhibition in the same cells (WARNER and SCOTT 1988). Also in line with this proposal was the later observation that sIgM-cross-linking of tolerance-susceptible splenic B cells from neonatal mice resulted in negative signaling through calcium elevation, while phorbol ester activation of PKC was stimulatory (YELLEN et al. 1991). This hypothesis has now received further support from the recent findings that PKC activation rescued Burkitt's lymphoma cells (KNOX et al. 1992), Ramos cells (VALENTINE and LICCIARDI 1992), and the immature BKS-2 lymphoma (MUTHUKKUMAR et al. 1993) from anti-Ig-induced death. Interestingly, spontaneous apoptosis of germinal center B cells (KNOX et al. 1992), sheep ileal Peyer's patch B cells (MOTYKA et al. 1993), and even resting splenic B cells (ILLERA et al. 1993) was prevented upon activation of PKC.

These studies suggest that insufficient PKC activation may be at least partially responsible both for antigen-induced apoptosis in immature B cells and for apoptotic death-by-default of mature B cells deprived of antigenic stimulation.

Since it has been shown that sustained PKC activation blocks the ability of anti-Ig antibodies to induce Ca^{2+} mobilization in B cells and in B lymphomas (BIJSTERBOSCH and KLAUS 1987; GOLD and DEFRANCO 1987; MIZUGUCHI et al. 1987), low level PKC activation might result in sustained intracellular Ca^{2+} elevation, which in turn would trigger the apoptosis program (McCONKEY et al. 1990). It should be stressed, however, that multiple PKC isoforms, which may differentially respond to antigenic stimulation, are expressed in B lymphocytes (HAGGERTY and MONROE 1994; TERAJIMA et al. 1992; TSUTSUMI et al. 1992b). Since pre-B cell lines, B lymphomas and splenic B cells express different PKC isoforms, which may vary in substrate specificity and activation requirements (MARQUEZ et al. 1992), whether or not PKC expression is developmentally regulated is an interesting issue regarding susceptibility of B cells to tolerance induction.

Much less attention has been paid to the cyclic AMP (cAMP) cascade over the past few years. However, it is generally believed that the cAMP/proteine kinase A pathway conveys an "off " signal which contributes to the regulation of "on" signals generated through antigen receptors or other surface receptors (KAMMER 1988). Physiological agents such as prostaglandin E_2 (PGE$_2$), which are known to elevate intracellular cAMP, are powerful negative regulators of B cell activation and differentiation (MUTHUSAMY and BONDADA 1993; PHIPPS et al. 1990; ROPER and PHIPPS 1992), and a key costimulatory function has been assigned to PGE$_2$ in antigen-induced B cell tolerance (PHIPPS et al. 1989; SCHAD and PHIPPS 1988). The contribution of cAMP to the regulation of apoptosis in lymphocytes is so far largely unknown. Recent reports have provided new evidence in this regard, although a consensus is not yet possible. For example, murine thymo-cytes were shown to be highly sensitive to the synergistic action of glucocorti-coids and cAMP-elevating agents, which resulted in enhanced apoptotic death, compared to thymocytes treated with glucocorticoids alone (McCONKEY et al. 1993). In contrast, cAMP analogs did not modulate glucocorticoid-induced death of T cell hybridoma, and TCR-induced apoptotic death was actually prevented by cAMP (LEE et al. 1993). While the latter report provides further support to the notion that glucocorticoid- and TCR-induced apoptotic pathways differ (KING and ASHWELL 1993), further investigations are needed to explain the reported dis-crepancies in the regulation of steroid action by cAMP.

With respect to the B lineage, germinal center (GC) B cells were shown to express higher amounts of cAMP than quiescent B cells, and a correlation was found between the unique propensity of GC B cells to undergo spontaneous apoptosis, and elevated levels of cAMP. Conversely, rescue from death of GC B cells afforded by phorbol esters and anti-CD40 antibodies was accompanied by decreased cAMP levels. In striking contrast, anti-CD40 treatment induced cAMP elevation in resting B lymphocytes (KNOX et al. 1993). Therefore, although cAMP is likely to be involved in the regulation of some apoptotic pathways in B lymphocytes, the differentiation/activation status of B cells may be of critical importance in this respect.

Full antigenic stimulation of B cells not only depends on the cascade of early biochemical events described above (first signal), but most often requires additional help from cytokines and cell contacts (second signal). Cross-talk between the BCR, cytokine receptors, and cell surface adhesion molecules is therefore likely to play a critical part in this process (CLARK and LEDBETTER 1994). Recent attention has been paid to B cell surface molecules such as CD40, which have the dual capacity of mediating activation of B cells themselves and of T cells through cell-to-cell ligand-receptor interactions. CD40 cross-linking by anti-CD40 antibodies or the T cell ligand for CD40, gp[39], delivers a comitogenic signal to B cells, prevents apoptotic death of GC B cells, and promotes immunoglobulin isotype switching (CALLARD et al. 1993; CLARK and LEDBETTER 1994; LEDERMAN et al. 1993; NOELLE et al. 1992). The observations that GC B cells die spontaneously unless rescued by anti-Ig and anti-CD40 antibodies (LIU et al. 1989), or recombinant gp[39] (HOLDER et al. 1993), is likely to reflect the physiological elimination of useless circulating B cells, i.e., those cells which are not stimulated through antigenic encounter and subsequent T-cell help. The CD40 molecule could play a wider role in the control of apoptotic death, since it has been shown that immature Burkitt's lymphoma cell lines (HOLDER et al. 1993; LEDERMAN et al. 1994; VALENTINE and LICCIARDI 1992) and the WEHI-231 B-cell line (TSUBATA et al. 1993) were rescued from anti-Ig-induced apoptosis through CD40. Even apoptotic death induced by hyper-cross-linking of mature splenic B cells was partially prevented by anti-CD40 antibodies and was totally blocked when anti-CD40 antibodies and IL-4 were used in combination (PARRY et al. 1994b). CD40 may also control additional apoptotic pathways, since, while interleukin-10 (IL-10) was shown to enhance spontaneous apoptosis in B-CLL cells, this effect could be prevented by anti-CD40 antibodies (FLUCKIGER et al. 1994). The anti-apoptotic potential of the CD40 molecule may seem paradoxical given its structural homology with the Fas receptor, which upon ligation induces apoptosis in a variety of cells including B cell lines (MAPARA et al. 1993; TRAUTH et al. 1989), although it may promote T cell activation under some circumstances (ALDERSON et al. 1993). Investigation of the signaling pathways downstream from these two receptors should help to clarify the situation. Recent studies have demonstrated that CD40 cross-linking on the surface of B cells triggers a PTK-dependent signaling cascade (KNOX et al. 1992; REN et al. 1994), which shares some of the characteristic features of the BCR-dependent cascade. Interestingly, while CD40 ligation induces phosphorylation of PLCγ2 and PI-3-kinase in Daudi B cell lines (REN et al. 1994), it does not stimulate the PLCγ-dependent elevation of IP3 and intracellular Ca^{2+} in GC B cells (KNOX et al. 1992). Though further experiments are needed in this field, the available evidence provided so far suggests that intracellular dialogue between CD40 and antigen receptors through partially overlapping signaling pathways may account for the reported ability of CD40 to potentiate antigenic stimulation of B Cells (WHEELER et al. 1993).

Appropriate expression of CD40 and CD80 receptors on B cells and of their respective counterreceptors on T cells ensures mutual cellular regulation, provided that the latter cells recognize MHC class II-bound peptides presented by the former cells. It has been suggested that expression of such ligand-receptor pairs is reciprocally regulated and that recognition of class II bound peptides by T cells may trigger or at least control this process (CLARK and LEDBETTER 1994). It is now widely accepted that MHC molecules are signal transducing receptors on B cells, and that antigen-primed B cells are positively stimulated through MHC ligation (WADE et al. 1993). An additional safety pathway has been now documented, which shows that, in striking contrast with activated B cells, MHC ligation on resting B cells results in apoptotic death through a cAMP-dependent pathway (NEWELL et al. 1993). Such a mechanism would ensure that only those B cells which received a specific first signal through sIg ligation should be allowed to receive help from antigen-specific T cells. Altogether, these findings provide compelling evidence that the immune system has developed multiple ways of controlling the efficiency of cognate and noncognate interactions.

In addition to cell-to-cell contact, soluble factors such as cytokines are essential for the activation and differentiation of B cells. Some of the biological effects of different cytokines may result from their ability to protect B cells from growth inhibition or apoptotic death (BAIXERAS et al. 1993). Although IL-2, IL-5, interferon (IFN)-α/β, tumor necrosis factor (TNF)-α/β, and more recently IL-10, were shown to be involved in this process, IL-4 has been the focus of several investigations which suggest that this Th2-specific lymphokine plays a key role in this respect. However, as it is often the case when the biological effects of cytokines are addressed, IL-4 may display proapoptotic properties, such as those recently evidenced in a murine B cell line (BISHOP et al. 1993) or in activated human monocytes (MANGAN et al. 1992). Nonetheless, an increasing amount of data has accumulated over the past few years which convincingly demonstrate that this lymphokine reverses the inhibitory effects of sIg ligation in B cells or B lymphomas (ALES-MARTINEZ et al. 1991; BRINES and KLAUS 1991, 1992, 1993; PARRY et al. 1994b; SCOTT et al. 1987), rescues splenic B cells and B-CLL cells from spontaneous apoptosis (DANCESCU et al. 1992; ILLERA et al. 1993), and counteracts the proapoptotic effects of IL-10 in B-CLL cells (FLUCKIGER et al. 1994). Investigation of the biochemical pathways which ensue IL-4 receptor ligation revealed that PKC and PTPase activations were rapidly induced and could perhaps control signals originating from sIg receptor ligation (HARADA et al. 1992; HARNETT et al. 1991; MIRE-SLUIS and THORPE 1991).

Most recent studies have now addressed the possibility that IL-10 may regulate apoptotic death in B cells. Interestingly, this lymphokine, which was shown to protect helper T cells from IL-2 deprivation-induced apoptotic death (TAGA et al. 1993), has now been found to prevent spontaneous death of GC B cells (LEVY and BROUET 1994). In striking contrast is the recent demonstration that IL-10 induces apoptotic death in B-CLL cells (FLUCKIGER et al. 1994). Such a discrepancy may reflect differential susceptibility of normal and malignant B cells to IL-10 treatment and may be of interest in the context of anti-tumor chemotherapy. Finally, since it has been previously reported that IL-10 does not afford protection from anti-Ig-induced B cell deletion (CUENDE et al. 1992), it appears likely that some but not all forms of apoptotic death are under the control of IL-10, although this topic clearly requires further investigations.

Once delivered to the cell, the multiple signals which originate from the antigen receptor, accessory receptors, and cytokine receptors must be integrated and conveyed to the nucleus in order to induce transcription of genes involved in the regulation of B cell proliferation and differentiation. Alternatively, these signals may be interpreted as to result in growth inhibition and ultimately cell death. Genetic control of such adverse pathways has received much attention over the past few years, with special regards to the lymphoid system. Although several genes have now been shown to be involved in this process in the immune system, expression and regulation of the c-*myc* and *bcl*-2 genes appear to play a key role in this respect (COHEN et al. 1992; EVAN and LITTLEWOOD 1993; GREEN et al. 1992; HIBNER and COUTINHO 1994; KING and ASHWELL 1993; KRAMMER et al. 1994; SCHWARTZ and OSBORNE 1993; WILLIAMS and SMITH 1993). The next section will focus on the recent developments in this field.

4 Genetic Controls of Apoptosis in B Lymphocytes

The c-*myc* oncogene has been classically involved in the control of cell proliferation. However, it has now become clear that the c-Myc protein can also induce apoptotic cell death under some circumstances (EVAN and LITTLEWOOD 1993). For example, high constitutive expression of c-*myc* in conjunction with a growth inhibitory signal is a potent inducer of apoptosis (ASKEW et al. 1991; EVAN et al. 1992; FANIDI et al. 1992). These observations may reflect the dilemma which faces a cell confronted with contradictory stimuli for proliferation and arrest. Thus, reception of an inhibitory signal during cell cycle progression would induce the cell to commit suicide (HIBNER et al. 1993; HIBNER and COUTINHO 1994; RUBIN et al. 1993). With respect to lymphocytes, recent observations suggested that the c-Myc protein may contribute to the regulation of cell proliferation vs death. TCR ligation-induced apoptosis in a T cell hybridoma was prevented by antisense inhibition of c-*myc* expression (SHI et al. 1992). In line with this study, was the recent report that some Burkitt's lymphoma cell lines could be rescued from spontaneous apoptotic death by antisense inhibition of c-*myc* expression. Moreover, protection by IFN-α from spontaneous apoptosis in these cell lines was shown to correlate with a decrease in c-Myc protein expression (MILNER et al. 1993). However, we have recently reported that prolonged expression of c-*myc* correlated with survival rather than apoptosis in anti-Ig-treated WEHI-231 B cells (HIBNER et al. 1993). Although somewhat surprising in view of the studies mentioned above, our observations have been recently supported by the finding that, while antisense c-*myc* oligonucleotide indeed prevented anti-Ig inhibition of WEHI-231 cells, such a treatment actually resulted in stabilization of c-*myc* mRNA and of c-*myc* protein expression (Fischer et al. 1994). Although such unexpected functions of an antisense oligonucleotide molecule may have to be further examined, together with our own observations these findings imply that sustained levels of c-Myc proteins are required to prevent anti-Ig-induced apoptosis in this immature B cell line. Thus, while *c-myc* is undoubtedly involved in the control of multiple forms of apoptosis, its precise function in these processes is unclear. In this respect, we believe that the early suggestion that appropriate temporal expression rather than absolute levels of *c-myc* could be relevant to growth regulation (KELLY et al. 1983) could now apply to the regulation of apoptotic death.

To date, the most intensively studied gene involved in programmed cell death regulation has been undisputably the *bcl*-2 gene. Bcl-2 prolongs cell survival and can be considered as an antidote to cell death (KORSMEYER 1992; REED 1994). Elucidation of the biochemical mechanisms by which the *bcl*-2 gene product prevents apoptosis may not be far off, since it has been recently shown that the Bcl-2 protein interferes with the generation of harmful reactive oxygen species (HOCKENBERY et al. 1993). The family of *bcl*-2 genes, which includes EBV and other virus *bcl*-2 homologs, has been growing from the recent discovery of human Bcl-2 related proteins, including Bcl-X and Bax proteins (REED 1994). A number of physiological and pathological situations are under the control of the Bcl-2 protein family in terms of cell death regulation. Bcl-2 is involved in regulating some of the survival pathways of developing B lymphocytes (KORSMEYER 1992). The topological distribution of Bcl-2 within secondary GCs is most instructive in this respect, demonstrating a close correlation between up-regulation of Bcl-2 and B cell survival and proliferation. Conversely, local down-regulation of Bcl-2 correlates with extensive death of those B cells which are not stimulated by signals preventing their entry into apoptosis (BONNEFOY et al. 1993; HOCKENBERRY et al. 1991; KNOX and GORDON 1993; LEVY and BROUET 1994; LIU et al. 1991). Bcl-2 up-regulation is also likely to favor the survival of peritoneal B cells (TSUBATA et al. 1994) and of IL-4 stimulated B-CLL malignant cell lines (DANCESCU et al. 1992). In line with these findings, APO-1/Fas-mediated apoptosis of B-CLL cells (MAPARA et al. 1993) or other cells (ITOH et al. 1993) correlates with down-regulation of Bcl-2.

However, recent studies have shown that Bcl-2 does not control all forms of apoptotic death in B lymphocytes. For example, although CD40 ligation rescued GC B cells from spontaneous apoptosis, and Burkitt's lymphoma cells from anti-Ig-induced apoptosis, none of these events could be significantly correlated with enhanced Bcl-2 expression (HOLDER et al. 1993). Moreover, anti-Ig-induced apoptosis in WEHI-231 immature B cells did not correlate with endogenous expression of Bcl-2 (GOTTSCHALK et al. 1994; HIBNER et al. 1993) nor was it prevented by overexpression of Bcl-2 (CUENDE et al. 1993), although a recent study suggested that Bcl-2 protein indeed partially protected this cell line from anti-Ig-induced apoptosis (KAMESAKI et al. 1994).

A series of recent experiments has provided some new clues, but also some uncertainty, about the implication of Bcl-2 in B cell development and responsiveness. For example, high expression of Bcl-2 was evidenced in pro-B cells and mature B cells, while down-regulation was found in pre-B and immature B cells (MERINO et al. 1994). These observations parallel the earlier report that transition from double negative to double positive thymocytes was accompanied by down-regulation of Bcl-2 and that peripheral T cells regain Bcl-2 expression (VEIS et al.

1993a). Such a dynamic regulation of Bcl-2 expression may explain differential susceptibility to environmental signals which control the development of lymphoid cells. However, immature self-reactive B cells from mice transgenic both for bcl-2 and an anti-erythrocyte autoantibody, were clonally deleted in the bone marrow in spite of high expression of the bcl-2 gene (NISITANI et al. 1993). High expression of Bcl-2 may therefore not be sufficient to override the strong inhibitory signals generated by extensive sIg-cross-linking by multivalent self-antigens. It should be therefore predicted that clonal deletion of B cells via moderate cross-linking of sIg receptors by soluble antigens should be prevented or delayed by Bcl-2 expression. Interestingly, this seems to be the case, since constitutive expression of the bcl-2 gene delayed cell death in chimeric mice transgenic for both soluble HEL and anti-HEL antibody (HARTLEY et al. 1993). Taken together, these experiments strongly suggest that Bcl-2 may play a critical role in some but not all steps of B cell ontogeny (STRASSER et al. 1994). However, such a view may be an oversimplification, since recent studies have shown that bcl-2 gene inactivation did not impede initial B and T cell development in young mice, although dramatically disturbing responsiveness of older animals (NAKAYAMA et al. 1993; VEIS et al. 1993b). Though it can be suggested that other members of the bcl-2 family replaced the defective bcl-2 gene in such genetically manipulated animals, it appears that characterization of the role of bcl-2 and related genes in lymphoid development and responsiveness is still in its infancy.

5 Conclusions

An impressive amount of data has accumulated over the last few years which have clarified some areas in the field of B cell activation and apoptosis, but, at the same time, further complications have been uncovered. Although we are probably far from integrating these observations in a conceptual model, future studies will certainly improve our knowledge in these fundamental aspects of immunology and cell biology. They should focus on determining how multiple intracellular signaling pathways are connected to each other and act in conjunction to dictate the fate of B cells. Biochemical, transgenic and gene targeting approaches will certainly allow us to critically define the molecular bases of such important processes. Imunologists should grasp the recent opportunities afforded by the discovery of new regulatory pathways in cell cycle control. Finally, pharmacologists should benefit from elucidation of the biochemical mechanisms which underlie the apoptotic pathways. This could hopefully, provide the future basis for treament of autoimmune or other immunological diseases.

References

Alderson MR, Armitage RJ, Maraskovsky E, Tough TW, Roux E, Schooley K, Ramsdell F, Lynch DH (1993) Fas transduces activation signals in normal human T-lymphocytes. J Exp Med 178: 2231–2235

Ales-Martinez JE, Warner GL, Scott DW (1988) Immunoglobulins D and M mediate signals that are qualitatively different in B cells with an immature phenotype. Proc Natl Acad Sci USA 85: 6919–6923

Ales-Martinez JE, Silver L, LoCascio N, Scott DW (1991) Lymphoma models for B-cell activation and tolerance. IX. Efficient reversal of anti-Ig-mediated growth inhibition by an activated TH2 clone. Cell Immunol 135: 402–409

Alexander D, Shiroo M, Robinson A, Biffen M, Shivnan E (1992) The role of CD45 in T-Cell activation: resolving the paradoxes. Immunol Today 13: 477–481

Askew D, Ashmun R, Simmons B, Cleveland J (1991) Constitutive c-myc expression in IL-3-dependent myeloid cell lines suppresses cycle arrest and accelerates apoptosis. Oncogene 6: 1915–1922

Bachmann MF, Rohrer UH, Kundig TM, Burki K, Hengartner H, Zinkernagel RM (1993) The influence of antigen organization on B-cell responsiveness. Science 262: 1448–1451

Baixeras E, Kroemer G, Cuende E, Marquez C, Bosca L, Martinez JEA, Martinez AC (1993) Signal transduction pathways involved in B-cell induction. Immunol Rev 132: 5–47

Benhamou LE, Cazenave P-A, Sarthou P (1990) Anti-immunoglobulins induce death by apoptosis in WEHI-231 B lymphoma cells. Eur J Immunol 20: 1405–1407

Benhamou LE, Watanabe T, Kitamura D, Cazenave P-A, Sarthou P (1994) Signaling properties of anti immunoglobulin resistant variants of WEHI-231. Eur J Immunol 24: 1993–1999

Bijsterbosch MK, Klaus GGB (1987) Tumor-promoting phorbol esters suppress receptor-stimulated inositol phospholipid degradation and Ca^{2+} mobilization in mouse lymphocytes, Eur J Immunol 17: 113–118

Bishop GA, Ramirez LM, Koretzky GA (1993) Growth inhibition of a B-cell clone mediated by ligation of IL-4 receptors or membrane IgM. J Immunol 150: 2565–2574

Bonnefoy JY, Henchoz S, Hardie D, Holder MJ, Gordon J (1993) A subset of anti-CD21 antibodies promote the rescue of germinal center B-cells from apoptosis. Eur J Immunol 23: 969–972

Bonnefoy-Berard N, Genestier L, Flacher M, Revillard JP (1994) The phosphoprotein phosphatase Calcineurin controls calcium-dependent apoptosis in B cell lines. Eur J Immunol 24: 325–329

Bossy D, Salamero J, Olive D, Fougereau M, Schiff C (1993) Structure, biosynthesis, and transduction properties of the human mu-psi-L complex: similar behavior of preB and intermediate preB-B cells in transducing ability. Int Immunol 5: 467–478

Brines RD, Klaus GGB (1991) Effects of anti-immunoglobulin antibodies, interleukin-4 and second messenger agonists on B-cells from neonatal mice. Int Immunol 3: 461–466

Brines RD, Klaus GGB (1992) Inhibition of lipopolysaccharide-induced activation of immature B-cells by anti-mu and anti-delta antibodies and its modulation by interleukin-4. Int Immunol 4: 765–771

Brines RD, Klaus GGB (1993) Polyclonal activation of immature B-cells by preactivated T-cells: the role of IL-4 and CD40-ligand. Int Immunol 5: 1445–1450

Brink R, Goodnow CC, Crosbie J, Adams E, Eris J, Mason DY, Hartley SB, Basten A (1992) Immunoglobulin-M and immunoglobulin-D antigen receptors are both capable of mediating lymphocyte-B activation, deletion, or anergy after interaction with specific antigen. J Exp Med 176: 991–1005

Brouns GS, Devries E, Vannoesel CJM, Mason DY, Vanlier RAW, Borst J (1993) The structure of the mu/pseudo light chain complex on human pre-B cells is consistent with a function in signal transduction. Eur J Immunol 23: 1088–1097

Brunswick M, Finkelman FD, Highet PF, Inman JK, Dintzis HM, Mond JJ (1988) Picogram quantities of anti-Ig antibodies coupled to dextran induce B cell proliferation. J Immunol 140: 3364–3372

Callard RE, Armitage RJ, Fanslow WC, Spriggs MK (1993) CD40 ligand and its role in X-linked hyper-IgM syndrome. Immunol Today 14: 559–564

Cambier JC, Bedzyk W, Campbell K, Chien N, Friedrich J, Harwood A, Jensen W, Pleiman C, Clark MR (1993) The B-cell antigen receptor: structure and function of primary, secondary, tertiary and quaternary components. Immunol Rev 132: 85–106

Carsetti R, Kohler G, Lamers MC (1993) A role for immunoglobulin-D: interference with tolerance induction. Eur J Immunol 23: 168–178

Chang TL, Capraro G, Kleinman RE, Abbas AK (1991) Anergy in immature lymphocytes-B: differential responses to receptor-mediated stimulation and T-helper cells. J Immunol 147: 750–756

Clark EA, Ledbetter JA (1994) How B and T cells talk to each other. Nature 367: 425–428

Cohen JJ, Duke RC, Fadok VA, Sellins KS (1992) Apoptosis and programmed cell death in immunity. Annu Rev Immunol 10: 267–293

Cooke MP, Heath AW, Shokat KM, Zeng YJ, Finkelman FD, Linsley PS, Howard M, Goodnow CC (1994) Immunoglobulin signal transduction guides the specificity of B cell-T cell interactions and is blocked in tolerant self-reactive B cells. J Exp Med 179: 425–438

Cuende E, Kroemer G, Alonso JM, Nemazee D, Martinez C, Ales-Martinez JE (1992) Inability of IL-2 and IL-10 to counteract B-cell clonal deletion. Cell Immunol 142: 94–102

Cuende E, Ales-Martinez JE, Ding LY, Gonzalez-Garcia M, Martinez-A. C, Nunez G (1993) Programmed cell death by bcl-2-dependent and independent mechanisms in B-lymphoma cells. EMBO J 12: 1555–1560

Dancescu M, Rubiotrujillo M, Biron G, Bron D, Delespesse G, Sarfati M (1992) Interleukin-4 protects chronic lymphocytic leukemic B-cells from death by apoptosis and upregulates Bcl-2 expression. J Exp Med 176: 1319–1326

Desiderio S (1994) The B cell antigen receptor in B-cell development. Curr Opin Immunol 6: 248–256

Evan GI, Littlewood TD (1993) The role of c-myc in cell growth. Curr Opin Genet Dev 3: 44–49

Evan GI, Wyllie AH, Gilbert CS, Littlewood TD, Land H, Brooks M, Waters CM, Penn LZ, Hancock DC (1992) Induction of apoptosis in fibroblasts by c-myc protein. Cell 69: 119–128

Fanidi A, Harrington EA, Evan GI (1992) Cooperative interaction between c-myc and bcl-2 proto-oncogenes. Nature 359: 554–556

Fearon DT (1993) The CD19-CR2 TAPA-1 complex, CD45 and signaling by the antigen receptor of B-lymphocytes. Curr Opin Immunol 5: 341–348

Fischer G, Kent SC, Joseph L, Green DR, Scott DW (1994) Lymphoma models for B cell activation and tolerance .10. Anti-mu-mediated growth arrest and apoptosis of murine B cell lymphomas is prevented by the stabilization of myc. J Exp Med 179: 221–228

Fluckiger AC, Durand I, Banchereau J (1994) Interleukin 10 induces apoptotic cell death of B-chronic lymphocytic leukemia cells. J Exp Med 179: 91–99

Fukuda T, Kitamura D, Taniuchi I, Maekawa Y, Benhamou LE, Sarthou P, Watanabe T (1994) Complementation by the HS1 protein of a mutant B cell line resistant to anti-IgM induced apoptosis (submitted)

Gaur A, Yao XR, Scott DW (1993) B-cell tolerance induction by cross-linking of membrane IgM, but not IgD, and synergy by cross-linking of both isotypes. J Immunol 150: 1663–1669

Gay D, Saunders T, Camper S, Weigert M (1993) Receptor editing: an approach by autoreactive B-cells to escape tolerance. J Exp Med 177: 999–1008

Gold MR, DeFranco AL (1987) Phorbol esters and dioctanoylglycerol block anti-IgM-stimulated phosphoinositide hydrolysis in the murine B cell lymphoma WEHI-231. J Immunol 138: 868–876

Goodnow CC (1992) Transgenic mice and analysis of B-cell tolerance. Annu Rev Immunol 10: 489–518

Goodnow CC, Crosbie J, Adelstein S, Lavoie TB, Smith Gill SJ, Brink RA, Pritchard-Briscoe H, Wotherspoon JS, Loblay RH, Raphael K, Trent RJ, Basten A (1988) Altered immunoglobulin expression and functional silencing of self-reactive B lymphocytes in transgenic mice. Nature 334: 676–682

Gottschalk AR, McShan CL, Merino R, Nunez G, Quintans J (1994) Physiological cell death in B lymphocytes. 1. Differential susceptibility of WEHI-231 sublines to anti-Ig induced physiological cell death and lack of correlation with Bcl-2 expression. Int Immunol 6: 121–130

Green DR, Bissonnette RP, Glynn JM, Shi Y (1992) Activation-induced apoptosis in lymphoid systems. Semin Immunol 4: 379–388

Haggerty HG, Monroe JG (1994) Mutant of the WEHI-231 B lymphocyte line that is resistant to phorbol esters is still sensitive to antigen receptor-mediated growth inhibition. Cell Immunol 154: 166–180

Haggerty HG, Wechsler RJ, Lentz VM, Monroe JG (1993) Endogenous expression of delta on the surface of WEHI-231 characterization of its expression and signaling properties. J Immunol 151: 4681–4693

Harada N, Yang G, Miyajima A, Howard M (1992) Identification of an essential region for growth signal transduction in the cytoplasmic domain of the human interleukin-4 receptor. J Biol Chem 267: 22752–22758

Harnett MM, Holman MJ, Klaus GGB (1991) IL-4 promotes anti-Ig-mediated protein kinase-C translocation and reverses phorbol ester-mediated protein kinase-C down-regulation in murine B cells. J Immunol 147: 3831–3836

Hartley SB, Crosbie J, Brink R, Kantor AB, Basten A, Goodnow CC (1991) Elimination from peripheral lymphoid tissues of self-reactive lymphocytes-B recognizing membrane-bound antigens. Nature 353: 765–769

Hartley SB, Cooke MP, Fulcher DA, Harris AW, Cory S, Basten A, Goodnow CC (1993) Elimination of self-reactive lymphocytes-B proceeds in 2 stages. Arrested development and cell death. Cell 72: 325 335

Hasbold J, Klaus GGB (1990) Anti-immunoglobulin antibodies induce apoptosis in immature B lymphomas. Eur J Immunol 20: 1685–1690

Hibner U, Coutinho A (1994) Signal antinomy: a mechanism for apoptosis induction. Cell Death Differ 1: 33–37

Hibner U, Benhamou LE, Haury M, Cazenave PA, Sarthou P (1993) Signaling of programmed cell death induction in WEHI-231 B lymphoma cells. Eur J Immunol 23: 2821–2825

Hockenberry DM, Zutter M, Hickley W, Nahm M, Korsmeyer SJ (1991) Bcl-2 protein is topologically restricted in tissues characterized by apoptotic cell death. Proc Natl Acad Sci USA 88: 6961–6965

Hockenbery DM, Oltvai ZN, Yin XM, Milliman CL, Korsmeyer SJ (1993) Bcl-2 functions in an antioxidant pathway to prevent apoptosis. Cell 75: 241–251

Holder MJ, Wang H, Milner AE, Casamayor M, Armitage R, Spriggs MK, Fanslow WC, MacLennan ICM, Gregory CD, Gordon J (1993) Suppression of apoptosis in normal and neoplastic human B-lymphocytes by CD40 ligand is independent of Bcl-2 induction. Eur J Immunol 23: 2368–2371

Illera VA, Perandones CE, Stunz LL, Mower DA, Ashman RF (1993) Apoptosis in splenic lymphocytes-B: regulation by protein kinase-C and IL-4. J Immunol 151: 2965–2973

Ishihara K, Wood WJ, Wall R, Sakaguchi N, Michnoff C, Tucker PW, Kincade PW (1993) Multiple B29 containing complexes on murine B-lymphocytes: common and stage restricted Ig-associated polypeptide chains. J Immunol 150: 2253–2262

Itoh N, Tsujimoto Y, Nagata S (1993) Effect of bcl-2 on Fas antigen-mediated cell death. J Immunol 151: 621–627

Jongstra J, Misener V (1993) Developmental maturation of the B-cell antigen receptor. Immunol Rev 132: 107–123

Justement LB, Campbell KS, Chien NC, Cambier JC (1991) Regulation of B-cell antigen receptor signal transduction and phosphorylation by CD45. Science 252: 1839–1842

Kamesaki H, Zwiebel JA, Reed JC, Cossman J (1994) Role of Bcl 2 and IL-5 in the regulation of anti-IgM induced growth arrest and apoptosis in immature B cell lines—a cooperative regulation model for B cell clonal deletion. J Immunol 152: 3294–3305

Kammer GM (1988) The adenylate cyclase-cAMP-protein kinase A pathway and regulation of the immune response. Immunol Today 9: 222–229

Kanazashi S, Hata D, Ishigami T, Jung EY, Shintaku N, Sumimoto S, Heike T, Katamura K, Mayumi M (1994) Induction of phosphatidylinositol turnover and Egr-1 messenger RNA expression by crosslinking of surface IgM and IgD in the human B-cell line-B104. Mol Immunol 31: 21–30

Kelly K, Cochran BH, Stiles CD, Leder P (1983) Cell-specific regulation of the c-myc gene by lymphocyte mitogens and platelet-derived growth factors. Cell 35: 603–610

King LB, Ashwell JD (1993) Signaling for death of lymphoid cells. Curr Opin Immunol 51: 368–373

Kishihara K, Penninger J, Wallace VA, Kundig TM, Kawai K, Wakeham A, Timms E, Pfeffer K, Ohashi PS, Thomas ML, Furlonger C, Paige CJ, Mak TW (1993) Normal B-lymphocyte development but impaired T-cell maturation in CD45-exon6 protein tyrosine phosphatase-deficient mice. Cell 74: 143–156

Kitamura D, Kaneko H, Miyagoe Y, Ariyasu T, Watanabe T (1989) Isolation and characterization of a novel human gene expressed specifically in the cells of hematopoietic lineage. Nucleic Acids Res 17: 9367–9379

Knox KA, Gordon J (1993) Protein tyrosine phosphorylation is mandatory for CD40-mediated rescue of germinal center B-cells from apoptosis. Eur J Immunol 23: 2578–2584

Knox KA, Finney M, Milner AE, Gregory CD, Wakelam MJO, Michell RH, Gordon J (1992) 2nd-messenger pathways involved in the regulation of survival in germinal centre B cells and in Burkitt lymphoma lines. Int J Cancer 52: 959–966

Knox KA, Johnson CD, Gordon J (1993) Distribution of cAMP in secondary follicles and its expression in B-cell apoptosis and CD40-mediated survival. Int Immunol 5: 1085–1091

Korsmeyer SJ (1992) Bcl-2—A repressor of lymphocyte death. Immunol Today 13: 285–288

Krammer PH, Behrmann I, Daniel P, Dhein J, Debatin K-M (1994) Regulation of apoptosis in the immune system. Curr Opin Immunol 6: 279–289

Lederman S, Yellin MJ, Covey LR, Cleary AM, Callard R, Chess L (1993) Nonantigen signals for B-cell growth and differentiation to antibody secretion. Curr Opin Immunol 5: 439–444

Lederman S, Yellin MJ, Cleary AM, Pernis A, Inghirami G, Cohn LE, Covey LR, Lee JJ, Rothman P, Chess L (1994) T-Bam/CD40-L on helper T lymphocytes augments lymphokine-induced B cell Ig isotype switch recombination and rescues B cells from programmed cell death. J Immunol 152: 2163–2171

Lee MR, Liou ML, Liou ML, Yang YF, Lai MZ (1993) cAMP analogs prevent activation-induced apoptosis of T-cell hybridomas. J Immunol 151: 5208–5217

Levy Y, Brouet JC (1994) Interleukin-10 prevents spontaneous death of germinal center B cells by induction of the Bcl-2 protein. J Clin Invest 93: 424–428

Liou LB, Warner GL, Scott DW (1992) Can peritoneal B-cells be rendered unresponsive. Int Immunol 4: 15–21

Liu YJ, Joshua DE, Williams GT, Smith CA, Gordon J, MacLennan ICM (1989) Mechanism of antigen-driven selection in germinal centres. Nature 342: 929–931

Liu YJ, Mason DY, Johnson GD, Abbot S, Gregory CD, Hardie DL, Gordon J, MacLennan ICM (1991) Germinal center cells express bcl-2 protein after activation by signals which prevent their entry into apoptosis. Eur J Immunol 21: 1905–1910

Mangan DF, Robertson B, Wahl SM (1992) IL-4 enhances programmed cell death (apoptosis) in stimulated human monocytes. J Immunol 148: 1812–1816

Mapara MY, Bargou R, Zugck C, Dohner H, Ustaoglu F, Jonker RR, Krammer PH, Dorken B (1993) APO-1 mediated apoptosis or proliferation in human chronic B-lymphocytic leukemia. Correlation with bcl-2 oncogene expression. Eur J Immunol 23: 702–708

Marquez C, Martinez C, Kroemer G, Bosca L (1992) Protein kinase-C isoenzymes display differential affinity for phorbol esters. Analysis of phorbol ester receptors in B-cell differentiation. J Immunol 149: 2560–2568

Matsuo T, Nomura J, Kuwahara K, Igarashi H, Inui S, Hamaguchi M, Kimoto M, Sakaguchi N (1993) Cross-linking of B-cell receptor-related MB-1 molecule induces protein tyrosine phosphorylation in early B-lineage cells. J Immunol 150: 3766–3775

McConkey DJ, Orrenius S, Jondal M (1990) Cellular signalling in programmed cell death (apoptosis). Immunol Today 11: 120–121

McConkey DJ, Aguilar-Santelises M, Hartzell P, Eriksson I, Mollstedt H, Orrenius S, Jondal M (1991) Induction of DNA fragmentation in chronic B-lymphocytic leukemia cells. J Immunol 146: 1072–1076

McConkey DJ, Orrenius S, Okret S, Jondal M (1993) Cyclic AMP potentiates glucocorticoid-induced endogenous endonuclease activation in thymocytes. FASEB J 7: 580–585

Melchers F, Karasuyama H, Haasner D, Bauer S, Kudo A, Sakaguchi N, Jameson B, Rolink A (1993) The surrogate light chain in B-cell development. Immunol Today 14: 60–68

Merino R, Ding LY, Veis DJ, Korsmeyer SJ, Nunez G (1994) Developmental regulation of the Bcl-2 protein and susceptibility to cell death in B-lymphocytes. EMBO J 13: 683–691

Metcalf ES, Klinman NR (1977) In vitro tolerance of bone marrow cells: a marker for B cell maturation. J Immunol 118: 2111–2116

Metzger H (1992) Transmembrane signaling: the joy of aggregation. J Immunol 149: 1477–1487

Milner AE, Grand RJA, Waters CM, Gregory CD (1993) Apoptosis in Burkitt lymphoma cells is driven by c-myc. Oncogene 8: 3385–3391

Mire-Sluis AR, Thorpe R (1991) Interleukin-4 proliferative signal transduction involves the activation of a tyrosine-specific phosphatase and the dephosphorylation of an 80-kDa protein. J Biol Chem 266: 18113–18118

Misener V, Downey GP, Jongstra J (1991) The immunoglobulin light chain related protein lambda 5 is expressed on the surface of mouse pre-B-cell lines and can function as a signal transducing molecule. Int Immunol 3: 1129–1136

Mizuguchi J, Yong-Yong J, Nakabayaschi H, Huang K-P, Beaven MA, Chused T, Paul WE (1987) Protein kinase C activation blocks anti-IgM-mediated signaling in BAL 17 B lymphoma cells. J Immunol 139: 1054–1059

Mond JJ, Feuerstein N, Finkelman FD, Huang F, Huang K-P. Dennis G (1987) B-lymphocyte activation mediated by anti-immunoglobulin antibody in the absence of protein kinase C. Proc Natl Acad Sci USA 84: 8588–8592

Mongini PKA, Blessinger CA, Dalton JP, Seki T (1992) Differential effects of cyclosporin A on diverse B cell activation phenomena triggered by crosslinking of membrane IgM. Cell Immunol 140: 478–494

Morris DL, Rothstein TL (1993) Abnormal transcription factor induction through the surface immunoglobulin-M receptor of lymphocytes-B 1. J Exp Med 177: 857–861

Motyka B, Griebel PJ, Reynolds JD (1993) Agents that activate protein kinase C rescue sheep ileal Peyer's patch B-cells from apoptosis. Eur J Immunol 23: 1314–1321

Murakami M, Tsubata T, Okamoto M, Shimizu A, Kumagai S, Imura H, Honjo T (1992) Antigen-induced apoptotic death of Ly-1 B cells responsible for autoimmune disease in transgenic mice. Nature 357: 77–80

Muthukkumar S, Udhayakumar V, Bondada S (1993) Elevation of cytosolic calcium is sufficient to induce growth inhibition in a B-cell lymphoma. Eur J Immunol 23: 2419–2426

Muthusamy N, Bondada S (1993) Differential regulation of surface immunoglobulin and Lyb2 mediated B-cell activation 2. cAMP dependent (Prostaglandin-E2) and independent (IFN-gamma) mechanisms of regulation of B-lymphocyte activation. Int Immunol 5: 949–956

Nakamura T, Sekar MC, Kubagawa H, Cooper MD (1993) Signal transduction in human B cells initiated via Ig beta ligation. Int Immunol 5: 1309–1315

Nakayama K, Nakayama K, Negishi I, Kuida K, Shinkai Y, Louie MC, Fields LE, Lucas PJ, Stewart V, Alt FW, Loh DY (1993) Disappearance of the lymphoid system in Bcl-2 homozygous mutant chimeric mice. Science 261: 1584–1588

Nemazee DA (1993) Promotion and prevention of autoimmunity by B-lymphocytes. Curr Opin Immunol 5: 866–872

Nemazee DA, Bürki K (1989a) Clonal deletion of B lymphocytes in a transgenic mouse bearing anti-MHC class I antibody genes. Nature 337: 562–566

Nemazee DA, Bürki K (1989b) Clonal deletion of autoreactive B lymphocytes in bone marrow chimeras. Proc Natl Acad Sci USA 86: 8039–8043

Newell MK, Vandemall J, Beard KS, Freed JH (1993) Ligation of major histocompatibility complex Class-II molecules mediates apoptotic cell death in resting B-lymphocytes. Proc Natl Acad Sci USA 90: 10459–10463

Nisitani S, Tsubata T, Murakami M, Okamoto M, Honjo T (1993) The bcl-2 gene product inhibits clonal deletion of self-reactive B-lymphocytes in the periphery but not in the bone marrow. J Exp Med 178: 1247–1254

Noelle RJ, Ledbetter JA, Aruffo A (1992) CD40 and its ligand, an essential ligand-receptor pair for thymus-dependent B-cell activation. Immunol Today 13: 431–433

Nossal GJV (1983) Cellular mechanisms of immunological tolerance. Annu Rev Immunol 1: 33–62

Nossal GJV (1994) Negative selection of lymphocytes. Cell 76: 229–239

Nossal GJV, Pike BL (1975) Evidence for the clonal abortion theory of B lymphocyte tolerance. J Exp Med 141: 904–917

Nossal GJV, Pike BL (1980) Clonal anergy: persistence in tolerant mice of antigen-binding B lymphocytes incapable of responding to antigen or mitogen. Proc Natl Acad Sci USA 87: 1602–1606

Ogimoto M, Katagiri T, Mashima K, Hasegawa K, Mizuno K, Yakura H (1994) Negative regulation of apoptotic death in immature B cells by CD45. Int Immunol 6: 647–654

Parker DC (1975) Stimulation of mouse lymphocytes by insoluble anti-mouse immunoglobulin. Nature 258:361–363

Parry SL, Holman MJ, Hasbold J, Klaus GGB (1994a) Plastic-immobilized anti-mu or anti-delta antibodies induce apoptosis in mature murine B lymphocytes. Eur J Immunol 24: 974–979

Parry SL, Hasbold J, Holman M, Klaus GGB (1994b) Hypercross-linking surface IgM or IgD receptors on mature B cells induces apoptosis that is reversed by costimulation with IL-4 and anti-CD40. J Immunol 152: 2821–2829

Phipps RP, Lee D, Schad V, Warner CL (1989) E-series prostaglandins are potent growth inhibtors for some B lymphomas. Eur J Immunol 19: 995–1001

Phipps RP, Roper RL, Stein SH (1990) Regulation of B-cell tolerance and triggering by macrophages and lymphoid dendritic cells. Immunol Rev 117: 135–158

Pure E, Vitetta E (1980) Induction of murine B cell proliferation by insolubilized anti-immunoglobulins. J Immunol 125: 1240–1242

Radic MZ, Erikson J, Litwin S, Weigert M (1993) B-lymphocytes may escape tolerance by revising their antigen receptors. J Exp Med 177: 1165–1173

Reed JC (1994) Bcl-2 and the regulation of programmed cell death. J Cell Biol 124: 1–6

Ron CL, Morio T, Fu SM, Goha RS (1994) Signal transduction via CD40 involves activation of Lyn kinase and phosphatidylinositol 3-kinase, and phosphorylation ot phospholipase C gamma 2. J Exp Med 179: 673–680

Reth M (1992) Antigen receptors on B lymphocytes. Annu Rev Immunol 10: 97–121

Reth M (1994) B cell antigen receptors. Curr Opin Immunol 6: 3–8

Rolink A, Melchers F (1991) Molecular and cellular origins of B lymphocyte diversity. Cell 66: 1081–1094

Roper RL, Phipps RP (1992) Prostaglandin-E2 and cAMP inhibit lymphocyte-B activation and simultaneously promote IgE and IgG1 synthesis. J Immunol 149: 2984–2991

Rubin LL, Philpott KL, Brooks SF (1993) The cell cycle and cell death. Curr Biol 3: 391–394

Sarthou P, Henry-Toulmé N, Cazenave P-A (1989) Membrane IgM cross-linking is not coupled to protein kinase C translocation in WEHI-231 B lymphoma cells. Eur J Immunol 19: 1247–1252

Schad VC, Phipps RP (1988) Two signals are required for accessory cells to induce B cell un-responsiveness. Tolerogenic Ig and prostaglandin. J Immunol 141: 79–84

Schreiber SL, Crabtree GR (1992) The mechanism of action of cyclosporin A and FK506. Immunol Today 13: 136–142

Schwartz LM, Osborne BA (1993) Programmed cell death, apoptosis and killer genes. Immunol Today 14: 582–590

Schwartz RS, Stollar BD (1994) Heavy-chain directed B-cell maturation: continuous clonal selection beginning at the pre-B cell stage. Immunol Today 15: 27–32

Scott DW, O'Garra A, Warren D, Klaus GGB (1987) Lymphoma models for B cell activation and tolerance-VI. Reversal of anti-Ig mediated negative signaling by T cell-derived lymphokines. J immunol 139: 3924–3929

Shi Y, Glynn JM, Guilbert LJ, Cotter TG, Bissonnette RP, Green DR (1992) Role for c-myc inactivation-induced apoptotic cell death in T cell hybridoma. Science 257: 212–214

Strasser A, Harris AW, Corcoran LM, Cory S (1994) Bcl-2 expression promotes B-lymphoid but not T-lymphoid development in scid mice. Nature 368: 457–460

Taga K, Cherney B, Tosato G (1993) IL-10 inhibits apoptotic cell death in human T cells starved of IL-2. Int Immunol 5: 1599–1608

Terajima J, Tsutsumi A, Freiremoar J, Cherwinski HM, Ransom JT (1992) Evidence for clonal heterogeneity of the expression of 6 protein kinase C isoforms in murine-B and lymphocytes T. Cell Immunol 142: 197–206

Tiegs SL, Russell DM, Nemazee D (1993) Receptor-editing in self-reactive bone marrow B cells, J Exp Med 177: 1009–1020

Tisch R, Roifman CM, Hozumi N (1988) Functional differences between immunoglobulins M and D expressed on the surface of an immature B-cell line. Proc Natl Acad Sci USA 85: 6914–6918

Trauth BC, Klas C, Peters AMJ, Matzku S, Möller P, Falk W, Debatin K-M, Krammer PH (1989) Monoclonal antibody-mediated tumor regression by induction of apoptosis. Science 245: 301–305

Trump BF, Berezesky IK (1992) The role of cytosolic Ca^{2+} in cell injury, necrosis and apoptosis. Curr Opin Cell Biol 4: 227–232

Tsubata T, Wu J, Honjo T (1993) B-cell apoptosis induced by antigen receptor crosslinking is blocked by a T-cell signal through CD40. Nature 364: 645–648

Tsubata T, Murakami M, Honjo T (1994) Antigen-receptor cross-linking induces peritoneal B-cell apoptosis in normal but not autoimmunity-prone mice. Curr Biol 4: 8–17

Tsutsumi A, Terajima J, Jung WM, Ransom J (1992a) Surface mu heavy chain expressed on pre B lymphomas transduces Ca^{2+} signals but fails to cause growth arrest of pre-B lymphomas. Cell Immunol 139: 44–57

Tsutsumi A, Freiremoar J, Ransom JT (1992b) Transient down-regulation of PKC-zeta RNA following crosslinking of membrane IgM on WEHI-231 B-lymphoma cells. Cell Immunol 142: 303–312

Udhayakumar V, Kumar L, Subbarao B (1991a) The influence of avidity on signaling murine lymphocytes-B with monoclonal anti-IgM antibodies—Effects on B-cell proliferation versus growth inhibition (tolerance) of an immature B-cell lymphoma. J Immunol 146: 4120–4129

Udhayakumar V, Muthukkumar S, Subbarao B (1991b) Cyclosporin-A blocks surface IgM-mediated growth inhibition in an immature B-lymphoma, BKS-2. Eur J Immunol 21: 2605–2608

Valentine MA, Licciardi KA (1992) Rescue from anti-IgM-induced programmed cell death by the B-cell surface proteins CD20 and CD40. Eur J Immunol 22: 3141–3148

Van Endert PM, Moldenhauer G (1992) Inhibitory and stimulatory signaling via immunoglobulin receptors. Dichotomous responses elicited in clonal B-cell populations. Eur J Immunol 22: 1229–1235

Veis DJ, Sentman CL, Bach EA, Korsmeyer SJ (1993a) Expression of the Bcl-2 protein in murine and human thymocytes and in peripheral T-lymphocytes. J Immunol 151: 2546–2554

Veis DJ, Sorenson CM, Shutter JR, Korsmeyer SJ (1993b) Bcl-2-deficient mice demonstrate fulminant lymphoid apoptosis, polycystic kidneys, and hypopigmented hair. Cell 75: 229–240

Wade WF, Davoust J, Salamero J, Andre P, Watts TH, Cambier JC (1993) Structural compartmentalization of MHC Class-II signaling function. Immunol Today 14: 539–546

Warner GL, Scott DW (1988) Lymphoma models for B-cell activation and tolerance-VII. Pathways in anti-Ig-mediated growth inhibition and its reversal. Cell Immunol 115: 195–203

Wheeler K, Pound JD, Gordon J, Jefferis R (1993) Engagement of CD40 lowers the threshold for activation of resting B-cells via antigen receptor. Eur J Immunol 23: 1165–1168

Williams GT, Smith CA (1993) Molecular regulation of apoptosis: genetic controls on cell death. Cell 74: 777–779

Yamanashi Y, Okada M, Semba T, Yamori T, Umemori H, Tsunasawa S, Toyoshima K, Kitamura D, Watanabe T, Yamamoto T (1993) Identification of HS 1 protein as a major substrate of proteintyrosine kinase(s) upon B-cell antigen receptor-mediated signaling. Proc Natl Acad Sci USA 90: 3631–3635

Yao XR, Scott DW (1993) Antisense oligodeoxynucleotides to the blk tyrosine kinase prevent anti-mu chain-mediated growth inhibition and apoptosis in a B-cell lymphoma. Proc Natl Acad Sci USA 90: 7946–7950

Yellen AJ, Glenn W, Sukhatme VP, Cao XM, Monroe JG (1991) Signaling through surface IgM in tolerance-susceptible immature murine B-lymphocytes. Developmentally regulated differences in transmembrane signaling in splenic B-cells from adult and neonatal mice. J Immunol 146: 1446–1454

Yellen-Shaw A, Monroe JG (1992) Differential responsiveness of immature-stage and mature-stage murine B cells to anti-IgM reflects both FcR-dependent and FcR-independent mechanisms. Cell Immunol 145: 339–350

Characterization of Signals Leading to Clonal Expansion or to Cell Death During Lymphocyte B Cell Activation

L. Bosca[1], C. Stauber[2], S. Hortelano[1], E. Baixeras[2], and C. Martinez-A.[2]

1 Introduction

The immune system is endowed with a multitude of different mechanisms to eliminate, paralyze or neutralize T and B lymphocytes expressing self-reactive antigen receptors that might endanger the individual's life. The ability of both types of lymphocytes to recognize and react to different stimuli is a learning process that occurs during lymphocyte differentiation, and the mechanisms implicated in self-tolerance intervene at determined control points following developmental criteria. B and T lymphocyte differentiation from committed precursor cells into antibody-secreting plasma cells or effector T cells proceeds through multiple steps that are defined by changes in the expression pattern of lineage specific genes (Moller 1994).

Antigen-independent stages of B cell maturation take place in the bone marrow, where lymphoid precursors commit to the B lineage and subsequently differentiate into surface IgM+ B cells. This differentiation process includes the transition from pro-B cells (stage at which the Ig genes are in germline configuration) towards pre-B cells (where VDJ recombination of the Ig variable region in the heavy chain locus generates υ chains that are expressed on the cell surface in association with the surrogate light chains, VpreB and 15. Finally,

[1] Instituto de Bioquimica, CSIC, Universidad Complutense, 28040 Madrid, Spain
[2] Centro Nacional de Biolecnologia CSIC, Universidad Autonoma, Campus de Cantoblanco, 28049 Madrid, Spain

rearrangements of the κ or λ chains take place and the IgM receptor is displayed on the surface of immature B cells. The transition from immature to mature B cells is accompanied by Ig heavy chain class switching. Later, mature B cells migrate out from bone marrow into the periphery (spleen and lymph nodes), where an antigen-dependent phase of B cell development takes place (MOLLER 1994).

During development, B cells are exposed to intense selection within the bone marrow so that potentially autorreactive cells are induced to undergo programmed cell death (PCD) while nonself-reactive lymphocytes are exported to the periphery. These observations raise interesting questions as to what determines these very different responses and what is the nature of the signaling pathways involved in the maturation process (BAIXERAS et al. 1993).

Identification of the mechanisms that control B cell survival at different stages of proliferation and differentiation of the precursors is crucial for the understanding of B cell biology. The protective effect over apoptosis exerted through CD40 signaling of B cells activated upon antigen-receptor cross-linking is illustrative (CLARK and LEDBETTER 1994; TSUBATA et al. 1993) CD40 is expressed in both pre-B cells and mature B cells, and activation through this molecule prevents apoptosis not only in circulating B cells but also in immature B cell lines such as WEHI-231 cells (CLARK and LEDBETTER 1994; TSUBATA et al. 1993). CD40 is closely related to the tumor necrosis factor receptor, whereas its natural ligand (CD40L) is structurally related to the family of tumor necrosis factor α molecules. CD40L is expressed on the cell surface of activated T cells, but not on resting T cells, therefore providing costimulatory signals in the process of B cell-T cell interaction (CLARK and LEDBETTER 1994; JENKINS and JOHNSON 1993; TSUBATA et al. 1993).

2 B Cell Activation

Mice that cannot rearrange variable region genes such as SCID mice or mice homozygous for disrupted RAG.1- / - or RAG.2- / - lack mature B cells. This, together with the restoration of B cell maturation in RAG.2 deficient mice by transgenic BCR, constitutes compelling evidence that B cells must display functional antigen receptors in order to complete differentiation (LOFFERT 1994; MÖLLER 1994). The ability of surface υ polypeptide to mediate these effects may depend on associated membrane proteins, which together with the υ polypeptide constitutes the B cell receptor complex. Surface immunoglobulin receptor facilitates differentiation by transducting intracellular signals after binding to ligands. However, binding of antigenic structures to BCR does not lead to differentiation or clonal expansion, rather it drives B cells into apoptosis. Only upon appropriate combinations of signals delivered by activated T cells and/or macrophages will B-lymphocytes initiate antibody production and isotype switching (BAIXERAS et al. 1993).

Escape from PCD occurs when specific combinations of surface molecules are ligated in concert with surface immunoglobulin. Many studies using soluble blocking antibodies, hybrid antibodies and antibodies bound to plastic have identified numerous molecules that may contribute to lymphocyte activation and are classified as costimulatory molecules (KRAMMER et al. 1994; KROEMER and MARTINEZ-A 1994). A few of these molecules act together with the antigen-specific signal to prevent induction of anergy in the responding cell. They included the B cell surface molecules B7/BB1, heat stable antigen, CD20, CD40 and CD2 (BAIXERAS et al. 1993). The transduction of signals involved in costimulatory interactions has been shown to influence the expression of early genes such as c-myc, c-myb or B-myb (GOLAY et al. 1992). Also, the levels of various transcription factors are controlled by these interactions; for example AP-1 expression is sustained after costimulation using protein kinase C and cyclic AMP-dependent protein kinase pathways (RINCON 1993).

Recently we have derived a system where purified splenic B cells, when confronted with nominal soluble antigen, undergo extensive PCD (GENARO and BOSCA 1993). In contrast, when the same antigen is presented on the surface of either B or T cells, the responding B cells undergo extensive clonal expansion. We have characterized the differential signals that will drive the B cells either into proliferation or PCD, This system can be extensively studied in B cells from υ/κ transgenic mice specific for H-2K^k haplotype that, upon stimulation with purified soluble MHC-1 alloantigen of the K^k specificity, initiates a rapid process that ends in PCD (BAIXERAS et al. 1993).

To date, two classes of PCD inhibitory costimuli have been well characterized in B cells: those mediated by soluble mediators, including cytokines, and those received via cell surface receptors like CD40 and CD2. Hereby we review the role that a soluble mediator, nitric oxide, a cell surface receptor CD2 and protein kinase C (PKC) plays in preventing apoptosis in B cells.

3 Nitric Oxide as a Mediator in the Immune System: Implications in Autoimmunity

Nitric oxide (NO) constitutes an important signaling molecule in a variety of cell systems, including the immune system (NATHAN 1992). NO was unexpectedly discovered as the molecule responsible for the vasodilation produced by acetyl-choline and other neurotransmitters in the presence of endothelial cells (FURCHGOTT 1988; MONCADA 1992; MONCADA et al. 1991). The NO generating system is now well identified but characterization of its biological and patho-physiological role is still in progress (BREDT and SNYDER 1992; HOFFMAN et al. 1990; MONCADA et al. 1991). At present, NO is considered an important intra- and intercellular regulatory molecule exhibiting functions as diverse as vasodilation neural communication, host defense and immunoregulation. NO is synthesized by many different cell types such as neurons, endothelial cells and monocytes, although the regulatory mechanism controlling its synthesis varies in different tissues (BREDT et al. 1990, LOWENSTEIN and SNYDER 1992; MONCADA et al. 1991; MONCADA 1992).

NO synthase, the enzyme involved in the production of NO from molecular O_2 and arginine, belongs to a growing family of isoenzymes all sharing structural and functional homology and conserving the same chemical reaction. By cDNA analysis at least four isoforms have been identified in mammalian tissues. They are encoded by at least three distinct genes, which in turn define the main characteristics of NO synthase's synthesis (KNOWLES and MONCADA 1992; LOWENSTEIN and SNYDER 1992): two genes encode the constitutively expressed enzymes, and another gene is responsible for expression of the inducible form of NO synthase. Neural and endothelial cells express both forms of the constitutive enzyme, which requires Ca^{2+} and calmodulin to be active. The cytokine-inducible isoenzyme is expressed in several cell types including macrophages, monocytes and hepatocytes and is induced upon stimulation with a wide array of cytokines, e.g., interferon-γ (IFN-γ), tumor necrosis factor-α and endotoxins (lipopoly-saccharide, LPS). This form is Ca^{2+} and calmodulin independent (BILLIAR et al. 1990; HAUSCHILDT et al. 1990; MARLETTA et al. 1988), and its activity is mainly controlled by transcriptional mechanisms and by substrate (arginine) availability (ALBINA et al. 1993; LYONS et al. 1992; XIE et al. 1993).

Since NO is a gaseous substance it acts not only on the agonist-stimulated producing cell (i.e., macrophages or dendritic cells) but also, through a diffusion process, may exert its physiological action over neighboring cells. This mode of action defines a new type of intercellular communication mechanism in which the synthesis of second messengers by the responding cell is achieved in the absence of additional transmembrane signaling events required to perceive extracellular messages (BREDT and SNYDER 1989,1992). This type of communication is especially important in the immune system where intercellular cognate recognition may provide an additional way to promote cell contact and to perceive the release of this messenger (KNOWLES and MONCADA 1992; KNOWLES et al. 1989; PALMER 1993; PALMER et al. 1987).

Expression of the Ca^{2+}-independent, cytokine-inducible NO synthase has been described in various cell types, in addition to macrophages (STUEHR et al. 1991; STUEHR and MARLETTA 1985); however, the current view, that the enzyme induced by cytokines is the same isotype in all tissues, is doubtful. In fact, using different mice strains we have identified at least three different species of mRNA probably generated by differential splicing (M. Velasco et al., unpublished observations). Furthermore, the NO synthase induced in interleukin-1-stimulated human hepatocytes exhibits an important degree of Ca^{2+}/calmodulin dependence, in contrast to the independence displayed by the enzyme expressed in macrophages (GELLER et al. 1993a,b). Finally, it is also possible that all forms of NO synthase so far characterized require Ca^{2+} and calmodulin to be active, but the interactions of these cofactors with the enzyme exhibits a broad range of affinity. In the case of the inducible isoenzyme, the affinity for calmodulin is so high that the enzyme appears as an oligomer with calmodulin tightly bound as a subunit.

Macrophage inducible NO synthase is by far the most well characterized isoenzyme among the cytokine-inducible forms, both from the biological and chemical points of view. The complex regulatory mechanism implicated in the control of its expression by cytokines and endotoxins has also been extensively studied (DING 1990; MARLETTA et al. 1988; GELLER 1993b: NATHAN 1992). The ability of different macrophage-like cell lines such as RAW 264.7 to release NO after activation allows detailed study of the mechanisms of response to combinations of cytokines and endotoxins. Thus, an extensive and complex relationship between individual factors, acting synergistically in most cases, has been revealed. Specifically, combinations of IFN-γ, INF-α and LPS produced one of the highest inductions of the enzyme in macrophages, in agreement with its role in host defense (LIEW et al. 1991; LYONS et al. 1992).

Interestingly, in addition to the short-term regulation, the main difference between the constitutive and inducible enzyme activities is the amount of NO released, quantitatively more important in cells expressing the inducible enzyme. The biological role of the NO released by these cells is more difficult to understand than that of the NO which is constitutively produced. NO, in addition to promoting the activation of guanylate cyclase, inhibits enzymes (aconitase, ribonucleotide reductase, ADP-ribosylation of proteins (DRAPIER and HIBBS 1986; LEPOIVREM et al. 1990; BRUNE and LAPETINA 1989), metabolic pathways (mitochondrial respiration, DNA synthesis in some types of cells; (GARG and HASSID 1989; GRANGER et al. 1980), and, presumably through these actions, participates in a vast array of processes, includinig host defense, autoimmunity and rejection of engrafted tissues (HOFFMAN et al. 1990; MCCARTNEY et al. 1993, LANGREHR et al. 1992; WEINBERG et al. 1994).

As previously indicated, one quantitatively important source of NO in the immune system is activated macrophages. NO plays an important role in antimicrobial immunity and in nonseptic inflammatory reactions (LOWENSTEIN et al. 1994; MONCADA and HIGGS 1993). Upon macrophage activation with LPS and IFN-γ, NO synthase induction is maximal and large amounts of NO are released. In contrast, glucocorticoids and Th2 cytokines such as interleukin-4 (IL-4), IL-10 and IL-13 inhibit NO synthase expression (MONCADA 1992; NATHAN 1992). In this way, a cross-modulation between Th1 cells, by increasing NO synthetase expression and the NO generating system, and Th2 cells, by inhibiting NO synthase expression, seems to operate in macrophages. This situation is of physiopathological relevance since it is possible that this pathway is functional in the response of the host in cases such as leishmaniasis and other parasitic pathologies in which the Th1/Th2 ratio is critical (LIEW et al. 1991).

An antitumoricidal activity for NO has been reported for various cell types (JONATHAN et al. 1994). NO is also involved in cell proliferation, the effect depending on the nature of the target cell. When murine splenic cells are activated with concanavalin A (ConA) or LPS in the presence of macrophages, proliferation is supressed due to the NO released by macrophages. The synthesis of IFN-γ by T cells seems to play a prominent role in NO synthase expression in macrophages, since the presence of anti-IFN-γ antibodies blocks NO production and prevents the antiproliferative role of macrophages on T cells (ALBINA et al. 1991).

A role for NO in the pathogenesis of spontaneous murine autoimmune disease has been reported for MRL-lpr/lpr (LANGREHR et al. 1992; WEINBERG et al. 1994). This strain of mice exhibits spontaneous autoimmune diseases involving lymphadenopathy, production of autoantibodies, arthritis, nephritis and other inflammatory dysfunctions. These animals have elevated plasma levels of IFN-γ, TNF-α and IL-1 and IL-6. At the molecular and genetic levels, part of the dysfunctions in MRL-lpr/lpr animals are due to a mutation in the Fas/Apo.1 gene (WATANABE-FUKUNAGA et al. 1992; WATSON et al. 1992). Characteristic of these animals is the presence of high levels of nitrites and nitrates in the blood and secretion of large amounts of these metabolites through the urine. By treating the mice with aminoguanidine or N-nitroarginie, two NO synthase inhibitors, some of the pathological symptoms associated with NO production (i.e., arthritis) significantly decrease. The high NO synthesis in MRL-lpr/lpr animals has been attributed to elevated levels of inducible NO synthase in various tissues (ISCHIROPOULOS et al. 1992). It is interesting to mention that macrophages, when activated under physiological conditions, release arginase, thereby reducing the substrate concentration required for NO synthesis. However, whether this is also true for MRL-lpr/lpr mice remains to be established (ALBINA et al. 1993; LYONS et al. 1992; XIE et al. 1993). Indeed, in addition to NO the generation of other oxygen reactive species is increased and they may participate in the peculiar pathogenesis of disease in these animals. For instance, the simultaneous presence of NO and H_2O_2 may produce peroxinitrites, a derivative of both reactive molecules which has been proposed to play a relevant role in the development of the disease in lpr mice (ISCHIROPOULOS et al. 1992).

4 Role of Nitric Oxide in B Cell Deletion

In contrast to activated macrophages, NO synthase is only poorly induced in B cells. Nevertheless, in LPS-activated B cells, NO synthase is significantly induced a few hours after stimulation and NO is released into the medium (Hortelano et al., unpublished observations).The role of NO in B cell function can be easily studied with the help of substances that intracellularly release NO as result of the activation of cellular esterases. Ex vivo purified B cells, in the absence of stimulation, after 4 h in vitro, initiate a series of changes that lead to PCD. Under these conditions, micromolar concentrations of NO block apoptosis. The release of NO induced in B cells by chemical donors is accompanied by an increase of intracellular levels of cyclic GMP, which is a good indicator for the presence of NO and which in turn may act as an additional second messenger (TSOU et al. 1993;

BREDT and SNYDER 1989). Moreover, the inhibition of apoptosis by NO is also observed when B cells are stimulated with aggregated antigens in the absence of the costimulatory signals (see above). A possible mechanism underlying protection from apoptosis by NO might involve elevated levels of Bcl-2 (GENARO et al. 1994, KELSOE and ZHENG 1993, MERINO et al. 1994, MOLLER 1992: NUÑEZ et al. 1990b). In fact, both in naive B cells and in B cells activated with soluble MHC-I alloantigen, the mRNA and protein levels of Bcl-2 are maintained over a long period of time (4–8 h) (KELSOE and ZHENG 1993; GENARO et al. 1994; MOLLER 1992).

The protective role of NO against PCD in B cells contrasts with the NO-dependent induction of apoptosis observed in other cell types such as macrophages (ALBINA et al. 1993; MERINO et al. 1994; NUÑEZ et al. 1990a), suggesting the existence of cell-specific pathways for the response to NO. Such a bivalent role of NO in macrophages and B cells is not unique, because in neurons both a neuroprotective and neurodestructive effects of NO have been reported. In this case the complex behavior has been explained on the basis of the different redox states of NO once it is released into the cytoplasm of the cell (LIPTON et al. 1993). Moreover, it is also possible that the involvement of other signaling molecules can influence regulation of the expression of the inducible form of NO synthase, as reported for the complex dual stimulation of macrophages with traces of LPS and IFN (BOGDAN et al. 1993). In conclusion, the generation of low but sustained amounts of NO may prolong the survival of B cells in secondary lymphoid organs.

5 Protein Kinase C Activation Prevents Apoptosis in B Lymphocytes

The role of protein kinase C (PKC) in B cell activation through the antigen receptor or by bacterial products (LPS and lipoproteins) is still unclear (BAIXERAS et al. 1993; MARQUEZ et al. 1992). The initial events after activation by antigens, LPS or synthetic lipopeptides, involve tyrosine phosphorylation, which in turn may deliver second messengers that activate PKC (DONG et al. 1993). However, if the temporal pattern of signal transduction is altered (i.e., cells are stimulated with phorbol esters), a complete blockage in B cell triggering is obtained. These data support the view of PKC as a modulatory step in the signaling process and suggest that only specific isoforms of PKC might participate in each specific activatory pathway (MARQUEZ et al. 1992). In addition to a modulatory role of PKC in signaling through the B cell receptor, it is possible that other receptor-operated interactions may cause PKC activation. In this regard, it cannot be excluded that some costimulatory signals may involve PKC activation in their mechanism (CLARK and LANE 1991; CLARK and LEDBETTER 1994; PARKER 1993; TSUBATA et al. 1993).

When B cells from mice carrying a υ/κ transgene specific for the haplotype K^k of MHC-I were used to study antigen-dependent B cell apoptosis, a high percentage of B cells recognized the solubilized alloantigen (GENARO et al. 1994; KELSOE and ZHENG 1993; MOLLER 1992). This recognition results in the release of early signals that are qualitatively identical to those obtained after stimulation with intact allogeneic cells. These signals involve a rapid increase in tyrosine phosphorylation via shared protein tyrosine kinases (among them Lyn, Fyn and Blk), and activation of a phosphoinositide-specific phospholipase C, which produces an increase in the inositol trisphosphate and diacylglycerol pools, resulting in Ca^{2+} mobilization and activation of some isoforms of PKC (GENARO and BOSCA 1993; KELSOE and ZHENG 1993; MOLLER 1992; GENARO et al. 1994). However, the signals obtained using solubilized alloantigen are quantitatively different from those elicited using intact allogeneic cells.

Since early signals only provide a partial view of the commitment to a biological response, it is useful to follow the proliferation of the cells after antigenic stimulation. In this case, only B lymphocytes activated with intact allogeneic cells proliferate, whereas those stimulated with solubilized alloantigen initiate an abortive signaling which results in cell death by apoptosis. Therefore, an additional (costimulatory) signal released through intercellular contact is required to achieve proliferation of the responder cells. Pretreatment of the responder cells with phorbol esters, pharmacological activators of PKC, is sufficient to provide the positive signal for survival upon interaction with the solubilized alloantigen. Thus, phorbol esters convert an apoptotic signal into a signal leading to a proliferative response.

6 CD2 Ligation Rescues B Cells
   from Programmed Cell Death

Binding of antigenic structures to the B cell receptor initiates responses as different as differentiation, clonal expansion or apoptosis. To escape the apoptotic pathway, the B cell requires additional signals triggered by other receptor interactions on the cell surface. It is the combination of signals delivered from surface Ig and other surface molecules that determines the outcome.

One possible candidate for such a coreceptor molecule that provides the costimulation to prevent cell death is CD2, a member of the immunoglobulin superfamily (MOINGEON et al. 1989a) In mice this receptor protein is expressed in all B and T cells and in Natural Killer (NK) cells (SEN et al. 1990; YAGITA et al. 1989). In humans, however, CD2 is not expressed in peripheral B cells, but only in a small fraction of bone marrow B cells and thymic B cells (MURAGUCHI et al. 1992; PUNNONEN and DE-VRIES 1993). Although the role of CD2 is well studied in T cells, little is known for B cells. In T cells, it exerts two main functions, which can be delineated to structurally distinct portions of the cytoplasmic domain (BIERER and HAHN 1993). As an adhesion molecule, it facilitates the interaction between T cells and antigen presenting cells (MOINGEON et al. 1989b). It also has a regulatory function in the antigen-specific response by the T cell receptor complex (MOINGEON et al. 1989a). Since CD2 is expressed at very early stages of differentiation, a possible role during B lymphopoiesis has been suggested (MURAGUCHI et al. 1992; SEN et al. 1990).

The role of CD2 in apoptosis has been studied in the mature mouse B cell line BAL-17, because it expresses surface IgM, IgD and CD2 molecules and high levels of the proto-oncogene *bcl*-2, which has been shown to inhibit cell death in many systems. Stimulation of CD2, either by cross-linking with an anti-CD2 antibody or by its physiological ligand sCD48, can rescue BAL-17 cells from apoptosis induced either by serum starvation or by increased free radical production in the presence of H_2O_2 (E. Baixeras, unpublished data). Thus, activation of CD2 can provide one signal to prevent apoptosis, suggesting that CD2 might be involved in delivering a costimulatory signal during the antigenic triggering of B cells and inhibiting the apoptotic pathway.

Ligation of CD2 induces tyrosine phosphorylation of at least two substrates with different kinetics (E. Baixeras, unpublished data). One of the substrates resembles the p56^{lck} kinase. In fact, both CD2 ligation and surface IgM cross-linking stimulate phosphorylation of the p56^{lck} kinase, whereas only the surface IgM cross-linking results in phosphorylation of the Lyn kinase. Furthermore, immunoprecipitations showed that CD2 associated with p56^{lck} kinase and this complex dissociates upon stimulation by either surface IgM or CD2 (E. Baixeras, unpublished data). This analysis clearly suggests a functional relationship between surface IgM and CD2 receptors, analogous to the observed interaction between T cell receptor-CD3 and CD2 in T cells. Moreover, our results, together with the finding that p56^{lck} kinase activity also increases in human T cells activated via CD2 (DANIELIAN et al. 1991,1993), indicate that p56^{lck} participates in signal transduction upon CD2 activation in both B and T cells.

CD2 is regulated during B cell differentiation and its expression coincides with Bcl-2 expression. We have also observed simultaneous high levels of expression of Bcl-2 and CD2 in B and T cell lines (E. Baixeras, unpublished data). Furthermore, Bcl-2 has been reported to associate with R-ras p23 (FERNANDEZ-SARABIA and BISCHOFF 1993). These results suggest that CD2 and Bcl-2 could cooperate in the prevention of apoptosis, either by direct interaction or via a common signal transduction pathway.

We can conclude that the CD2 molecule probably plays an essential role in modulation of the response to other distinct extracellular signals. According to our results and in analogy with the reported CD2-T cell receptor interactions, we propose that, in B cells, CD2 may interact with B cell receptor signaling via shared kinases such as p56^{lck}, thereby establishing a common signaling pathway for B cell receptor and CD2 when both molecules are coexpressed at the cell surface. How, if at all, this signal transduction is connected with the effect on stabilizing the cytoskeleton organization remains to be established.

Acknowledgements. We will like to thank Coral Bastos for the help in editing the manuscript. This work was supported by grants from Comision Interministerial de Ciencia y Tecnologia (PM 93/070, SAF 93-0123), Comunidad de Madrid, Fondo de Investigaciones Sanitarias and Pharmacia and by a fellowship from the European Molecular Biology Organization to C.S.

References

Albina J E, Abate JA, Henry W-J (1991) Nitric oxide production is required for murine resident peritoneal macrophages to supress mitogen-stimulated T cell proliferation. J Immunol 147: 144–148

Albina JE, Cui S, Mateo RB, Reichner JS (1993) Nitric-oxide mediated apoptosis in murine peritoneal macrophages. J Immunol 150: 5080–5085

Baixeras E, Kroemer G, Cuende E, Márquez C, Boscá L, Alés Martínez JE, Martinez-AC (1993) Signal transduction pathways involved in B cell induction. Immunol Rev 132: 5–47

Bierer BE, Hahn WC (1993) T cell adhesion, avidity regulation and signaling: a molecular analysis of CD2. Semin Immunol 5: 249–261

Billiar TR, Curran RD, Stuehr DJ, Stadler J, Simmons RL, Murray SR (1990) Inducible cytosolic enzyme activity for the production of nitrogen oxides from L-arginine in hepatocytes. Biochem Biophys Res Commun 168: 1034–1040

Bogdan C, Vodovotz Y, Paik J, Xie Q.-W, Nathan C (1993) Traces of bacterial lipopolysaccharide suppress IFN-induced nitric oxide synthase gene expression in primary mouse macrophages. J Immunol 151: 301–309

Bredt DS, Snyder SH (1989) Nitric oxide mediates glutamate-linked enhancement of cGMP levels in the corebellum. Proc Natl Acad Sci USA 86: 9030–9033

Bredt DS, Snyder SH (1992) Nitric oxide, a novel nouronal messenger. Neuron 8: 3–11

Bredt DS, Hwang PM, Snyder SH (1990) Localization of nitric oxide synthase indicating a neural role for nitric oxide. Nature 347: 768–770

Brune B, Lapetina EG (1989) Activation of a cytosotic ADP-ribosyltransferase by nitric oxide-generating agents. J Biol Chem 264: 8455–8458

Clark EA, Lane PJL (1991) Regulation of human B cell activation and adhesion.Annu Rev Immunol 9: 97–127

Clark EA, Ledbetter JA (1994) How B and T cells talk to each other. Nature 367: 425–428

Danielian S, Fagard R, Alcover A, Acuto O, Fischer S (1991) The tyrosine kinase activity of p56lck is increased in human T cells activated via CD2. Eur J Immunol 21: 1967–1970

Danielian S, Alcover A, Polissard L, Stefanescu M, Acuto O, Fisher S, Fagard R (1993) Both T cell receptor (TCR)-CD3 complex and CD2 increase the tyrosine kinase activity of p56lck. CD2 can mediate TCR-CD3 independent and CD45-dependent activation of p56lck. Eur J Immunol 22: 2915–2921

Ding AH, Nathan CF, Graycar J, Derynck R, Stueh DJ, Srimal S (1990) Macrophage deactivating factor and transforming growth factors 1, 2 and 3 inhibit induction on macrophage nitrogen oxide synthesis by IFN. J Immunol 145: 940–944

Dong Z, Qi X, Xie K, Fidler IJ (1993) Protein tyrosine kinase inhibitors decrease induction of nitric Oxoid synthase activity in lipopolysaccharide-responsive and lipopolysaccharide-nonresponsive murine macrophages. J Immunol 151: 2717–2724

Drapier JC, Hibbs JB Jr (1986) Murine cytotoxic activated macrophages inhibit aconitase in tumor cells. Inhibition involves the iron-sulfur prosthetic group and is reversible. J Clin Invest 78: 790–795

Fernandez-Sarabia MJ, Bischoff JR (1993) Bcl-2 associates with the ras-related protein R-ras p23. Nature 366: 274–275

Furchgott RF (1988) Studies on relaxation of Rabbit aorta by sodium nitrite: the basis for the proposal that the acid activatable inhibitory factor from retractor penis is inorganic nitrite and the endothelium-derived relaxing factor is nitric oxide. In: Vanhoutte PM (ed) Mechanism of vasodilation. Raven, New York, p401

Garg UC, Hassid AS (1989) Nitric oxide-generating vasodilators and B-bromo-cyclic guanosine monophosphate inhibit mitogenesis and proliferation of cultured rat vascular smooth muscle cells. J Clin Invest 83: 1774–1777

Geller DA, Lowenstein CJ, Shapiro RA, Nussler AK, Di Silvio M, Wang SC, Nakayama DK, Simmons RL, Snyder SH, Billiar TR (1993a) Molecular cloning and expression of inducible nitric oxide synthase from human hepatocytes. Proc Natl Acad Sci USA 90: 3491–3495

Geller DA, Nussler AK, Di Silvio M, Lowenstein CJ, Shapiro RA, Wang SC, Simmons RL, Billiar TR (1993b) Cytokines, endotoxin, and glucocorticoids regulate the expression of inducible nitric oxide synthase in hepatocytes. Proc Natl Acad Sci USA 90: 522–526

Genaro AM and Boscá L (1993) Early signals in alloantigen induced B cell proliferation. J Immunol 151: 1832–1843

Genaro AM, Gonzálo JA, Boscá L, Martínez-A. C (1994) CD2 occupancy prevents apoptosis in murine B lymphocytes by upregulating Bcl-2 expression. Eur J Immunol 24: 2515–2521

Golay J, Cusmano G, Introma M (1992) Independent regulation of c-myc, b-myb and c-myb gene expression by inducers and inhibitors of proliferation in human B lymphocytes. J Immunol 149: 300–308

Granger DL, Taintor RR, Cook JL, Hibbs JB (1980) Injury in neoplastic cells by murine macrophages leads to inhibition of mitochondrial respiration. J Clin Invest 65: 357

Hauschildt S, Luckhoff A, Mulsch A, Kohler J, Bessler W, Busse RS (1990) Induction and activity of NO syntase in bone marrow derived macrophages are independent of Ca^{2+}. Biochem J 270: 351

Hoffman RA, Langrehr JM, Billiar TR, Curran RD, Simmons RL (1990) Alloantigen-induced activation of rat splenocytes is regulated by oxidative metabolism of L-arginiene. J Immunol 145: 2220–2226

Ischiropoulos H, Zhu L, Beckman JS (1992) Peroxynitrite formation from macrophage-derived nitric oxide. Arch Biochem Biophys 298: 446–451

Jenkins MK, Johnson JC (1993) Molecules involved in T-cell costimulation. Curr Opin Immunol 5: 361–367

Jonathan S, Reichner JS, Mateo RB, Albina JE (1994) Activated murine macrophages induce apoptosis in tumor cells through nitric oxide-dependent or -independent mechanisms. Cancer Res 54: 2462–2467

Kelsoe G, Zheng B (1993) Sites of B-cell activation in vivo. Curr Opin Biol 5: 418–422

Knowles RG, Moncada S (1992) Nitric oxide as a signal in blood vessels. Trends Biochem Sci 17: 399–402

Knowles RG, Palacios M, Palmer RMJ, Moncada S (1989) Formation of nitric oxide from L-arginine in the central nervous system: a transduction mechanism for stimulation of the soluble guanylate cyclase. Proc Natl Acad Sci USA 86: 5159–5162

Krammer P, Behrman I, Daniel P, Dhein J, Debatin K-M (1994) Regulation of apoptosis in the immune system. Curr Opin Immunol 6: 279–289

Kroemer G, Martinez-AC (1994) Pharmacological inhibition of programmed cell death. Immunol Today 15: 235–242

Langrehr JM, Murase N, Markus PM, Cai X, Neuhaus P, Schraut W, Simmons RL, Hoffman RA (1992) Nitric oxide production in host-versus-graft and graft-versus-host reactions in the rat. J Clin Invest 90: 679–683

Lepoivre M, Chenais B, Yapo A, Lemaire G, Thelander L, Tenu J-P (1990) Alterations of ribonucleotide reductase activity following induction of the nitrite-generating pathway in adenocarcinoma cells. J Biol Chem 265: 14143–14149

Liew FY, Li Y, Severn A, Millott S, Schmidt J, Slater M, Moncada S (1991) A possible novel pathway of regulation by murine T helper type-2 (Th2) cells of a Th1 cell activity via the modulation of the induction of nitric oxide synthase on macrophages. Eur J Pharmacol 21: 3009–3014

Lipton SA, Choi YB, Pan ZH, Lei SZ, Chen HS, Sucher NJ, Loscalzo J, Singel DJ, Stamler JS (1993) A redox-based mechanism for the neuroprotective and neurodestructive effects of nitric oxide an related nitrosocompounds. Nature 364: 626–632

Loffert D, Schaal S, Erlich A, Hardy RR, Zon Y-R, Muller W, Rajewsky K (1994) Early B-cell development in the mouse: insights from mutations introduced by gene targeting. Immunol Rev 137: 135–172

Lowenstein CJ, Snyder SH (1992) Nitric oxide, a novel biological messenger. Cell 70: 705–707

Lowenstein CJ, Dinerman JL, Snyder SH (1994) Nitric Oxide: a physiologic Messenger. Ann Intern Med 120: 227–237

Lyons CR, Orloff GJ, Cunningham JM (1992) Molecular cloning and functional expression of a inducible NOS from a murine macrophage cell line. J Biol Chem 267: 6370–6374

Marletta MA, Yoon PS, Iyengar R, Leaf CD, Wishnok JS (1988) Macrophage oxidation of L-arginine to nitrite and nitrate: Nitric oxide is an intermediate. Biochemistry 27: 8706–8711

Marquez C, Martínez-AC, Kroemer G, Boscá L (1992) Protein kinase C isoenzymes display differential affinity for phorbol esters. Analysis of phorbol ester receptors in B cell differentiation. J Immunol 149: 2560–2568

McCartney F, Allen N, Mizel DE, Albina JE, Xie Q-W, Nathan CF, Wahl SM (1993) Supression of arthritis by an inhibitor of nitric oxide synthase. J Exp Med 178: 749–754

Merino R, Ding L, Veis DJ, Korsmeyer SJ, Nuñez G (1994) Developmental regulation of the bcl-2 protein and susceptibility to cell death in B lymphocytes. EMBO J 23: 683–691

Moingeon P, Chang H, Sayre PH, Clayton LK, Alcover A, Gardner P, Reinherz EL (1989a) The structural biology of CD2. Immunol Rev 111: 111–114

Moingeon P, Chang HC, Wallner BP, Stebbins C, Frey AZ, Reinherz EL (1989b) CD2 mediated adhesion facilitates T lymphocyte antigen recognition function. Nature 339: 312

Möller G (1992) Cytokines in infectious disease. Immunol Rev 126: 5–178

Möller G (ed) (1994) B-cell differentiation. Immunol Rev 137: 5–229

Moncada S (1992) The L-arginine: nitric oxide pathway. Acta Physiol Scand 145: 201–227

Moncada S, Higgs A (1993) The L-arginine-nitric oxide pathway. N Engl J Med 329: 2002–2012

Moncada S, Palmer RMJ, Higgs EA(1991) Nitric oxide: physiology, pathophysiology and pharmacology. Pharmacol Rev 43: 109–143

Muraguchi A, Kawamura N, Hori A, Horii Y, Ichigi Y, Kimoto M, Kishimoto T (1992) Expression of the CD2 molecule on human B lymphoid progenitors. Int Immunol 4: 841–849

Nathan C (1992) Nitric oxide as a secretory product of mammalian cells. FASEB J 6: 3051–3064

Nuñez G, London L, Hockenbery D, Alexander M, McKearn JP, Korsmeyer SJ (1990a) Deregulated Bcl-2 gene expression selectively prolongs survival of growth factor-deprived hematopoietic cell lines. J Immunol 144: 3602–3610

Nuñez G, London, Hockenbery D, Alexander M, McKearn JP, Korsmeyer SJ (1990b) Deregulated Bcl-2 gene expression selectively prolongs survival of growth factor-deprived hematopoietic cell lines. J Immunol 144: 3602–3610

Palmer RMJ (1993) The discovery of nitric oxide in the vessel wall. Arch Surg 128: 396–401

Palmer RMJ, Ferridge AG, Moncada S (1987) Nitric oxide accounts for the biological activity of endothelium-derived relaxing factor. Nature 327: 524–526

Parker DC (1993) T cell dependent B cell activation. Annu Rev Immunol 11: 331–360

Punnonen J, de-Vries JE (1993) Characterization of a novel CD^{2+} human thymic B cell subset. J Immunol 151: 100–110

Rincon M, Tugores A, Landazuri MA, Lopez-Botet M (1993) Costimulation of cAMP and protein kinase C pathways inhibits the CD3-dependent T cell activation and leads to a persistent expression of the AP-1 transcription factor. Cell Immunol 149: 343–356

Sen J, Rosenberg N, Burakoff S (1990) Expression and ontogeny of CD2 on murine B cells. J Immunol 144: 2925–2930

Stuehr DJ, Marletta MA (1985) Mammaliam nitrate biosynthesis: mouse macrophages produce nitrite and nitrate in response to Escherichia coli lipopolysaccharide. Proc Natl Acad Sci USA 82: 7738–7742

Stuehr DJ, Cho JJ, Kwon NS, Nathan C (1991) Purification and characterization of the cytokine-induced macrophage nitric oxide synthase: a FAD- and FMN containing flavoprotein. Proc Natl Acad Sci USA 88: 7773–7777

Tson K, Snyder GL, Greengard P (1993) Nitric oxide/cAMP pathway stimulates phosphorylation of DARPP-32, a dopamine- and cAMP-regulated phosphoprotein, in the substantianigra. Proc Natl Acad Sci USA 90: 3462–3465

Tsubata T, Wu J, Honjo T (1993) B cell apoptosis induced by antigen receptor crosslinking is blocked by a T cell signal through CD40. Nature 364: 645–648

Watanabe-Fukunaga R, Brannan Cl, Copeland NG, Jenkins NA, Nagata S (1992) Lymphoproliferation disorder in mice explained by defects in Fas antigen that mediates apoptosis. Nature 356: 314–317

Watson ML, Rao JK, Gilkeson GS, Ruiz P, Eicher EM, Pisetsky DS, Matsuzawa A, Rochell JM, Seldin MF (1992) Genetic analysis of MRL/lpr mice: relationship of the Fas apoptosis gene to didease manifestation and renal modifying loci. J Exp Med 176: 1645–1656

Weinberg JB, Granger DL, Pisetsky DS, Seldin MF, Misukonis MA, Mason SN, Rippen AM, Ruiz P, Wood ER, Gilkeson GS (1994) The role of nitric oxide in the pathogenesis of spontaneous murine autoimmune disease: Increased nitric oxide production and nitric oxide syntase expression in MRL lpr/lpr mice, and reduction of spontaneous glomerulonephritis and arthritis by orally administered N monomethyl-L-arginine. J Exp Med 179: 651–660

Xie Q-W, Whisnant R, Nathan C (1993) Promoter of the mouse gene encoding calcium independent nitric oxide synthase confers inducibility by interferon and bacterial lipopolysaccharide. J Exp Med 177: 1779–1784

Yagita H, Nakamura T, Karasuyama H, Okumura K (1989) Monoclonal antibodies specific for murine CD2 reveal its presence on B as well as T cells. Proc Natl Acad Sci USA 86: 645–649

Glucocorticoid-Induced Death of Immune Cells: Mechanisms of Action

J.W. Montague[1] and J.A. Cidlowski[1, 2]

1 Introduction

Cell death occurs in many physiological situations, including embryogenesis, differentiation, and metamorphosis. It is an important mechanism in maintaining homeostasis by providing a counterbalance to mitosis. There are at least two types of cell death that are now recognized: necrosis and apoptosis. In necrosis, the cell undergoes irreversible swelling and lysis in response to a variety of signals which are primarily nonphysiological; the plasma membrane disrupts and then spills the intracellular contents into the environment, resulting in activation of the immune response. Apoptosis, by contrast, is inherently "programmed" as part of the cellular processes, allowing the cell to die in response to a variety of signals without a deleterious effect on surrounding cells, i.e., it does not elicit an immune response. Details of the morphology of apoptosis have been extensively reported (Kerr et al. 1972; Arends and Wyllie 1991; Schwartzman and Cidlowski 1993a;

[1] Department of Biochemistry and Biophysics, University of North Carolina at Chapel Hill, Chapel Hill, NC 27599, USA

[2] Department of Physiology, CB # 7545, and Lineberger Comprehensive Cancer Institute, University of North Carolina at Chapel Hill, Chapel Hill, NC 27599, USA

COMPTON and CIDLOWSKI 1992), but briefly, apoptosis involves separation of the cell from its neighbors, condensing of the cytoplasm, condensing of the chromatin, which moves to the margins of the nuclear envelope, convolution of the plasma membrane, and, finally, blebbing off of apoptotic bodies, which contain various organelles and chromatin fragments.

The condensed chromatin of apoptosis is often associated with internucleo-somal cleavage, displaying DNA fragments that are multiples of 180–200 base pairs in size (KERR et al. 1972; WYLLIE 1980; WYLLIE et al. 1984; COHEN and DUKE 1984; ARENDS et al. 1990). When such fragmented DNA is electrophoresed through an agarose gel and stained with ethidium bromide, the cleavage products will form a "ladder" pattern (the "rungs" of which are composed of integer multiples of the nucleosomal-sized fragments), diagnostic of cells undergoing apoptosis. The endonuclease responsible for cleaving the DNA in the linker regions (between the nucleosomes) has been characterized in many lymphocytic cells as being calcium- and magnesium-dependent. Internucleosomal cleavage is always associated with apoptosis, but apoptosis is not strictly defined by inter-nucleosomal cleavage and may actually incorporate other types of DNA fragmentation. Recently, additional patterns of DNA fragmentation from dying or dead cells have been reported to occur in nonlymphocytic cell lines (OBERHAMMER et al. 1992,1993). It is not clear if these cells are actually undergoing apoptosis or a novel form of programmed cell death. Additional indicators, such as specific morphology or the requirement of energy, as in the case of apoptosis, are mandatory to help properly categorize a certain death response.

Apoptosis occurs often in the immune system. Cell death helps shape the immune system as it matures by deleting autoreactive T cells (SMITH et al. 1989) and unreacted B cells (HASBOLD and KLAUS 1990). Immune cells can also be stimulated to undergo apoptosis with growth factor withdrawal (NIETO et al. 1990; WILLIAMS et al. 1990) or by glucocorticoid administration (COMPTON and CIDLOWSKI 1992; WYLLIE 1980; COHEN and DUKE 1984). The apoptotic effect of glucocorticoids on the different types of immune cells is well documented, although most of the studies concerning the effect of glucocorticoids on immune cells involve immature thymocytes. These can be easily isolated in large numbers from rats or mice, which provides a convenient and adaptable model for the study of programmed cell death, or they can be studied as a specific cell line, such as the S49 mouse thymoma cell line.

In this review, we will cover the topic of glucocorticoid-induced death of immune cells, keeping in mind that, despite intensive study, the events that lead to glucocorticoid-stimulated cell death are still poorly understood. Some general metabolic and genetic responses during steroid hormone treatment in immune cells will be discussed, followed by several apoptotic-specific effects of glucocorticoids, such as gene induction and calcium fluxes. Since the studies included in these sections were performed mostly on mature and immature T cells, we included a separate section for apoptosis in B cells. This is followed by a discussion of Bcl-2 and its role in preventing glucocorticoid-induced apoptosis. Finally, we demonstrate how these different effects of glucocorticoids can be integrated by a repressor model of nuclease activation that carries out the apoptotic process.

2 General Effects of Glucocorticoids on Metabolism and Gene Regulation

Glucocorticoid effects in immune cells had been studied long before the seminal apoptotic paper was published by KERR et al. in 1972. For several decades prior, glucocorticoids were known to elicit a lytic response in immune cells. During this time there was much research concerning altered metabolic responses following glucocorticoid treatment. For example, thymocyte and lymphocyte nuclei exposed to hormone were described as displaying a pyknotic phenotype, with abnormal chromatin arrangements and nuclear edema, followed by the eventual dissolution of the nuclear membrane and karyolysis (DOUGHERTY and WHITE 1945; BURTON et al. 1967; WHITFIELD et al. 1968; COWAN and SORENSEN 1964). Interestingly, WHITFIELD et al. noted in 1968 that the effect of cortisol on lymphocyte nuclei was identical to lymphocytic response to irradiation—both of these insults have been subsequently demonstrated to induce apoptosis. One well-studied example of altered metabolic effects is the decrease in glucose uptake in both lymphocytic and nonlymphocytic cells (MUNCK and LEUNG 1977). Another example is the decrease of amino acid transport (MORITA and MUNCK 1964) and nucleoside accumulation (MAKMAN et al. 1968), as well as decreased protein and nucleic acid biosynthesis (NORDEEN and YOUNG 1976). ATP production (MAKMAN et al. 1971) and RNA polymerase activity (BELL and BORTHWICK 1975) are also diminished following addition of hormone. Thus, glucocorticoids have been shown to inhibit several anabolic processes. Our laboratory has shown that glucocorticoids are also capable of stimulating catabolic processes, so that protein and RNA degradation is actually enhanced (MACDONALD and CIDLOWSKI 1982; MACDONALD et al. 1980; CIDLOWSKI 1982). Indeed, glucocorticoids are capable of stimulating the activity of several hydrolytic enzymes in lymphocytes, including acid phosphatase (CLARKE and WILLS 1978), two serine hydrolases (MACDONALD and CIDLOWSKI 1981), ribonuclease (AMBELLAN and HOLLANDER 1966; WIERNIK and MACLEOD 1965; MASHBURN et al. 1969), and deoxyribonuclease (WIERNIK and MACLEOD 1965).

In addition to altering metabolic processes in immune cells, glucocorticoids can, when complexed with an activated receptor, induce or inhibit specific genes resulting in measurable changes of a number of mRNAs (ROUILLER et al. 1988; MACDONALD and GOLDFINE 1988; HIRATA 1981; BLACKWELL et al. 1980; COLBERT and YOUNG 1986; BURNSTEIN et al. 1990; EASTMAN-REKS and VEDECKIS 1986; BARBOUR et al. 1988). Obviously, glucocorticoids exert a wide range of actions over varied genes and proteins. What role any or all of these actions may play in glucocorticoid-induced death of immune cells is a question currently under extensive investigation.

3 Glucocorticoid-Induced Apoptotic Genes

Glucocorticoid-induced apoptosis of some immune cells has been shown to require protein synthesis (WYLLIE et al. 1984; COHEN and DUKE 1984; COMPTON et al. 1988; McCONKEY et al. 1990), although the apoptotic nuclease is constitutively expressed (GAIDO and CIDLOWSKI 1991; VANDERBILT et al. 1982). Such a requirement was suggested by several groups (WYLLIE et al. 1984; COHEN and DUKE 1984) who showed that the RNA and protein synthesis blockers actinomycin D and cyclo-hexamide, respectively, could prevent steroid-induced death of thymocytes. This result implies that the activated glucocorticoid receptor can induce programmed cell death-specific genes. Meanwhile, because glucocorticoids have also been shown to inhibit protein synthesis, the inhibitory action of the glucocorticoid receptor may also be important in shutting off certain proteins so that apoptosis may progress. Indeed, in certain cell types (S49.1, HL-60, U 937, Mol t4, Daudi, MRC-5, Raji, K 562) use of inhibitors of protein and mRNA synthesis alone was enough to induce apoptosis (MARTIN et al. 1990; CARON-LESLIE and CIDLOWSKI 1994). CARON-LESLIE and CIDLOWSKI proposed that the inhibition of protein synthesis does not directly cause apoptosis, but rather is part of a cascade of events which are dependent on protein inhibition to lead to apoptotic death. The necessary inter-play between gene induction and gene inhibition during glucocorticoid-induced apoptosis has not yet been defined, but such data can be unified by a repressor model in which key genes and proteins that mediate the apoptotic process are kept in check through posttranslational modification, association with inhibitors, or as inactive precursors.

Studies of genes expressed specifically in apoptotic immune cells have addressed the issue of the requirement of synthesizing proteins in the death process. HARRIGAN et al. (1989) isolated and characterized 11 genes induced in glucocorticoid-treated WEHI-7TG cells, the majority of which showed an increase of message within 0.5–1 h after dexamethasone treatment. This work was continued by BAUGHMAN et al. (1991), who reported two more glucocorticoid-regulated genes from WEHI-7TG cells. Seven of these 13 clones have been identified (BAUGHMAN et al. 1992), two are repressed in response to hormone, the remaining 11 are induced. The types of induced proteins that were identified include chondroitin sulfate proteoglycan core protein, mitochondrial PO_4 carrier protein, immunoglobulin-related glycoprotein-70, Lupus-Graves antigen, a G-protein-coupled receptor, and calmodulin. Interestingly, calmodulin gene expres-sion was also shown by DOWD et al. (1991) to be induced by glucocorticoid treatment of WEHI7.2 lymphocytes, supporting the theory that Ca^{2+}-calmodulin-dependent enzymes are involved in the cell death process (McCONKEY et al. 1989a,b). These calmodulin results reflect the possibility that Ca^{2+} plays an important role in glucocorticoid-induced apoptosis, a subject which is discussed in greater detail in the following section. Two other mRNAs associated with programmed cell death were described by OWENS et al. (1991). The protein encoded by the clone RP-2 has an α-helical domain and a membrane-spanning region, which suggest it is an integral membrane protein, while the protein associated with RP-8 has a zinc finger domain, suggestive of DNA binding activity. A heat labile factor, thought to be a Ca^{2+} pore, produced 60 min following methylprednisolone treatment of thymocytes appears to be another protein necessary for the apoptotic process (MCCONKEY et al. 1989). Further work is required to determine exactly how all of these glucocorticoid-induced genes and proteins play a role in carrying out the process of apoptosis. Some proteins may help in activating the apoptotic nuclease, while others, such as proteases or transglutaminases, may contribute to the other aspects of cell death like the formation of apoptotic bodies.

4 Effect of Glucocorticoid Administration on Ca^{2+} Levels

The role Ca^{2+} may play, if any, in glucocorticoid-induced apoptosis of lymphocytes is still under dispute. The following section will discuss data concerning whether or not glucocorticoid treatment results in an increase in Ca^{2+} levels which then initiate the apoptotic process. Most of the work reported has been performed on immature rat thymocytes or thymoma cell lines, although some data include other lymphoid tissues.

Early work by KAISER and EDELMAN (1977) demonstrates a Ca^{2+} requirement for glucocorticoid-induced death of rat thymocytes. When comparing thymocytes treated with steroid to thymocytes treated with the Ca^{2+} ionophore A23187, the authors noted that the two treatments had similar effects, namely, cytolysis and inhibition of uridine metabolism. When Ca^{2+} was removed from the media, the cells showed decreased sensitivity to hormone-induced death. The authors then tested lymph node lymphocytes (KAISER and EDELMAN 1978) to determine whether this "Ca^{2+} effect" was specific only for thymocytes or all lymphocytes in general. Although both cell types displayed sensitivity to hormone and A23187, the lymph node lymphocytes did not appear to require Ca^{2+} for this hormone effect. The difference in Ca^{2+} requirements for the two cell types can be trivially explained by concluding there is a differential sensitivity to Ca^{2+} between thymocytes and lymph node lymphocytes. Interestingly, however, more recent data from several different groups are revealing conflicting results as to the role of Ca^{2+} just in hormone-induced death of thymocytes.

McCONKEY et al. (1989a) provided evidence that a sustained increase in cytosolic Ca^{2+} concentration resulted from treatment of thymocytes with methylprednisolone and preceded DNA degradation. In these studies, the Ca^{2+} level increased eightfold over a 2 h time period. The resulting DNA degradation could be blocked by the addition of RU486, suggesting the glucocorticoid receptor is directly involved. A decrease in DNA degradation in the presence of the intracellular Ca^{2+} buffer quin-2 demonstrated the requirement of an elevated Ca^{2+} concentration for endonuclease activation. This buffering effect could be overcome by the addition of Ca^{2+} ionophore, resulting in restoration of endonuclease activity. This group also demonstrated a direct correlation between Ca^{2+} concentration and amount of DNA fragmentation (McCONKEY et al. 1989b). In a similar study, the effect of A23187 on S49.1 (mouse thymoma) cells was shown to result in DNA fragmentation (CARON-LESLIE and CIDLOWSKI 1991), again suggesting a role for Ca^{2+} in hormone-induced thymocyte apoptosis.

In direct contrast to this idea is research indicating that Ca^{2+} is not required for glucocorticoid-induced apoptosis. NICHOLSON and YOUNG (1979) used nuclear fragility, determined by the inability of nuclei to withstand cellular lysis caused by a hypotonic shock, as an early indicator of glucocorticoid effect on P1798 lymphosarcoma cells. Although there was an increase in Ca^{2+} uptake after these lymphoid cells were treated with hormone, no correlation could be made with the increase in nuclear fragility. In an attempt to correlate these data with those of McCONKEY et al., who showed that an elevated Ca^{2+} level was part of the glucocorticoid-induced death process of thymocytes, ISEKI et al. (1993) obtained rat (and mouse) thymocytes and looked for glucocorticoid-inspired changes in Ca^{2+} flux. With fura-2 as a fluorescent Ca^{2+} indicator, Ca^{2+} levels were measured up to 15 min after treatment, during which time a glucocorticoid-induced increase in intracellular Ca^{2+} levels was not observed. The differences in experimental techniques (Ca^{2+} indicators and time of incubation) prevents a true comparison of the two results, however.

The human T cell leukemic cell line CEM-C7 also displays sensitivity to glucocorticoids with a different Ca^{2+} response. For example, these cells do not appear to require extracellular Ca^{2+} to mediate the DNA fragmentation (ALNEMRI and LITWACK 1990). Likewise, BANSAL et al. (1990) were able to show that DNA degradation occurred in CEM-C7 cells after dexamethasone treatment in a Ca^{2+}-free media; however, they indicate that this DNA does not have the characteristic ladder pattern, which suggests they are looking at DNA from necrotic cells that are dying in the absence of Ca^{2+}; therefore the CEM-C7 response to glucocorticoids may not be apoptotic. Perhaps since CEM-C7 cells are a transformed cell line they may have lost a component of the normal apoptotic pathway so the differences observed may result from cell-specific responses to glucocorticoids. For example, thymocytes readily take up extracellular Ca^{2+} to activate an endogenous nuclease in response to hormone, whereas CEM-C7 cells apparently contain a non-Ca^{2+}/Mg^{2+}-dependent nuclease. Additionally, DNA fragmentation occurs much more rapidly following hormone treatment in thymocytes than in CEM-C7 cells (BANSAL et al. 1990). Another possibility that needs to be considered is that subtle Ca^{2+} level changes, resulting from the release of internal stores of Ca^{2+}, are responsible for the glucocorticoid effect (BANSAL et al. 1990). This possibility is explored by LAM et al. (1993), who present evidence that glucocorticoids release Ca^{2+} from intracellular stores in W7MG1 mouse lymphoma cells, thus allowing for continuation of Ca^{2+}-requiring mechanisms during apoptosis, even in the absence of extracellular Ca^{2+}. Unlike immature thymocytes, which show a significant increase in intracellular Ca^{2+} levels after glucocorticoid treatment, the W7MG1 cells show only a slight increase in cytosolic Ca^{2+} levels 4 h after addition of dexamethasone. However, when ionomycin, which promotes

Ca^{2+} uptake and releases Ca^{2+} from several internal sources, and thapsigargin, which releases Ca^{2+} only from the ER, were used to probe for determining the effect of glucocorticoids on internal Ca^{2+} release, there was a significant decrease in the levels of mobilizable Ca^{2+} released from organelle storage sites in the hormone-treated cells as compared to control cells. Thapsigargin treatment revealed that not only are levels of mobilizable Ca^{2+} from ER significantly reduced after incubation in dexamethasone, but that thapsigargin treatment alone result- ed in a dose-dependent decrease of cell growth and viability and an increase in DNA degradation in the typical apoptotic ladder pattern. These data imply that release of Ca^{2+} from the ER may be an important step in glucocorticoid-induced apoptosis.

The different conclusions as to the role of Ca^{2+} emphasize the complexity of steroid-induced death in immune cells. Due to the amount of evidence of a Ca^{2+}/ Mg^{2+}-dependent apoptotic nuclease (WYLLIE 1980; COHEN and DUKE 1984; ARENDS et al. 1990), it is highly probable that glucocorticoid treatment does result in increased Ca^{2+} levels, whether by an influx of extracellular Ca^{2+} or a release of Ca^{2+} from internal stores; however, comparable studies between thymocytes and CEM-C7 cells need to be performed to ascertain if there are cell-specific mecha- nisms for glucocorticoid-induced apoptosis.

5 Actions of Ca^{2+}-Binding Proteins in Glucocorticoid-Induced Apoptosis

Studies on the effect of Ca^{2+}-binding proteins in glucocorticoid-induced apoptosis complement the results described above, suggesting that Ca^{2+} does play an important role during the process in thymocytes, although perhaps not in CEM- C7 cells. Calmodulin, a Ca^{2+}-binding protein involved in a variety of cellular events, such as division, motility and contractility, demonstrates increased mRNA levels following addition of dexamethasone to WEHI7.2 cells (DOWD et al. 1991). The rise appears to be a result of an increase in transcription of calmodulin mRNA rather than a result of message stability. Inhibitors of calmodulin action, such as calmidazolium and trifluoperazine, blocked steroid-induced DNA degradation in thymocytes (McCONKEY et al. 1989b; ISEKI et al. 1993) and cell death in WEHI7.2 cells (DOWD et al. 1991); however, this effect was not observed in CEM-C7 cells (BANSAL et al. 1990). Stable expression of the Ca^{2+}-binding protein calbindin-D_{28K} in WEHI7.2 cells provided interesting results that point to a Ca^{2+} requirement for glucocorticoid-induced death (DOWD et al. 1992) of those cells. Calbindin-D_{28K} can bind five or six Ca^{2+} ions with high affinity (LEATHERS et al. 1990). WEHI7.2 cells containing overexpressed calbindin-D_{28K} were not as susceptible to dexa- methasone-induced cell death. This protective effect correlated with the relative levels of overexpression—the greater the concentration of calbindin-D_{28K}, the greater the resistance to dexamethasone-induced death. A similar protective effect was noted when cells were treated with the Ca^{2+} ionophore A23187, indicating the anti-apoptotic effect may be a result of sequestering Ca^{2+} ions. Thus, for most examples, Ca^{2+} ions appear to be a result of glucocorticoid treatment of immune cells and are necessary to create the phenotypic character-istics of apoptosis.

6 Glucocorticoid-Induced Apoptosis of B Cells

B cells are also capable of undergoing apoptosis, whether in response to being placed in culture, as in the case of germinal centers of secondary lymphoid organs (HOLDER et al. 1992), or being treated with steroid hormone. The disorder chronic lymphocytic leukemia of B cell type (B-CLL) is characterized by small, immature resting B lymphocytes accumulating in the periphery. Treatment involves gluco-corticoid administration, which results in a decrease of these peripheral lym-phocytes. The mechanism of this response is not understood, but probably involves apoptosis. Previous work had shown that a significant fraction of B-CLL B cells would undergo apoptosis spontaneously when placed in culture (COLLINS et al. 1989). These facts, plus the knowledge of the effect of glucocorticoids on immature thymocytes, prompted McCONKEY et al. (1991) to compare the effect of methylprednisolone on B-CLL cells to normal peripheral blood lymphocytes. The data showed that methylprednisolone treatment resulted in increased DNA fragmentation, the formation of apoptotic ladders, and decreased viability in the B-CLL cells as compared to the normal peripheral blood lymphocytes. Additional-ly, the hormone-treated B-CLL cells showed an increase in cytosolic Ca^{2+} levels 2 h after addition of hormone, an effect that was blocked by both RU486 and cyclohexamide. Thus, the leukemic B cells display a sensitivity to hormone similar to that of the immature thymocytes. Other studies of hormone-induced apoptosis in B cells include studies of the protein Bcl-2.

7 Bcl-2 Blocks Glucocorticoid-Induced Death
of Immune Cells

bcl-2 (B-cell leukemia/lymphoma-2 gene) is a proto-oncogene first identified by its association with B cell malignancies (REED 1994). The concentration of the 26 kDa Bcl-2 protein is highly regulated during maturation of the B cell (MERINO et al. 1994); however, if a t(14:18) chromosomal translocation occurs, which places the bcl-2 gene under control of immunoglobulin heavy chain enhancer elements, the protein is expressed at much higher levels and apoptotic death of these cells is blocked. This anti-programmed cell death effect of Bcl-2 was first noted in the case of interleukin-3 (IL-3) withdrawal from immature pre-B cells (VAUX et al. 1988; HOCKENBERRY et al. 1990). Although these cells did not die, they also did not proliferate. They appeared to be stuck at the G_0 phase of the cell cycle (HOCKENBERRY et al. 1990). Therefore, the cancerous effect of Bcl-2 may result from a lowering of the rate of cell death without an increase in the rate of cell growth (REED 1994). We (CARON-LESLIE et al. 1994) have shown that glucocorticoids can inhibit protein synthesis in Bcl-2 expressing S49 cells without activating apoptosis. Such data dissociate the growth inhibition effect from the apoptotic effect of glucocorticoids.

Reports of anti-apoptotic activity of Bcl-2 led researchers to further explore the protective effects of Bcl-2 with other inducers of apoptosis, namely glucocorticoids. ALNEMRI et al. (1992) showed that B cells expressing high levels of the Bcl-2 protein did not have the decrease in viability after triamcinolone acetonide treatment as compared to a similar strain that contained much lower levels of Bcl-2. These resistant cells also did not have internucleosomally cleaved DNA, nor did they proliferate, as seen in the case of IL-3 withdrawal, suggesting they were in G_0. Information about how Bcl-2 might play a role in B cell development was provided when MERINO et al. (1994) demonstrated that pro-B cells (the least developed) and mature B cells contained high levels of Bcl-2, and pre-B cells and immature B cells contained low levels of bcl-2. When cells at the various stages of development are treated with dexamethasone, the pre-B and the immature B cells are much more susceptible to the death-inducing effects of the hormone than the pro-B and mature B cells, Thus, the amount of glucocorticoid-induced apoptosis appears to depend on the developmental stage of that cell, which is also correlated with the levels of Bcl-2 protein. This developmental effect is also seen with thymocytes. The mature medullary thymocytes are positive for Bcl-2 while the immature cells in the cortex are negative (PEZZELLA et al. 1990; HOCKENBERRY et al. 1991). Consequently, dexamethasone treatment almost completely eliminates the immature CD4+ CD8+ cells without affecting the mature thymocytes. Our laboratory (OLDENBERG and CIDLOWSKI 1994) has noted that mature thymocytes contain the same level of glucocorticoid receptors as immature cells, therefore, the decreased response to hormone cannot be attributed to diminished glucocorticoid receptor levels. The deleterious glucocorticoid effect on immature thymocytes is overcome in transgenic mice containing a bcl-2 vector expressed in the thymus, although negative selection of the transgenic thymocytes still occurs (SENTMAN et al. 1991). Additionally, the bcl-2 gene was transfected into S49.1 and WEHI7.2 cells, a thymoma and lymphoma cell line, respectively, and was shown to enhance resistance to dexamethasone-induced death and DNA fragmentation in both cases (MIYASHITA and REED 1992), although it did not prevent inhibition of proliferation induced by glucocorticoids. A similar protective effect was noted for transgenic mice expressing Bcl-2 in B cells (MERINO et al. 1994). Thus, Bcl-2 appears to confer resistance to glucocorticoid-induced death in several types of immune cells, but does not block the proliferation inhibition effect of glucocorticoids. Additionally, we (CARON-LESLIE and CIDLOWSKI 1994) observed that expression of Bcl-2 protein in S49 cells prevents dexamethasone-induced apoptosis of these cells, but does not prevent cyclohexamide- or A23187-induced apoptosis, indicating that Bcl-2 somehow interferes with the signal leading to apoptosis rather than interfering with the apoptotic process itself.

8 The Repressor Model of Nuclease Activation

Our research has explored the mechanisms of glucocorticoid-induced apoptosis of thymocytes from rat thymic or thymoma cell lines. We have incorporated these data with those of others studying apoptosis to propose that apoptosis is a repressed phenotype that can be activated by addition of glucocorticoids. For this situation we propose that all normal cells express the inherent genes necessary to carry out the apoptotic process. The resulting proteins are kept in an inactive state by mechanisms such as inhibitors, inactive precursors, or posttranslational modification. This model (Fig. 1) is based on our current understanding of apoptotic activation in thymocytes and immune cells with similar responses to glucocorticoids and accounts for the effects of glucocorticoid treatment on protein turnover and Ca^{2+} levels as well as nuclease activation and DNA cleavage.

Activation of the apoptotic nuclease is a committed step to programmed cell death and is therefore critical in regulating apoptosis in immune cells. This

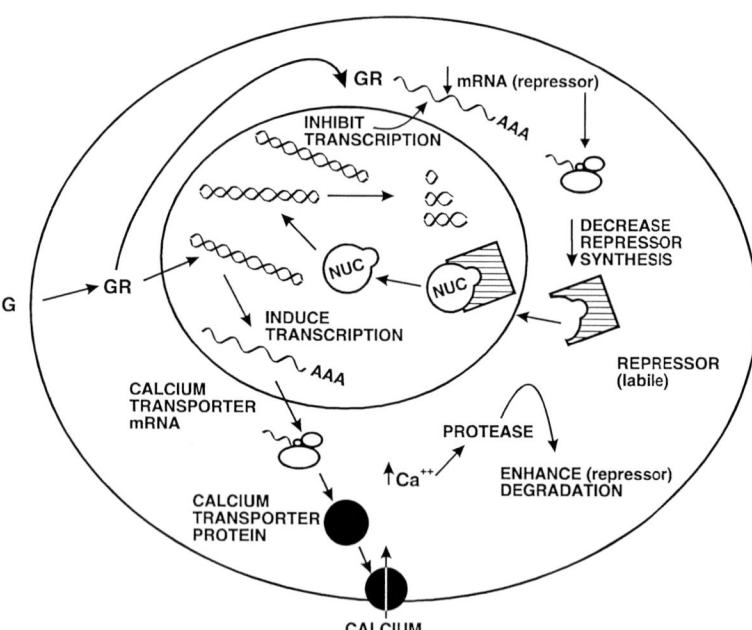

Fig. 1. Repressor model for glucocorticoid-induced apoptosis activation in rat thymocytes. Glucocorticoid (*G*)-induced apoptosis is initiated directly through the glucocorticoid receptor (*GR*). Cellular responses to the activated receptors by induction or inhibition of specific genes may account for the activation of the apoptotic process. The evidence for a nuclease (*NUC*) that is constitutively expressed in all cells is incorporated into our hypothesis that apoptotic nuclease activity is inhibited by a labile repressor protein. Glucocorticoid treatment could allow for activation of this nuclease by decreasing the repressor protein MRNA levels. Glucocorticoids may concomitantly act to eliminate the repressor protein by stimulating its degradation. This degradation could occur with the induction of a specific protease or with the induction of a Ca^{2+} transporter protein that increases Ca^{2+} levels, resulting in activation of a protease. Any of these actions could result in release of an active nuclease

nuclease was originally thought to be induced by glucocorticoids (COMPTON and CIDLOWSKI 1987), based on the inhibitory effect of RNA and protein synthesis blockers (WYLLIE et al. 1984; COHEN and DUKE 1984), but there is no evidence for an increase in either levels of nuclease protein or mRNA. More recent work is suggestive of a constitutively expressed nuclease (SCHWARTZMAN and CIDLOWSKI 1993b; GAIDO and CIDLOWSKI 1991; NIKONOVA et al. 1993). Such a nuclease would require inhibition until the proper apoptotic signal was received. Several lines of evidence suggest that the nuclease exists in a complex with a repressor protein. When nuclear extracts from dexamethasone-treated rat thymocytes were assayed for nuclease activity, a low molecular weight (18 kDa) nuclease was identified (NUC18) (GAIDO and CIDLOWSKI 1991). In control cells, those not treated with dexamethasone, nuclease activity was apparent at a much higher molecular weight (approximately 100 kDa), indicating NUC18 is part of a complex that separates after glucocorticoid treatment. Additionally, thymocyte nuclear extracts from similar control cells were capable of internucleosomal DNA degradation only after being passed over a sucrose gradient or gel filtration column (SCHWARTZMAN and CIDLOWSKI 1993b). This again suggests that the nuclease is associated with other proteins that are capable of preventing any activity until the nuclease is separated from the complex, either by specific glucocorticoid action or by a physical means, such as a sucrose gradient.

The inhibition of glucocorticoid-induced apoptosis by the antagonist RU486 (MCCONKEY et al. 1989; CARON-LESLIE and CIDLOWSKI 1991; COMPTON and CIDLOWSKI 1986) indicates that programmed cell death is initiated directly through the glucocorticoid receptor. The activated glucocorticoid receptor has several responses, discussed in this review, that, singly or in concert, may account for induction of apoptosis. We hypothesize that glucocorticoids may inhibit the transcription of mRNA that encodes a repressor protein, thus resulting in increased levels of unrepressed nuclease. This proposal is supported by the fact that, in some cases, blocking protein synthesis alone is enough to cause apoptosis (MARTIN et al. 1990; CARON-LESLIE and CIDLOWSKI 1994). This suggests that the repressor protein is labile and requires constant synthesis to keep the nuclease in check. In addition to decreasing repressor mRNA levels, glucocorticoids may increase transcription of Ca^{2+} transporter mRNA, such as a Ca^{2+} pore protein (MCCONKEY et al. 1989a), The newly synthesized Ca^{2+} transporter protein could contribute to the observed Ca^{2+} influx, which could then activate a protease that degrades the repressor protein. Or, glucocorticoids could directly increase transcription of a protease mRNA, which could specifically degrade the repressor protein. Interestingly, recent reports implicate the cysteine protease interleukin-1β converting enzyme (ICE) in the apoptotic process (GAGLIARDINI et al. 1994; MIURA et al. 1993). Any of the proposed mechanisms would result in increased levels of active nuclease, capable of cleaving the DNA in the characteristic internucleosomal pattern of apoptosis.

If the nuclease is constitutively expressed, then why do we observe inhibition of glucocorticoid-induced apoptosis in the presence of RNA and protein synthesis inhibitors? There are several explanations. First, the production of a protein involved in regulating the level of repressor protein may be decreased, thus interfering with the normal effect of glucocorticoids. Second, the inhibited proteins may regulate Ca^{2+} influx, which plays a critical role in glucocorticoid-induced apoptosis. Third, the inhibitors may block production of the protease necessary to release the repressor from the apoptotic nuclease. Thus, inhibitors of RNA and protein synthesis could block glucocorticoid-induced apoptosis through several pathways.

Constitutive expression of the apoptotic nuclease could provide several advantages. For example, the constant presence of a repressed nuclease would allow the cell to quickly initiate apoptosis, because de novo synthesis of the nuclease would not be necessary. Also, the metabolic demands on the cell would be reduced and would require less energy because the substrate and the enzyme are colocalized, preventing the need to export RNA and import protein, actions that would become increasingly difficult in a dying cell.

The importance of apoptosis in the maintenance of many different systems is becoming increasingly apparent. It is crucial, therefore, to more clearly define the mechanisms of this fascinating process. We have begun this task by proposing the repressor model, which is based on current knowledge of apoptosis in the immune system. The immune system is a great resource for studying the intricacies of apoptosis. As detailed in this review, even one apoptotic signal given to one cell type (e.g., glucocorticoid treatment of immature thymocytes) results in a multifaceted response (e.g., Ca^{2+} flux, DNA degradation). Knowing how these apoptotic responses interact to result in the death of a cell will provide great insight into this essential component of life.

Acknowledgement.   This work was supported by NIH DK 32078.

References

Alnemri E, Litwack G (1990) Activation of internucleosomal DNA cleavage in human CEM lymphocytes by glucocorticoid and novobiocin. J Biol Chem 265: 17323–17333

Alnemri E, Fernandes T, Haldar S, Croce C, Litwack G (1992) Involvement of bcl-2 in glucocorticoid-induced apoptosis of human pre-B-leukemias. Cancer Res 52: 491–495

Ambellan E, Hollander V (1966) The role of ribonuclease in regression of lymphosarcoma P1798. Cancer Res 26: 903–908

Arends M, Wyllie A (1991) Apoptosis: mechanisms and roles in pathology. Int Rev Exp Pathol 32: 223–251

Arends M, Morris R, Wyllie A (1990) Apoptosis. The role of the endonuclease. Am J Pathol 136: 593–608

Bansal N, Houle A, Melnykovych G (1990) Dexamethasone-induced killing of neoplastic cells of lymphoid derivation: lack of early calcium involvement. J Cell Physiol 143: 105–109

Barbour K, Berger S, Berger F, Thompson E (1988) Glucocorticoid regulation of the genes encoding thymidine kinase, thymidylate synthase, and ornithine decarboxylase in P1798 cells. Mol Endocrinol 2: 78–84

Baughman G, Harrigan M, Campbell N, Nurrish S, Bourgeois S (1991) Genes newly identified as regulated by glucocorticoids in murine thymocytes. Mol Endocrinol 5: 637–644

Baughman G, Lesley J, Trotter J, Hyman R, Bourgeois S (1992) Tcl-30, A new T cell-specific gene expressed in immature glucocorticoid-sensitive thymocytes. J Immunol 149: 1488–1496

Bell P, Borthwick N (1975) Glucocorticoid effects on DNA-dependent RNA polymerase activity of rat thymocytes. J Steroid Biochem 7: 1147–1150

Blackwell G, Carnuccio R, DiRosa M, Flower R, Parente L, Persico P (1980) Macrocortin: a polypeptide causing the anti-phospholipase effect of glucocorticoids. Nature 287: 147–149

Burnstein K, Jewell C, Cidlowski J (1990) Human glucocorticoid receptor cDNA contains sequences sufficient for receptor down-regulation. J Biol Chem 265: 7248–7291

Burton A, Storr J, Dunn W (1967) Cytolytic action of corticosteroids on thymus and lymphoma cells in vitro. Can J Biochem 45: 289–297

Caron-Leslie L, Cidlowski J (1994) Evaluation of the role of protein synthesis inhibition in apoptosis in glucocorticoid sensitive and glucocorticoid resistant S49 cells. Endocr J 2: 47–52

Caron-Leslie L, Cidlowski J (1991) Similar actions of glucocorticoids and calcium on the regulation of apoptosis in S49 cells. Mol Endocrinol 5: 1169–1179

Caron-Leslie L-A, Evans R, Cidlowski J (1994) (in press)

Cidlowski J (1982) Glucocorticoids stimulate ribonucleic acid degradation in isolated rat thymic lymphocytes in vitro Endocrinology 111: 184–190

Clarke C, Wills E (1978) The activation of lymphoid tissue lysosomal enzymes by steroid hormone. J Steroid Biochem 9: 135–139

Cohen J, Duke R (1984) Glucocorticoid activation of a calcium-dependent endonuclease in thymocyte nuclei leads to cell death. J Immunol 132: 38–42

Colbert R, Young D (1986) Glucocorticoid-induced messenger ribonucleic acids in rat thymic lymphocytes: rapid primary effects specific for glucocorticoids. Endocrinology 119: 2598–2605

Collins R, Vershuer L, Harmon B, Prentice R, Popt J, Kerr J (1989) Spontaneous programmed death (apoptosis) of B-chronic lymphocytic leukaemia cells following their culture in vivo. Br J Haematol 71: 343–350

Compton M, Cidlowski J (1986) Rapid in vivo effects of glucocorticoids on the integrity of rat lymphocyte genomic DNA. Endocrinology 118: 39–45

Compton M, Cidlowski J (1987) Identification of a glucocorticoid-induced nuclease in thymocytes. J Biol Chem 262: 8288–8292

Compton M, Cidlowski J (1992) Thymocyte Apoptosis: a model of programmed cell death. Trends Endocrinol Metab 3: 17–23

Compton M, Haskill J, Cidlowski J (1988) Analysis of glucocorticoid actions on rat thymocyte deoxyribonucleic acid by fluorescence-activated flow cytometry. Endocrinology 122: 2158–2164

Cowan W, Sorensen G (1964) Electron microscopic observations of acute thymic involution produced by hydrocortisone. Lab Invest 13: 353–370

Dougherty T, White A (1945) Functional alterations in lymphoid tissue induced by adrenal cortical secretion. Am J Anat 77: 81–116

Dowd D, MacDonald P, Komm B, Haussler M, Miesfeld R (1991) Evidence for early induction of calmodulin gene expression in lymphocytes undergoing glucocorticoid-mediated apoptosis. J Biol Chem 266: 18423–18426

Dowd D, MacDonald P, Komm B, Haussler M, Miesfeld R (1992) Stable expression of the calbindin-D28K complementary DNA interferes with the apoptotic pathway in lymphocytes. Mol Endocrinol 6: 1843–1848

Eastman-Reks S, Vedeckis W (1986) Glucocorticoid inhibition of c-myc, c-myb, and c-Ki-ras expression in a mouse lymphoma cell line. Cancer Res 46: 2457–2462

Gagliardini V, Fernandez P-A, Lee R, Drexler H, Rotello R, Fishman M, Yuan J (1994) Prevention of vertebrate neuronal death by the $crmA$ gene. Science 263: 826–828

Gaido M, Cidlowski J (1991) Identification, purification, and characterization of a calcium-dependent endonuclease (NUC18) from apoptotic rat thymocytes. J Biol Chem 266: 18580–18585

Harrigan M, Baughman G, Campbell N, Bourgeois S (1989) Isolation and characterization of glucocorticoid- and cyclic AMP-induced genes in T lymphocytes. Mol Cell Biol 9: 3438–3446

Hasbold J, Klaus G (1990) Anti-immunoglobulin antibodies induce apoptosis in immature B cell lymphomas. Eur J Immunol 20: 1685–1690

Hirata F (1981) The regulation of lipomodulin, a phospholipase inhibitory protein, in rabbit neutrophils by phosphorylation. J Biol Chem 256: 7730–7733

Hockenberry D, Nunez G, Milliman C, Schreiber R, Korsmeyer S (1990) Bcl-2 is an inner mitochondrial membrane protein that blocks programmed cell death. Nature 348: 334–336

Hockenberry D, Zutter M, Hickey W, Nahm N, Korsmeyer S (1991) Bcl-2 protein is topographically restricted in tissues characterized by apoptotic cell death. Proc Natl Acad Sci USA 88: 6961–6965

Holder M, Knox K, Gordon J (1992) Factors modifying survival pathways of germinal center B cells. Glucocorticoids and transforming growth factor-b, but not cyclosporin A or anti-CD19, block surface immunoglobulin mediated rescue from apoptosis. Eur J Immunol 22: 2725–2728

Iseki R, Kudo Y, Iwata M (1993) Early mobilization of Ca^{2+} is not required for glucocorticoid-induced apoptosis in thymocytes. J Immunol 151: 5198–5207

Kaiser N, Edelman I (1977) Calcium dependence of glucocorticoid-induced lymphocytolysis. Proc Natl Acad Sci USA 74: 638–642

Kaiser N, Edelman I (1978) Further studies on the role of calcium in glucocorticoid-induced lymphocytolysis. Endocrinology 103: 936–942

Kerr J, Wyllie A, Currie A (1972) Apoptosis: a basic biological phenomenon with wide ranging implications in tissue kinetics. Br J Cancer 26: 239–257

Lam M, Dubyak G, Distelhorst C (1993) Effect of glucocorticosteroid treatment on intracellular calcium homeostasis in mouse lymphoma cells, Mol Endocrinol 7: 686–693

Leathers V, Linse S, Forsen S, Norman A (1990) Calbindin-D_{28K} a 1α, 25-dihydroxyvitamin D3 induced calcium-binding protein, binds five or six Ca^{2+} ions with high affinity. J Biol Chem 265: 9838–9841

MacDonald R, Cidlowski J (1981) Glucocorticoid regulation of two serine hydrolases in rat splenic lymphocytes in vitro. Biochim Biophys Acta 678: 18–26

MacDonald R, Cidlowski J (1982) Glucocorticoids inhibit precursor incorporation into protein in splenic lymphocytes by stimulating protein degradation and expanding intracellular amino acid pools. Biochim Biophys Acta 717: 236–247

MacDonald R, Martin T, Cidlowski J (1980) Glucocorticoids stimulate protein degradation in lymphocytes: a possible mechanism of steroid-induced cell death. Endocrinology 107: 1512–1524

Makman M, Dvorkin B, White A (1968) Influences of cortisol on the utilization of precursors of nucleic acids and protein by lymphoid cells in vitro. J Biol Chem 243: 1485–1497

Makman M, Dvorkin B, White A (1971) Evidence for induction by cortisol in vitro of a protein inhibitor of transport and phosphorylation processes in rat thymocytes. Proc Natl Acad Sci 68: 1269–1273

Martin S, Lennon S, Bonham A, Cotter T (1990) Induction of apoptosis (programmed cell death) in human leukemic HL-60 cells by inhibition of RNA or protein synthesis. J Imm 145: 1859–1867

Mashburn B, Freeman C. Hollander V (1969) Effect of in vitro glucocorticoid treatment on acid ribonuclease activity in P1798 lymphosarcoma cells. Proc Soc Exp Biol Bed 131: 108–111

McConkey D, Nicotera P, Hartzell P, Bellomo G, Wyllie A, Orrenius S (1989a) Glucocorticoids activate a suicide process in thymocytes through an elevation of cytosolic Ca^{2+} concentration. Arch Bioc Biop 269: 365–370

McConkey D, Hartzell P, Nicotera P. Orrenius S (1989b) Calcium-activated DNA fragmentation kills immature thymocytes. FASEB J 3: 1843–1849

McConkey D, Hatzell P, Orrenius S (1990) Rapid turnover of endogenous endonuclease activity in thymocytes: effects of inhibitors of macromolecular synthesis. Arch Biochim Biophys 278: 284–287

McConkey D, Aguilar-Santelises M, Hartzell P, Eriksson I, Mellstedt H, Orrenius S, Jondal M (1991) Induction of DNA fragmentation in chronic B-lymphocytic leukemia cells. J Immunol 146: 1072–1076

McDonald A, Goldfine I (1988) Glucocorticoid regulation of insulin receptor gene transcription in IM-9 cultured lymphocytes. J Clin Invest 81: 499–504

Merino R, Ding L, Veis D, Korsmeyer S, Nunez G (1994) Developmental regulation of the Bcl-2 protein and susceptibility to cell death in B lymphocytes. EMBO J 13: 683–691

Miura M, Zhu H, Rotello R, Hartwieg E, Yuan J (1993) Induction of apoptosis in fibroblasts by IL-1-beta converting enzyme, a mammalian homolog of the C. elegans cell death gene ced-3. Cell 75: 653–660

Miyashita T, Reed J (1992) bcl-2 gene transfer increases relative resistance of S49.1 and WEHI7.2 lymphoid cells to cell death and DNA fragmentation induced by glucocorticoids and multiple chemotherapeutic drugs. Cancer Res 52: 5407–5411

Morita Y, Munck A (1964) Effects of glucocorticoids in vivo and in vitro on net glucose uptake and amino acid incorporation in rat thymus cells. Biochim Biophys Acta 93: 150–157

Munck A, Leung K (1977) Glucocorticoid receptors and mechanisms of action, In: Pasqualini J (ed) Receptors and mechanisms of action of steroid hormones. Decker, New York, pp 311–397

Nicholson M, Young D (1979) Independence of the lethal actions of glucocorticoids on lymphoid cells from possible hormone effects of calcium uptake. J Supramol Struct 10: 165–174

Nieto M, Gonzalez A, Lopez Rivas A, Diaz-Espada K, Gambon F (1990) IL-2 protects against anti-CD3-induced cell death in human medullary thymocytes. J Immunol 145:1364–1368

Nikonova L, Beletsky I, Umansky S (1993) Properties of some nuclear nucleases of rat thymocytes and their changes in radiation-induced apoptosis. Eur J Biochem 215: 893–901

Nordeen S, Young D (1976) Glucocorticoid action on rat thymic lymphocytes: experiments utilizing adenosine to support cellular metabolism lead to a reassessment of catabolic hormone action. J Biol Chem 251: 7295–7303

Oberhammer F, Fritsch G, Pavelka M, Froschl G, Tiefenbacher R, Purchio T, Schulte-Hermann R (1992) Induction of apoptosis in cultured hepatocytes and in the regressing liver by transforming growth factor-b1 occurs without activation of an endonuclease. Toxicol Lett 64/65: 701–704

Oberhammer F, Wi)son J, Dive C, Morris I, Hickman J, Wakeling A, Walker P, Sikorska M (1993)

Apoptotic death in epithelial cells: cleavage of DNA to 300 and/or 500 kb fragments prior to or in the absence of internucleosomal fragmentation. EMBO J 12: 3679–3684

Oldenburg N, Cidlowski J (1994) (in preparation)

Owens G, Hahn W, Cohen J (1991) Identification of mRNAs associated with programmed cell death in immature thymocytes. Mol Cell Biochem 11: 4177–4188

Pezzella F, Tse A, Cordell J, Pulford K, Gutter K, Mason R (1990) Expression of the bcl-2 oncogene protein is not specific for the 14;18 chromosome translocation. Am J Pathol 137: 225–232

Reed J (1994) BCl-2 and the regulation of programmed cell death. J Cell Biol 124: 1–6

Rouiller D, McKeon C, Taylor S, Gorden P (1988) Hormonal regulation of insulin receptor gene expression. J Biol Chem 263: 13185–13190

Schwartzman R, Cidlowski J (1993a) Apoptosis: the biochemistry and molecular biology of programmed cell death. Endocr Rev 14: 133–151

Schwartzman R, Cidlowski J (1993b) Mechanism of tissue-specific induction of internucleosomal deoxyribonucleic acid cleavage activity and apoptosis by glucocorticoids. Endocrinology 133: 591–599

Sentman C, Shutter J, Hockenberry D, Kanagawa O, Korsmeyer S (1991) bcl-2 inhibits multiple forms of apoptosis but not negative selection in thymocytes. Cell 67: 879–888

Smith C, Williams G, Kingston R, Jenkinson E, Owen J (1989) Anitbodies to CD3/T-cell receptor complex induce death by apoptosis in immature T cells in thymic cultures. Nature 337: 181–184

Vanderbilt J, Bloom K, Anderson J (1982) Endogenous nuclease. J Biol Chem 257: 13009–13017

Vaux D, Cory S, Adams J (1988) Bcl-2 gene promotes haemopoietic cell survival and cooperates with c-myc to immortalize pre-B cells. Nature 335: 440–442

Whitfield J, Perris A, Youdale T (1968) Destruction of the nuclear morphology of thymic lymphocytes by the corticosteroid cortisol. Exp Cell Res 52: 349–362

Wiernik P, MacLeod R (1965) The effect of a single large dose of 9 α-fluoroprednisolone on nucleo depolymerase activity and nucleic acid content of the rat thymus. Acta Endocrinol (Copenh) 49: 138–144

Williams G, Smith C, Spooncer E, Dexter T, Taylor D (1990) Haemopoietic colony stimulating factors promote cell survival by suppressing apoptosis. Nature 343: 76–79

Wyllie A (1980) Glucocorticoid-induced thymocyte apoptosis is associated with endogenous nuclease activation. Nature 284: 555–556

Wyllie A, Morris R, Smith A, Dunlop D (1964) Chromatin cleavage in apoptosis: association with condensed chromatin morphology and dependence on macromolecular synthesis. J Pathol 142: 67–77

Thymocyte Apoptosis
by Glucocorticoids and cAMP

M. Jondal[1], Y. Xue[1], D.J. McConkey[2], and S. Okret[3]

1 Negative Selection in the Thymus

The T cell receptor (TCR) repertoire is known to arise from a series of random genetic recombinational events, similar to the generation of immunoglobulin diversity, which give rise to antigen binding heterodimers capable of responding to a wide spectrum of MHC-presented peptides. However, in this differentiation process only a small number of cells reach the mature state, as most are deleted by a process called negative selection (VON BOEHMER et al. 1989; JANEWAY et al. 1992). In the mouse around 50×10^6 cells/day are formed in the thymus; of these, only approximately $1-2 \times 10^6$ mature to CD4+ helper and CD8+ cytotoxic T cells. Negative selection can be a consequence of either lack of self-MHC recognition or high avidity recognition of dominant self-peptides (ASHTON-RICKARDT et al. 1994; JANEWAY et al. 1992; SEBZDA et al. 1994).

This deletion process targets the cortical TCR[Lo] CD4+CD8+ subpopulation of thymocytes and is mediated by TCR engagement. Negative selection can be mimicked using activating anti-TCR antibodies or bacterial superantigens, resulting in DNA fragmentation and cell death typical of apoptosis (SMITH et al. 1989; SHI et al. 1989; McCONKEY et al. 1989). The relevance of these observations has been

---

[1] Microbiology and Tumor Biology Center, Karolinska Institute, 17177 Stockholm, Sweden
[2] Department of Cell Biology, MD Anderson Cancer Center, 1515 Holcombe Boulevard, Houston, TX 77030, USA
[3] Department of Medical Nutrition, Karolinska Institute, Novum 14157 Huddinge, Sweden

confirmed in studies utilizing mice transgenic for a particular TCR, in which administration of specific antigenic peptides elicits an identical apoptotic response (MURPHY et al. 1990; MAMALAKI et al. 1992).

Paradoxically, the TCR on CD4$^+$CD8$^+$ thymocytes also appear to be required to promote differentiation into functionally mature CD4$^+$CD8$^-$ or CD4$^-$CD8$^+$ thymocytes. Thus, one of the greatest challanges for immunlogists at present involves defining the molecular mechanisms underlying this dual signaling function of the TCR. One proposal is that positive selection is promoted by moderate avidity TCR interactions, whereas high avidity interactions lead to apoptosis. Strong support for this model has recently emerged from studies with transgenic thymocytes in organ culture, in which low concentrations of peptides promote maturation while high concentrations are lethal to the cells (ASHTON-RICKARDT et al. 1994; SEBZDA et al. 1994; HOGQUIST et al. 1994). Additionally, it appears that independent signal transduction pathways contribute to the outcome of TCR engagement. For example, it has been shown that TCR triggering is fairly inefficient at promoting apoptosis in vitro but requires an additional signal that can be contributed by Thy-1 (NAKASHIMA et al. 1991), CD28 (PUNT et al. 1994) or CD4/CD8 (McConkey et al., manuscript submitted). In addition, second signals provided by steroid hormones (ZACHARCHUK et al. 1990; IWATA and colleagues, this volume) or protein kinase C activation (McCONKEY et al. 1989) may inhibit TCR-mediated apoptosis. Thus, both TCR avidity differences and the presence or absence of parallel signaling pathways are likely to influence positive and negative selection.

A second observation that remains to be explained concerns why apoptosis is fairly readily observed in CD4$^+$CD8$^+$ thymocytes exposed to diverse stimuli, whereas in their immediate precursors (CD4$^-$CD8$^-$ cells) and in mature thymocytes it is not. Compelling evidence has recently emerged from several laboratories to suggest that developmental regulation of the expression of the *bcl*-2 oncogene is involved (GRATIOT-DEANS et al. 1993; VEIS et al. 1993; MOORE et al. 1994; ANDJELIC et al. 1993). Levels of Bcl-2 protein are relatively high in the apoptosis-resistant immature and mature compartments, while the level in the apoptosis-sensitive CD4$^+$CD8$^+$ cells are low. The important question that remains is how these fluctuations are promoted within the thymic microenvironment.

2 Glucocorticoids and the Thymus

Early findings demonstrated that stress-induced involution of the thymus is due to adrenal glucocorticoid (GC) release (SELYE 1936; INGLE 1940). Involution was caused by a process involving the typical morphological changes and DNA cleavage of apoptosis (WYLLIE et al. 1980). One crucial event in the induction of apoptosis is the activation of a Ca^{2+}-dependent endonuclease which cleaves DNA into nucleosome sized fragments. Several defined endonucleases have been implicated (GAIDO and CIDLOWSKI 1991; PEITSCH et al. 1993; BARRY and EASTMAN

1993). The endonuclease is constitutively present in isolated nuclei and can be activated by Ca^{2+} and Mg^{2+} treatment (COMPTON and CIDLOWSKI 1992). GC induced thymocyte apoptosis requires protein synthesis and RNA transcription and is associated with an increase in cytosolic Ca^{2+} (COMPTON and CIDLOWSKI 1992).

Thymocytes respond to GC by apoptosis both in vitro and in vivo. In vivo, the immature $CD4^+/CD8^+$ thymocyte fraction is rapidly killed whereas both the precurser population ($TCR^-/CD4^-/CD8^-$) and mature thymocytes ($CD4^+$ or $CD8^+$) are comparatively resistant (S. Chow, personal communication). Peripheral T cells are resistant to GC but become sensitive upon activation (KABELITZ 1993). Inflammatory mediators such as interleukins, interferon (IFN) and tumor necrosis factor (TNF) provoke the release of corticotropin releasing factor (CRF) and adrenocorticotropic hormone (ACTH) from hypothalamus and the pituary gland, respectively (BATEMAN et al. 1989; BESEDOVSKY et al. 1991). ACTH-induced adrenal GC release is an important negative feedback loop to prevent overactivation in the peripheral immune system (GONZALO et al. 1993).

Using anti-CD3 monoclonal antibodies as a model for negative selection, we had found that pretreatment of mice with a GC receptor antagonist (RU486, Roussel-Uclaf) protected immature $CD4^+/CD8^+$ thymocytes from an apoptotic reaction (JONDAL et al. 1993). More recently, using peptide treatment of TCR transgenic mice to induce apoptosis in immature thymocytes (MURPHY et al. 1990), we have found a similar protective effect (manuscript in preparation). Interestingly, the same thymocyte subpopulation was also protected from apoptosis induced by a cAMP-inducing drug, (N-ethyl)-carboxamide-adenosine (NECA), agonistic for adenosine A2 receptors. However, the involvment of adrenal GC release in the effects of anti-CD3 monoclonal antibodies, TCR binding peptides and NECA needs to be addressed in further studies with adrenalectomized mice to distinguish a possible role for increased corticosterone levels in blood. Still, adrenalectomy would not influence steroid synthesis within the thymic gland itself (VACCHIO et al. 1994) (see below).

In rats, the level of circulating GC is low during the first week of life, at a time when the thymic gland is large and active. In a recent intriguing report, VACCHIO et al. (1994) have identified GC production within the thymic gland itself. Enzymes involved in the GC synthetic pathway were shown to be present in radioresistant thymic epithelial cells which produced pregnenolone and deoxycorticorsterone in vitro.

Recent findings implicate certain proteases with specificity for aspartate residues, such as IL-1β converting enzyme (ICE), and the lytic proteins granzyme B and fragmentin-2 in the induction of apoptosis (JACOBSON and EVAN 1994; SHI et al. 1994). These proteins are homologous with the ced-3 gene product in C. elegans, necessary for apoptosis in certain defined cell populations. Downstream, these proteases may dysregulate the cyclin/cdc kinase system to initiate apoptosis (SHI et al. 1994). How, or if, GC and the activated glucocorticoid receptor (GR) are related to these effector pathways is presently not known.

Thus, GC is known to affect both immature $CD4^+/D8^+$ thymocytes and to control the peripheral immune system and GC may actually be synthesized within a subcompartment of the gland. These circumstances indicate that endogenous GC may participate as one important regulator of normal thymic differentiation, as earlier suggested (ZAZCHARACHUK et al. 1990; IWATA et al. 1991).

3 The Glucocorticoid Receptor

Glucocorticoids excert their effects cells via a specific receptor. It is generally believed that GCs, being lipophilic in nature, enter target cells by passive diffusion (BALLARD 1979; GIORGI and STEIN 1981), although in some systems there appear to be evidence in support of an active, energy driven transport mechanism (RAO 1981; SPINDLER et al. 1991; ALLÉRA and WILDT 1992). In the resting state, the GR exists as a large multiprotein heteromeric complex that contains one molecule of GR and a 90 kDa heat shock protein (HSP) dimer, HSP 56/59 and HSP 70. Upon binding of the hormone, the complex undergoes a process termed transformation, which results in the release of the free ligand-bound GR (for review, see PRATT 1993). This process permits the GR to translocate to the nucleus, where its biological effects are manifested through its ability to regulate the expression of a network of genes in a tissue-specific manner (BEATO 1988, 1991; LUCAS and GRANNER 1992). This is accomplished by the interaction of the GR with specific DNA sequences termed glucocorticoid response elements (GREs), which most often lies in the promoter region of regulated genes. The sequence of most GREs is partially palindromic in nature, and the GR-GRE complex contains a dimer of GR, with one molecule of GR contacting each half of the palindrome (TSAI et al. 1988; WRANGE et al. 1989). Once bound to the GREs the GR modulates (induces or represses) the activity of the target promoter (for reviews, see BEATO 1988, 1991).

The GR, like the other members of the nuclear receptor superfamily, has a conserved domain structure, each domain harboring distinct and independent functions. The most highly conserved domain is the central DNA-binding domain, while the NH_2-terminal domains are the most variable both in size and sequence. The NH_2-terminal domain of the GR harbors the major transactivation capacity. Weaker transactivation capacity is found in the 5'end of the COOH-terminal domain. The importance of that region for transactivation of target genes in vivo has been demonstrated by the inability of mutated GRs that lack this NH_2-terminal transactivation domain to induce apoptosis (DIEKEN et al. 1990). The COOH-terminal hormone binding domain is not only responsible for ligand binding but also for interaction with hsp90 and possibly for dimerization (for reviews, see WAHLI and MARTINEZ 1991; GRONENMEYER 1992).

The GR has been shown to be a phosphoprotein. Phosphorylated sites are present in the untransformed receptor and hyperphosphorylation is induced following ligand binding (ORTI et al. 1992). However, the exact function of phosphorylation/dephosphorylation in the GRs mechanism of action is still unclear. The main sites for phosphorylation has been localized to the NH_2-terminal transactivation domain, where six phosphoserines and one phosphothreonine were identified (BODWELL et al. 1991). This indicated a role for phosphorylation in transactivation. However, preliminary mutational studies have so far failed to identify a major role for the NH_2-terminal phosphorylation sites (MASON and HOUSLEY 1993).

No effect by cAMP or the protein phosphatase inhibitor okadaic acid on the phosphorylation of the GR has been observed (SOMERS and DEFRANCO 1992; MOYER et al. 1993). This was investigated since cAMP or okadaic acid in some experimental systems can enhance transcriptional stimulation by the GR (see below).

The GR induces transcription from GC target genes following binding to GREs. The location of these GREs with respect to the transcription start site can vary greatly, from within 200 base pairs, seen for example in the mouse mammary tumor virus (MMTV) gene (PAYVAR et al. 1983), to over 2 kilobases upstream, seen in the tyrosine aminotranserease (TAT) gene (JANTZEN et al. 1987). For many genes induced by GCs, it has now been shown that the GR does not act alone, but requires the presence of additional transcription factors. Although the GR can function alone when placed close to the TATA box in an artificial reporter gene, it is inactive when positioned further upstream (STRÄHLE et al. 1988). The inability of the GR to stimulate transcription from a distance alone can, however, be compensated for by other factors working in synergism manner with the receptor (STRÄHLE et al. 1988; SCHÜLE et al. 1988a,b). The GR has been demonstrated to cooperate with numerous transcription factors, including other steroid receptors (STRÄHLE et al. 1988; SCHÜLE et al. 1988; ANKENBAUER et al. 1988). Recently, full GC inducibility of the phosphoenolpyruvate carboxykinase gene was demonstrated to require the presence of a GRE and a basal promoter/cAMP response element which binds a cAMP response element binding protein (CREB). A direct protein–protein interaction between the GR and CREB could be detected, possibly explaining the functional cooperation between the two elements (IMAI et al. 1993).

4 Glucocorticoid Receptor and cAMP Signal Transduction Pathways

Signaling and transcriptional regulation through the second messenger cAMP occurs through a multistep process involving activation of protein kinase A (PKA) and subsequent phosphorylation of transcription factors such as CREB or activating transcription factors (ATFs) (GONZALES and MONTMINY 1989; HABNER 1990). The CREB/ATF family consists of a series of transcription factors that function through binding to the cAMP responsive element (CRE). Although each of the CREB/ ATF proteins bind CREs as homodimers, in some cases they may bind as heterodimers both within the CRFEB/ATF family or with members of the AP-1 transcription factor family, with different transcriptional effects as a result.

Several genes have been shown to be regulated by both GCs and cAMP. For example, GCs and cAMP synergistically activate transcription of genes encoding phosphoenolpyruvate carboxykinase, vasopressin, proenkaphalin and neuropeptide Y (IMAI et al. 1993; JOSHI and SEBOL 1991; VERBEEK et al. 1991; HIGUCHI et al. 1988). Similarly, cAMP and GCs synergistically activate transcription of several genes in the murine thymoma cell line WEHI-7 (HARRIGAN et al. 1989). Furthermore, GCs and cAMP synergistically trigger cell lysis in WEHI-7 and S49 lymphoma cells (VEDECKIS and BRADSLOW 1983; GRUOL et al. 1986). In contrast, the cAMP-mediated transcriptional activation of α-1 acidic glycoprotein gene transcription is repressed by GCs (STAUBER et al. 1992).

Recent studies have indicated that cAMP and/or PKA may play an important role in regulating signal transduction through the steroid hormone receptor superfamily (DENNER et al. 1990; POWER et al. 1991; RANGARAJAN et al. 1992; SOMERS and DEFRANCO 1992; NORDEEN et al. 1993: MOYER et al. 1993). For example, it has been shown that cAMP can, in certain cells and with some promoters, potentiate effects mediated by the GR. This can also be seen with transfected GC-regulated reporter genes which lack CREs (RANGARAJAN et al. 1992; SOMERS and DEFRANCO 1992; NORDEEN et al. 1993). This synergistic effect does not appear to involve the NH_2-terminal transactivation of COOH-terminal ligand binding domains of the GR but rather by enhanced DNA-binding activity for the GR to its cognate GRE (RANGARAJAN et al. 1992). No effect by cAMP/PKA on GR expression was observed. However, results from other groups have in other systems demonstrated that cAMP treatment leads to increased GR expression, which correlates with increased transcriptional activation of transfected or endogenous GC regulated genes (OIKARINEN et al. 1984; GRUOL et al. 1986; DONG et al. 1989). Thus, several mechanisms may be responsible for the enhancing activity. Although the exact mechanisms for the argumentation of GC responses by cAMP/PKA is unclear, it does not seem to involve direct GR phosphorylation (see above). Instead, phosphorylation of various components of the GR signal transduction pathway other than the GR may influence the transcriptional response. Phosphorylation may activate CREB or ATFS, which through a direct protein-protein interaction with the GR may influence transcriptional responses. In fact, a direct interaction between the GR and the CREB has been demonstrated in vitro, which might account for the synergistic activation of the phosphohenolpyruvate carboxykinase gene by GCs and cAMP seen in vivo (IMAI et al. 1993). However, the effect of cAMP on GR transcriptional activity may be very complex and also occur in cells lacking CREB (RANGARAJAN et al. 1992). Finally, the complexity of the system is demonstrated by the observation that the GC antagonist RU486 acquires agonistic properties when cells are treated with activators of PKA (NORDEEN et al. 1993).

5 cAMP Regulation of T Cell Activation and Apoptosis

Prostaglandins and pharmacologic agents that elevate cAMP are known to be relatively potent inhibitors of T cell activation, and previous work has shown that they are capable of blocking both early and late consequences of productive TCR engagement, such as phospholipase C activation and interleukin-2 production (PATEL et al. 1987; LERNER et al. 1988). Indeed, the existence of an interrelationship between the TCR and adenylate cyclase-regulated signal transduction pathways is strongly supported by the recent observation that the type I regulatory subunit and the catalytic subunit of cAMP-dependent protein kinase cocap and can be coimmunoprecipitated with the TCR following TCR triggering (SKÅLHEGG et al. 1994). When thymocytes and certain other lymphoid cell types are treated with prostaglandin E_2 or pharmacological agents that elevate cAMP, they undergo apoptosis (GRUOL et al. 1986; McCONKEY et al. 1990b; LEE et al. 1993; SUZUKI et al. 1991). cAMP-induced apoptosis has also been reported to be enhanced by TNF-α (KIZAKI et al. 1993). However, recent work by LEE et al. (1993) has shown that cAMP can antagonize T cell receptor-mediated apoptosis without affecting other relatively late molecular events induced by TCR triggering. This type of mutual inhibition by two apoptotic pathways is reminiscent of the effects of GCs or protein kinase C (PKC) on PKC-mediated apoptosis and may therefore contribute in some way to positive selection.

Importantly, moderate elevations in cAMP have also been reported to occur as a direct response to TCR triggering (LEDBETTER et al. 1986), and it is therefore possible that cAMP may have positive effects on TCR signal transduction function under some circumstances, particularly when potent PKC activation is also involved (PATEL et al. 1987).

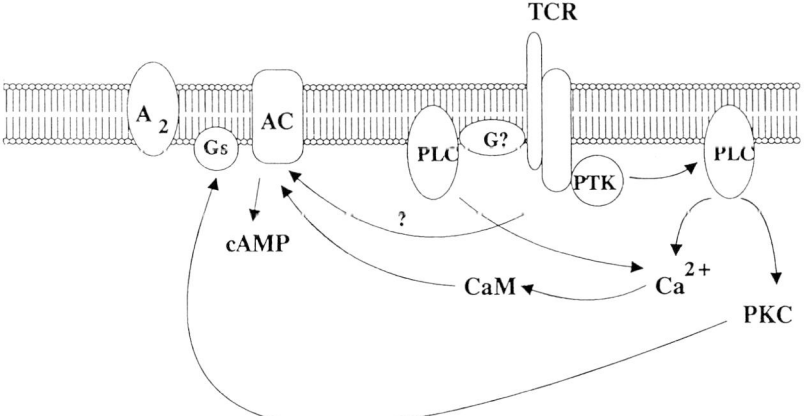

Fig. 1. Proposed pathways involved in the T cell receptor (TCR)-dependent potentiation of a cAMP response to the adenosine analogue NECA, using the Jurkat T cell line (KVANTA et al. 1989, 1990, 1991) *A₂*, adenosine A_2 receptor; *G*, G proteins; *AC*, adenylate cyclase; *PLC*, phospholipase C; *PTK*, protein tyrosine kinase; *CaM*, calmodulin

Using the human T cell line Jurkat we have found previously that stimulation of cells with anti-CD3 antibodies potentiates a cAMP response through adenosine A2 receptors (KVANTA et al. 1989,1990). Further work using the adenylate cyclase stimulator forskolin indicated that there are at least two different mechanisms involved in this receptor cross-talk, one of which depends on PKC (KVANTA et al. 1991) (Fig. 1).

In summary, moderate cAMP elevations, in combination with defined transductions signals, may have a positive effect on T cell activation whereas high elevations may have a negative effect including the direct or indirect (through GR) induction of apoptosis.

6 Potentiation of Glucocorticoid-Induced Apoptosis by cAMP

As discussion above, the capacity of cAMP to promote GR expression and function is well documented. Given that cAMP and GCs are each capable of inducing thymocyte apoptosis when administered individually, these observations beg the question of whether they might synergize when added together. Indeed, previous work by Bourgeois and coworkers has demonstrated that cAMP promotes GC-mediated cytolysis of T cells via a mechanism involving enhancement of GR function (GRUOL et al. 1986). Moreover, we have shown that cAMP potentiates GC-mediated apoptosis in thymocytes via a mechanism involving increased hormone binding that is independent of effects of cAMP on GR expression (McCONKEY et al. 1993). Also, a similar effect had been found in the T cell line CEM-C7 (M. Jondal, manuscript in preparation). Pretreatment of these cells with cAMP for 24 h increases both the apoptotic response to GC and hormone binding and both of these effects are dose-related. Potention of GC-induced apoptosis by cAMP can also be observed in leukemic cells from patients with chronic B lymphocytic leukemia (M. Aguilar-Santelises et al., submitted). Furthermore, we have also obtained evidence for the relevance of this phenomenon to thymocyte apoptosis in vivo in experiments using NECA, which appears to induce thymocyte apoptosis via elevations of cAMP. When GC and NECA are administered to mice together, efficient apoptosis can be observed at doses of the agents which alone are insufficient to promote an effect (JONDAL et al. 1993). Interestingly, the effects of high dose NECA are inhibited by the steroid receptor antagonist RU486, suggesting that the basal levels of steroid present in the circulation may be required for the effect. Together, these observations suggest that cAMP might promote GR function at subthreshold levels of steroid hormone.

7 Proposed Model

Negative selection in thymus and in the peripheral immune system may partly be dependent on the induction of apoptosis by endogenous GC. Cyclic AMP-dependent signaling may also be important for the induction of apoptosis in lymphoid cells, acting both independently and dependently of GC. It is not clear at present which level in the signaling chain, GC and cAMP interact. It might be either at the level of transcriptional regulatory proteins or at the gene regulation level.

It should be pointed out that thymocyte apoptosis may also occur independent of GC, as exemplified by in vitro experiments using either single cell suspensions or organ tissue cultures, However, in such experiments serum, containing GC, is often used and recent data suggest that the epithelial component within the thymus can produce its own GC (VACCHIO et al. 1994). It is likely, though, that an apoptotic reaction can be the consequence of the "unbalancing" of many different transduction signals which may act upon particular switch and effector molecules (MCCONKEY et al. 1990a). The hypothetical role for GC in thymic negative selection presented in this chapter should be looked upon as being one important component in a system that may have a considerable amount of redundancy.

In summary, we suggest that without recognition of self (lack of positive selection) thymocytes are vulnerable to apoptosis induction mediated by endogenous GC (Fig. 2). Medium activity TCR interaction with self (positive selection) would rescue thymocytes from apoptosis by some undefined intracellular signal, possibly involving PKC (MCCONKEY et al. 1990a). High affinity interaction with self would lead to apoptosis (negative selection) by intracellular signals associated with cAMP, including effects both independent and dependent on the GR.

If steroid receptors are of major importance in immunoregulation, that may also have some bearing on the well known difference in immune reactivity

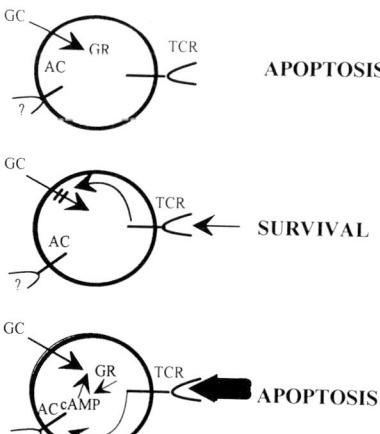

Fig. 2. The role of glucocorticoid (*GC*), glucocorticoid receptor (*GR*), cAMP and T cell receptor (*TCR*) in thymic selection. For further explanation, see text

between sexes (GROSSMAN 1989). The overall hormonal background may fine-tune the GR to a tighter negative immune regulation in males than in females. This, in turn, might be related to the higher occurrence of some autoimmune diseases in females (AHMED et al. 1985).

References

Alléra A, Wildt L (1992) Glucocorticoid-recognizing and -effector sites in rat liver plasma membrane. Kinetics of corticosterone uptake by isolated membrane vesicles-I. Binding and transport. J Steroid Biochem Mol Biol 42: 737–756

Andjelic S, Jain N, Nikolic-Zugic J (1993) Immature thymocytes become sensitive to calcium-mediated apoptosis with the onset of CD8, CD4 and the T cell receptor expression: a role for bcl-2? J Exp Med 178: 1745–1751

Ankenbauer W, Strähle U, Schütz G (1988) Synergistic action of glucacorticoid and estradiol responsive elements. Proc Natl Acad Sci USA 85: 7526–7530

Ahmed AS, Penhale WJ, Talal N (1985) Sex hormones, immune responses, and autoimmune diseases. Am J Pathol 121: 531–551

Ashton-Rickardt PG, Bandeira A, Delaney JR, van Kaer L, Pircher HP, Zinkernagel RM, Tonegawa S (1994) Evidence for a differential acidity model of T cell selection in the thymus. Cell 76(4): 593–596

Ballard PL (1979) Delivery and transport of glucocorticoids to target cells. In: Baxter JD, Rosseau GG (eds) Glucocorticoid hormone action. Monogr Endocrinol 12: 25–48

Barry MA, Eastman A (1993) Identification of deoxyribonuclease II as an endonuclease involved in apoptosis. Biochem Biophys 1: 440–450

Bateman A, Singh A, Kral T, Solomon S (1989) The immune-hypothalamic-pituitary-adrenal axis. Endocr Rev 10: 92–106

Beato M (1988) Gene regulation by steroid hormones. Cell 56: 335–344

Beato M (1991) Transcriptional control by nuclear receptors. FASEB J 5: 2044–2051

Besedovsky HO, del Rey A, Klusman I, Furukawa H, Monge-Arditi G, Kabiersch A (1991) Cytokines as modulators of the hypothalamus-pituitary-adrenal axis. J Steroid Biochem Mol Biol 40: 613–621

Bodwell JE, Orti E, Coull JM, Pappin DJC, Smith LI, Swift F (1991) Identification of phosphorylated sites in the mouse glucocorticoid receptor. J Biol Chem 266: 7549–7555

Compton MM, Cidlowski JA (1992) Thymocyte apoptosis. A model of programmed cell death. TEM 3(1): 17–23

Denner LA, Weigel NL, Maxwell BL, Schrader WT, O'Mally BW (1990) Regulation of progesterone receptor-mediated transcription by phosphorylation. Science 250: 1740–1743

Dieken ES, Meese EJ, Miesfeld RL (1990) nt' glucocorticoid receptor transcripts lack sequences encoding the amino-terminal transcriptional modulatory domain. Mol Cell Biol 10: 4574–4581

Dong Y, Aronsson M, Gustafsson J-Å, Okret S (1989) The mechanism of cAMP-induced glucocorticoid receptor expression. J Biol Chem 264: 13679–13683

Gaido ML, Cidlowski JA (1991) Indendification, purification and characterization of a calcium dependent endonuclease (NUC18) from apoptotic rat thymocytes. J Biol Chem 206(28) : 18580–18585

Giorgi EP, Stein WD (1981) The transport of steroids into animal cells in culture. Endocrinology 108: 688–697

Gonzales GA, Montminy MR (1989) Cyclic AMP stimulates somatostatin gene transcription by phosphorylation of CREB at serine 133. Cell 59: 675–680

Gonzalo JA, González-Garcia A, Martínez-A C, Kroemer G (1993) Glucocorticoid mediated control of the activation and clonal deletion of peripheral T cells in vivo. J Exp Med 177: 1239–1246

Gratiot-Deans J, Ding L, Turka LA, Nuñez G (1993) Bcl-2 proto-oncogene expression during human T cell development. Evidence for biphasic regulation. J Immunol 151: 83–91

Gronemeyer H (1992) Control of transcription activation by steroid hormone receptors. FASEB J 6: 2524–2529

Grossman C (1989) Possible underlying mechanisms of sexual dimprphism in the immune response, fact and hypothesis. J Steroid Biochem 34: 241–251

Gruol DJ, Campell N, Burgeois 5 (1986) Cyclic AMP-dependent protein kinase promotes glucocarticoid receptor function. J Biol Chem 261: 4909–4914

Habner JF (1990) Cyclic AMP response element binding proteins: a cornucopia of transcription factors. Mol Endocrinol 4: 1087–1094

Harrigan MT, Baughman G, Campbell F, Burgeois S (1989) Isolation and characterization of glucocorticoid- and cyclic AMP-induced genes in T lymphocytes. Mol Cell Biol 9: 3448–3456

Higuchi H, Yang H-YT, Sabol SL (1988) Rat neuropeptide Y precursor gene expression, J Biol Chem 263: 6288–6295

Hogquist KA, Jameson SC, Heath WR, Howard JL, Bean MJ, Carbone FR (1994) T Cell receptor antagonist peptides induce positive selection. Cell 76: 17–27

Imai E, Miner JN, Mitchell JA, Yamamoto KR, Granner DK (1993) Glucocorticoid receptor-cAMP response element-binding protein interaction and the response of the phosphoenolpyruvate carboxykinase gene to glucocorticoids. J Biol Chem 268: 5353–5356

Ingle DJ (1940) Effect of two steroid compounds on weight of thymus of adrenalectomized rats. Proc Soc Exp Biol Med 44: 174–175

Iwata M, Hanaoka S, Sato K (1991) Rescue of thymocytes and T cell hybridomas from glucocarticoid induced apoptosis by stimulation via the T cell receptor/CD3 complex: a possible in vitro model for positive selection of the T cell repertoire. Eur J Immunol 21: 643–648

Jacobson M, Evan GI (1994) Breaking the ICE. Curr Biol 4: 4–7

Janeway C, Rudensky A, Rath S, Murphy D (1992) It is easier for a camel to pass the needle's eye. Curr Biol 2(1): 26–28

Jantzen H-M, Strähle U, Gloss B, Stewart F. Schmid W. Boshart M, Mikicek R, Schütz G (1987) Cooperativity of glucocorticoid response elements located far upstream of the tyrosine aminotransferase gene. Cell 49: 29–38

Jondal M, Okret S, McConkey D (1993) Killing of immature CD4+CD8+ thymocytes in vivo by anti-CD3 or 5'-(N-ethyl)-carboxamido-adenosine is blocked by glucocorticoid receptor antagonist RU-486. Eur J Immunol 23: 1246–1250

Joshi J, Sabol SL (1991) Proenkephalin gene expression in C6 rat glioma cells: potentiation of cyclic adenosine-3 , 5'-monophosphate-dependent transcription by glucocorticoids. Mol Endocrinol 5: 1069–1080

Kabelitz D (1993) Induction of activation-driven death (apoptosis) in activated but not resting peripheral blood T cells. J Immunol 150(10): 4338–4345

Kizaki H, Nakada S, Ohnishi Y, Azuma Y, Mizuno Y, Tadakuma T (1993) Tumor necrosis factor-alpha enhances cAMP-induced programmed cell death in mouse thymocytes. Cytokine 5(4): 342–347

Kvanta A, Nordstedt C, Jondal M, Fredholm BB (1989) Activation of protein kinase C via the T cell receptor complex potentiates cyclic AMP responses in T cells. Naunyn Schmiedebergs Arch Pharmacol 340: 715–717

Kvanta A, Gerwins P, Jondal M, Fredholm BB (1990) Stimulation of T cells with OKT3 antibodies increases forskolin binding and cyclic AMP accumulation. Cell Signal 2: 461–470

Kvanta A, Jondal M, Fredholm BB (1991) CD3 dependent increase in cyclic AMP in human T-cells following stimulation of the CD2 receptor. Biochim Biophys Acta 1093: 178–183

Ledbetter JA, Parsons M, Martin PJ, Hansen JA, Rabinovitch PS, June CH (1986) Antibody binding to CD5 (Tp67) and TP44 T cell surface molecules: effects on cyclic nucieotides, cytoplasmic free calcium, and cAMP-mediated suppression. J Immunol 137: 3299–3305

Lee MR, Liou ML, Yand YF, Lai MZ (1993) cAMP analogs prevent activation-induced apoptosis of T cell hybridomas. J Immunol 151: 5208–5217

Lerner A, Jacobson B, Miller RA (1988) Cyclic AMP concentrations modulate both calcium flux and hydrolysis of phosphatidyinositot phosphates in mouse T lymphocytes. J Immunol 140: 936–940

Lucas PC, Granner DK (1992) Hormone response domains in gene transcription. Annu Rev Biochem 61: 1131–1173

Mamalaki C, Norton T, Tanaka Y, Townsend AR, Chandler P, Simpson F. Kioussis D (1992) Thymic depletion and peripheral activation of class I major histocompatibility complex-restricted T cells by soluble peptide in T cell receptor transgenic mice. Proc Natl Acad Sci USA 89: 11342–11346

Mason SA, Housley PR (1993) Site directed mutagenesis of the phosphorylation sites in the louse glucocorticoid receptor. J Biol Chem 268: 21501–21504

McConkey DJ, Hartzell P, Orrenius S, Jondal M (1989) Calcium-dependent killing of immature thymocytes by stimulation via the CD3/T cell receptor complex. J Immunol 143: 1801–1806

McConkey DJ, Orrenius S, Jondal M (1990a) Cellular signalling in programmed cell death (apoptosis). Immunol Today 11(4): 120–121

McConkey DJ, Orrenius S, Jondal M (1990b) Agents that elevate cAMP stimulate DNA fragmentation in thymocytes. J Immunol 145: 1227–1230

McConkey DJ, Orrenius S, Okret S, Jondal M (1993) Cyclig AMP potentiates glucocorticoid-induced endogenous endonuclease activation in thymocytes. FASEB J 7: 580–585

Moore NC, Anderson G, Williams GT, Owen JJT, Jenkinson EJ (1994) Developmental regulation of Bcl-2 expression in the thymus. Immunology 81: 115–119

Moyer ML, Borror KS, Bonat BJ, DeFranco D, Nordeen SK (1993) Modulation of cell signalling pathways can enhance or impair glucocortidoid-induced gene expression without altering the state of receptor phosphorylation. J Biol Chem 268: 22933–22940

Murphy KM, Heimberger KB, Loh DY (1990) Induction by antigen of intrathymic apoptosis of CD4+CD8+TCR^{LO} thymocytes in vivo. Science 250: 1720–1723

Nakashima I, Zhang YH, Rahman SM, Yoshida T, Isobe K, Ding LN, Iwamoto T, Hamaguchi M, Ikezawa H, Taguchi R (1991) Evidence of synergy between Thy-1 and CD3/TCR complex in signal delivery to murine thymocytes for cell death. J Immunol 147: 1153–1162

Nordeen SK, Bona BJ, Moyer ML (1993) Latent agonist activity of the steroid antagonist, RU486, is unmasked in cells treated with activators of protein kinase A. Mol Endocrinol 7: 731–742

Oikarinen J, Hämäläinen L, Oikarinen A (1984) Modulation of glucocorticoid receptor activity by cyclic nucleotides and its implications on the regulation of human skin fibroblast growth and protein synthesis. Biochim Biophys Acta 779: 158–165

Orti E, Bodwell JE, Munck A (1992) Phosphorylation of steroid hormone receptors. Endocr Rev 131: 105–128

Patel MD, Samelson LE, Klausner RD (1987) Multiple kinases and signal transduction Phosphorylation of the T cell antigen receptor complex. J Biol Chem 262: 5831–5838

Payvar F, deFranco D, Firestone GL, Edgar B, Wrange Ö, Okret S, Gustafsson J-A, Yamamoto KR (1983) Sequence specific binding of glucocorticoid receptor to MTV DNA at sites within and upstream of the transcribed region. Cell 35: 381–392

Peitsch MC, Polzar B, Stephan H, Crompton T, MacDonald HR, Mannherz HG, Tschopp J (1993) Characterization of the endogenous deoxyribonuclease involved in nuclear DNA degradation during apoptosis (programmed cell death). EMBO J 12: 371–377

Power RF, Mani SK, Codina J, Conneely OM, O'Malley BW (1991) Dopaminergic and ligand independent activation of steroid hormone receptors. Science 254: 1636–1639

Pratt WB (1993) Role of heat shock proteins in steroid hormone function. In: Parker MG (ed) Steroid hormone action. IRL Press, Oxford, pp 64–93 (Frontiers in molecular biology)

Punt JA, Osborne BA, Takahama Y, Sharrow SO, Singer A (1994) Negative selection of CD4+CD8+ thymocytes by T cell receptor-induced apoptosis requires a co-stimulatory signal that can be provided by CD28. J Exp Med 179: 709–713

Rangarajan PN, Umensono K, Evans RM (1992) Modulation of glucocorticoid receptor function by protein kinase A. Mol Endocrinol 6: 1451–1457

Rao GS (1981) Mode of entry of steroid and thyroid hormones into cells. Mol Cell Endocrinol 21: 97–108

Schüle R, Muller M, Kaltschmidt C, Renkawitz R (1988a) Many transcription factors interact synergistically with steroid receptors. Science 242: 1418–1420

Schüle R, Muller M, Otsuka-Murakami H, Renkawitz R (1988b) Cooperativity of the glucocorticoid receptor and the CACCC-box binding factor. Nature 332: 87–90

Sebzda E, Wallace VA, Mayer J, Yeung RSM, Mak TW, Oshashi PS (1994) Positive and negative thymocyte selection induced by different concentrations of a single peptide. Science 263: 1615–1618

Selye H (1936) Thymus and adrenals in the response of the organism to injuries and intoxicants. Br J Exp Pathol 17: 234–248

Shi Y, Sahai BM, Green DR (1989) Cyclosporin A inhibits activation-induced cell death in T cell hybridomas and thymocytes. Nature 339: 652–626

Shi Lianfa, Nishioka WK, Th'ng J, Bradbury EM, Litchfield DW, Greenberg AH (1994) Premature p34^{cdc2} activation required for apoptosis. Science 263: 1143–1145

Skålhegg BS, Taskén K, Hansson V, Hultfeldt HS, Jahnsen T, Lea T (1994) Location of cAMP-dependent protein kinase type I with the TCR-CD3 complex. Science 263: 84–87

Smith CA, Williams GT, Kingston R, Jonkinson EJ, Owen JT (1989) Antibodies to CD3/T cell receptor complex induce death by apoptosis in immature T cells in thymic cultures. Nature 337: 181–184

Somers JP, deFranco DB (1992) Effects of Okadaic acid, a protein phosphatase inhibitor, on glucocorticoid receptor-mediated enhancement. Mol Endocrinol 6(1): 26–34

Spindler KD, Kanuma K, Grossman D (1991) Uptake of corticosterone into isolated rat lever cells: possible involvment of NA+/K+-ATPase. J Steroid Biochem Mol Biol 38: 721–725

Stauber C, Altschmied J, Akerblom IE, Marron JL, Mellon PL (1992) Mutual cross-interference between glucocorticoid receptor and CREB inhibits transactivation in placental cells. New Biol 4: 527–540

Strähle U, Schmid W, Schütz G (1988) Synergistic action of the glucocorticoid receptor with transcription factors. EMBO J 7: 3389–3395

Suzuki K, Tadakuma T, Kizaki H (1991) Modulation of thymocyte apoptosis by isoproterenol and prostaglandin E2. Cell Immunol 134(1): 235–240

Tsai SY, Carlstedt-Duke J, Weigel NL, Dahlman K, Gustafsson J-Å, Tsai MR, O'Malley BW (1988) Molecular interactins of steroid hormone receptor with its enhancer element: evidence for receptor dimer formation. Cell 55: 361–369

Vacchio MS, Papadopoulos V, Ashwell JD (1994) Steroid production in the thymus: implications for thymocyte selection. J Exp Med (in press)

Vedeckis VV, Bradshaw HD Jr (1983) DNA fragmentation in S49 lymphoma cells killed with glucocorticoids and other agents. Mol Cell Endocrinol 30: 215–227

Veis JD, Sentman CL, Bach EA, Korsmeyer SJ (1993) Expression of the Bcl-2 protein in murine and human thymocytes and in peripheral T lymphocytes. J Immunol 151: 2546–2554

Verbeeck MAE, Sutanto N, DurbachJPH (1991) Regulation of vasopressin messenger RNA levels in the small cell lung carcinoma cell line GLC-8: interactions between glucocorticoids and second messengers. Mol Endocrinol 5: 795–801

von Boehmer H, Teh HS, Kisielow P (1989) The thymus selects the useful, neglects the useless and destroys the harmful. Immunol Today 10: 57–60

Wahli W, Martinez E (1991) Superfamily of steroid nuclear receptors: positive and negative regulators of gene expression. FASEB J 5: 2243–2249

Wrange Ö, Eriksson P, Perlmann T (1989) The purified activated glucocorticoid receptor is a homodimer. J Biol Chem 254: 5253–5259

Wyllie AH, Kerr JFR, Currie AR (1980) Cell death: the significance of apoptosis. Int Rev Cytol 68: 251–306

Zacharchuk CM, Mercep M, Chakraborti PK, Simons SS Jr, Ashwell JD (1990) Programmed T lymphocyte death. Cell activation and steroid-induced pathways are mutually antagonistic. J Immunol 145: 4037–4045

Regulation of Apoptosis via Steroid Receptors

M. Iwata

1 Introduction

Steroid hormones play essential roles in a variety of physiological processes including embryonic development, sexual differentiation and maturation, and metamorphosis. The homeostatic regulation of metabolism and cell turnover that determines tissue sizes and shapes are also under the influence of these hormones. Apoptosis is involved in many of these phenomena. The pharmacological or surgical manipulation of animals to change steroid levels often causes involution or enlargement of certain tissues partly through the enhancement or inhibition of apoptosis. For example, an elevation of blood glucocorticoid level, by an injection of glucocorticoids or by excessive stress, causes thymus involution due to apoptosis in cortical immature thymocytes (CLAMAN 1972). By contrast, adrenalectomy of mice causes not only depletion of glucocorticoids from the plasma but also a marked increase in the thymus size (SHORTMAN and JACKSON 1974). Glucocorticoid-induced apoptosis is dependent on the binding of glucocorticoids to glucocorticoid hormone receptors (GRs), which is also required in

Mitsubishi Kasei Institute of Life Sciences, 11, Minamiooya, Machida-shi, Tokyo, 194, Japan the general effects of steroids (DUVAL et al. 1984). The steroid receptors such as GR, mineralcorticoid receptors, progesterone receptors, androgen receptors, and estrogen receptors are members of a superfamily of ligand-inducible transcription factors. The steroid receptor superfamily also includes retinoic acid receptors, thyroid hormone receptors, vitamin D_3 receptors, ecdysone receptors, and COUP transcription factor. They are related to v-*erbA* oncogene.

2 Glucocorticoid-Induced Apoptosis in Thymocytes

Glucocorticoid-induced death in immature thymocytes is one of the classical examples of apoptosis, in which the typical morphological changes, such as chromatin condensation, nucleolar disruption, and cytoplasmic contraction, take place (WYLLIE et al. 1980). Without glucocorticoid binding, GRs exist mainly in the cytosol associating with proteins such as heat shock protein 90 (hsp90). The binding of glucocorticoids induces dissociation of GR from hsp90 and trans-location of the glucocorticoid-GR complex from the cytosol to the nucleus (PRATT et al. 1989). The complex acts as a translation regulatory factor, inducing or enhancing the expression of certain genes. Indeed, glucocorticoid-induced apop-tosis in thymocytes is inhibited by inhibitors of mRNA and protein synthesis. Thus, it is postulated that there is a "death gene(s)" that codes a protein(s) responsible for the induction of apoptosis. Some candidate genes, such as RP-2 and RP-8, have been cloned (OWENS et al. 1991; SCHWARTZ and OSBORNE 1993).

2.1 DNA Fragmentation and the Role of Ca^{2+}

The morphological changes in glucocorticoid-induced apoptosis are usually asso-ciated with endonuclease cleavage of DNA into oligonucleosomal fragments (WYLLIE 1980; WYLLIE et al. 1984). This DNA degradation is one of the early signs in most of the apoptotic processes. COHEN and DUKE (1984) have shown that DNA fragmentation in isolated nuclei of murine thymocytes is induced by Ca^{2+} and Mg^{2+}. They suggested that a Ca^{2+}, Mg^{2+}-dependent and Zn^{2+}-sensitive endonucle-ase is constitutively present in the nuclei of thymocytes, and that the protein for which synthesis is necessary for glucocorticoid-induced thymocyte death is not the endonuclease itself, but is in some way involved in its activation. SCHWARTZMAN and CIDLOWSKI (1993) detected intranucleosomal DNA cleavage activity in nuclear extracts of glucocorticoid-treated apoptotic rat thymocytes, but not in control thymocytes. In both cases, however, millimolar concentrations of Ca^{2+} and Mg^{2+} or Mn^{2+} were required for optimal DNA cleavage activity. Intracellular free Mg^{2+} concentrations in many cells range from 0.1–0.7 mM (PRESTON 1990), whereas intracellular free Ca^{2+} concentrations ($[Ca^{2+}]_i$) in normal and activated thymocytes are usually on the order of 0.1–1 μM (ISEKI et al. 1993). JONES et al. (1989) have shown that submicromolar concentrations of Ca^{2+} induce DNA fragmentation in rat liver nuclei in the presence of physiological levels of ATP and NAD^+. Thus, within cells that are in the process of apoptosis, it may be possible that endonuclease activation is induced by the combination of physiological concentrations of intracellular Ca^{2+}/Mg^{2+} and other components.

A sustained increase in $[Ca^{2+}]_i$ is considered to be essential for several apoptotic processes (ALLBRITTON et al. 1988; MCCONKEY et al. 1989; ISEKI et al. 1991; NICOTERA et al. 1992), but not for others (ALNEMRI and LITWACK 1990; BANSAL et al. 1990; MCCONKEY et al. 1990; SUZUKI et al. 1990; ISEKI et al. 1991; NICOTERA et al. 1992). It was reported that glucocorticoid-induced death in rat thymocytes was dependent on a sustained increase in $[Ca^{2+}]_i$ and was inhibited by depletion of extracellular Ca^{2+} with EGTA or buffering of intracellular Ca^{2+} with quin-2/AM (KAISER and EDELMAN 1977; MCCONKEY et al. 1989). However, NICHOLSON and YOUNG (1979) reported that it is unlikely that glucocorticoid-induced changes in Ca^{2+} uptake initiate the lethal actions of glucocorticoids. To resolve this problem, we employed microscopic fluorometry that enabled us to monitor real-time $[Ca^{2+}]_i$ on a single cell basis (ISEKI et al. 1993). The results indicated that dexamethasone (DEX), a potent synthetic glucocorticoid, does not induce an increase in $[Ca^{2+}]_i$ above the control level either in murine or rat thymocytes for at least 1 h after the start of the culture. We also found that DEX-induced apoptosis in both murine and rat thymocytes is not inhibited by EGTA. High concentrations (25 μM and over) of quin-2/AM inhibited DNA fragmentation, but failed to inhibit cytolysis. Furthermore, we found that a proper combination of the calcium ionophore ionomycin, and the protein kinase C activator phorbol 12-myristate 13-acetate (PMA), inhibits glucocorticoid-induced apoptosis (IWATA et al. 1993). Thus, we suggested that an early increase in $[Ca^{2+}]_i$ is neither induced by glucocorticoids nor responsible for glucocorticoid-induced apoptosis in thymocytes (ISEKI et al. 1993). By fluorocytometric analysis, DECKERS et al. (1993) detected an elevation of $[Ca^{2+}]_i$ in methylprednisolone-treated murine thymocytes 3–6 h after addition of the glucocorticoid and suggested that the elevation of $[Ca^{2+}]_i$ is not involved in the induction of the apoptosis. Thus, $[Ca^{2+}]_i$ may increase somehow after the cells are committed to apoptosis. The measurement of $[Ca^{2+}]_i$ in most of the experiments depends on the loading of the cells with a fluorescent Ca^{2+} chelating agent. The loading, however, often disturbs the cell's functions and sometimes results in cell death. A more improved monitoring system for $[Ca^{2+}]_i$ is required for determining the role of $[Ca^{2+}]_i$ in the later stage of glucocarticoid-induced apoptosis.

Extensive DNA strand breaks caused by oxidative stresses induce the activation of poly(ADP-ribose) synthetase in cells. The enzyme utilizes NAD as substrate and depletes NAD and ATP from the cells. Inhibitors of poly(ADP-ribose) synthetase prevent oxidant-induced cell lysis probably by preventing the depletion of NAD and ATP, but they do not prevent the DNA strand breakage (SCHRAUFSTATTER et al. 1986). Similarly, 3-aminobenzamide, a potent inhibitor of poly(ADP-ribose) synthetase, effectively prevents glucocorticoid-induced thymocyte lysis, but it does not prevent the DNA strand breakage (HOSHINO et al. 1993). Thus, ATP depletion caused by glucocorticoids, through the induction of

DNA fragmentation and the activation of poly(ADP-ribose) synthetase, may result in the inhibition of Ca^{2+}-ATPases and an increase in $[Ca^{2+}]_i$.

Originally apoptosis was defined morphologically, but DNA fragmentation has been often used as a major indicator of apaptosis. COHEN et al. (1992), however, dissociated some of the key morphological changes of apoptosis, such as heterochromatin condensation, from internucleosomal DNA fragmentation by treating thymocytes with glucocorticoid and Zn^{2+}. The dissociation was further confirmed in isolated liver nuclei treated with Ca^{2+} and Mg^{2+} in the presence of Zn^{2+} (SUN et al. 1994). BROWN et al. (1993) found that Zn^{2+} inhibits cleavage of DNA into oligonucleosomal fragments but does not prevent the cleavage of DNA into high molecular weight fragments. Thus, key enzymes other than the Ca^{2+}/Mg^{2+}-dependent endonuclease appear to be involved at the earliest stages of induction of apoptosis by glucocorticoids.

It has been suggested that in CEM-C7 human lymphocytes glucocorticoid-induced DNA cleavage and cytolysis do not involve a Ca^{2+}-requiring mechanism (ALNEMRI and LITWACK 1990; BANSAL et al. 1990). In T cell hybridomas, we have shown that glucocorticoid-induced apoptosis does not accompany early mobilization of $[Ca^{2+}]_i$, and that the apoptosis is not inhibited by EGTA but is inhibited by ionomycin with or without PMA (ISEKI et al. 1991).

2.2 Involvement of Ca^{2+}-Independent Protein Kinase C

Glucocorticoid-induced apoptosis in murine thymocytes appears to be dependent on protein kinase C (PKC), since PKC inhibitors inhibit glucocorticoid-induced DNA fragmentation and cytolysis in murine thymocytes (OJEDA et al. 1990; IWATA et al. 1994). PKC is a family of closely related enzymes, consisting of Ca^{2+}-dependent (PKC-α, $-\beta$I, -β II, and -γ) and Ca^{2+}-independent (PKC-δ, -ϵ, -η (L), -θ, -ζ, and -λ) isozymes. We found that glucocorticoid selectively induces an increase in Ca^{2+}-independent PKC activity in the particulate fraction of immature thymocytes but not in that of mature T cells. The increase and the apoptosis was inhibited by actinomycin D, cycloheximide, or the GR antagonist RU 38486, Immunoblotting studies revealed the selective translocation of PKC-ϵ from the cytosolic fraction to the particulate fraction upon glucocorticoid treatment. Thus, glucocorticoid-induced apoptosis in immature thymocytes appears to involve GR-mediated activation of PKC-ϵ through de novo synthesis of macromolecules (IWATA et al. 1994).

Protein dephosphorylation may be also a essential step for glucocorticoid-induced apoptosis. We found that okadaic acid, a potent inhibitor of protein phosphatase 1 and 2A, inhibits glucocorticoid-induced apoptosis in T cell hybridomas (OHOKA et al. 1993). The okadaic acid-sensitive step appeared to be after the translocation of GR and the expression of the genes controlled by glucocorticoid response elements. However, the effect of okadaic acid on murine thymocyte apoptosis was hard to assess, as it inhibited glucocorticoid-induced DNA fragmentation but enhanced cytolysis in thymocytes.

3 The Role of Glucocorticoids in Thymic Selection

The major population of glucocorticoid-sensitive thymocytes is immature and double positive (CD4+CD8+). These cells constitute approximately 80% of the total thymocytes (HUGO et al. 1991). It is known that the vast majority of double positive cells is destined to die within the thymus after a short life-span (EGERTON et al. 1990), whereas some of these cells appear to survive and differentiate into single-positive cells and to be exported from the thymus. The T cell repertoire is molded by thymic selection that is based on the regulation of apoptosis in each T cell clone at its double positive stage. The peak concentrations of glucocorticoid hormones (0.1–1 μM) in the plasma of a normal mouse or rat can induce death in its double positive thymocytes in vitro (WYLLIE 1980; COHEN and DUKE 1984; IWATA et al. 1991). Therefore, it appears that immature T cell clones which are positively selected should be protected from glucocorticoid-induced death. It is likely that the clonal selection in the thymus is dependent on the affinity or avidity of the T cell receptors (TCRs) to self MHC-encoded molecules with self antigens. Indeed, in TCR-transgenic mice, TCR antagonist peptides, or low concentrations of the antigen peptide recognized by the transgenic TCR in combination with MHC molecules, can mediate positive selection, whereas high concentrations of the antigen peptide result in thymocyte deletion (negative selection; ALLEN 1994).

3.1 Positive Selection of Thymocytes

We have previously found that cross-linking of TCR/CD3 molecules with a specific antibody at a proper concentration rescued normal mouse thymocytes from glucocorticoid-induced apoptosis in vitro (IWATA et al. 1991). Thus, we have proposed a hypothesis that positive selection of T cell clones is based on the inhibition of glucocorticoid-induced apoptosis in thymocytes by a proper TCR/CD3-mediated signal (IWATA et al. 1991). Depending on our hypothesis, it may be possible to analyze the positive selection signals in vitro.

CARRERA et al. (1992) have suggested that signal transduction through TCR and other molecules is involved in positive selection. It is evident that not only TCR engagement but also other molecular interactions are required for effective cell-cell interaction and signaling during T cell ontogeny. CD4, CD8, and LFA-1/ICAM-1 molecules are known to play particularly important roles in development and/or selection of thymocytes (MACDONALD et al. 1988; RAMSDELL and FOWLKES 1989; FINE and KRUISBEEK 1991). We found that the inhibitory effect of anti-CD3 on glucocorticoid-induced death was significantly enhanced or stabilized by costimulation via LFA-1 (Fig. 1), while, with anti-CD3 alone, the extent of the inhibition and the optimal dose of anti-CD3 for inhibition varied from experiment to experiment, as we described before (IWATA et al. 1991). Costimulation via CD4 or CD8 may also enhance or modify TCR/CD3-mediated signals. PUNT et al. (1993) have shown that treatment of the fetal thymus with a combination of anti-TCR-β and anti-CD4 antibodies dominantly induced CD4+CD8− thymocytes.

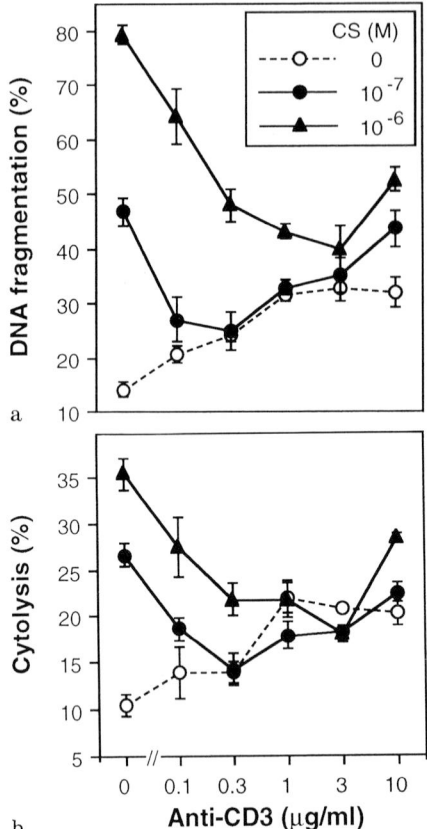

Fig. 1a, b. The combination of anti-LFA-1 and anti-CD3 inhibits glucocorticoid-induced DNA fragmentation and cytolysis in murine thymocytes. BALB/c thymocytes were cultured in plastic plates that had been coated with 1 µg/ml anti-CD11a (M17/4.2) together with various concentrations of anti-CD3 (145–2C11). Medium alone (*open circles*), 10^{-7} *M* corticosterone (*CS; closed circles*), or 10^{-6} *M* CS (*closed triangles*) was added 2 h after the start of the culture. After further culture for 16 h, **a** DNA fragmentation and **b** cytolysis were assessed as described (IWATA et al. 1991)

The protective effect of the TCR/CD3-mediated stimulation was mimicked by a combination of ionomycin and PMA, but ionomycin or PMA alone failed to inhibit glucocorticoid-induced DNA fragmentation and cytolysis (IWATA et al. 1993). Thus, a PKC isozyme(s) other than PKC-ε may be involved in the protective signal. In murine T cell hybridomas, we (IWATA et al. 1991) and ZACHARCHUK et al. (1990) independently found that TCR/CD3-mediated stimulation and glucocorticoids are mutually antagonistic in the induction of apoptosis. Our subsequent study (ISEKI et al. 1991) suggested that the TCR/CD3-mediated stimulation in T cell hybridomas involves an elevation of $[Ca^{2+}]_i$ and the activation of PKC. The combination of ionomycin and PMA mimicked the effect of the TCR/CD3-mediated stimulation. SHI et al. (1992) suggested that inhibition of the activation-induced apoptosis may be due to the inhibition of c-*myc* expression by glucocorticoid. However, the precise mechanism remains to be elucidated.

3.2 Negative Selection of Thymocytes

As an in vivo model of negative selection, some groups observed the death of immature thymocytes in mice after an injection of an anti-CD3 or anti-TCR monoclonal antibody. JONDAL et al. (1993) found that the GR antagonist RU 38486 inhibited anti-CD3-induced death in double positive thymocytes, suggesting that ondogenous glucocorticoid is involved in negative selection. However, it may be necessary to consider some other possibilities. For example, the injection of anti-CD3 induces polyclonal activation of mature T cells and the systemic production of various lymphokines, which disturb various systems in the body (FERRAN et al. 1990; ALEGRE et al. 1990). Polyclonal activation of T cells by an injection of anti-CD3 or *Staphylococcus aureus* enterotoxin B (SEB), a superantigen, appears to induce an increase in blood glucocorticoid levels (GONZALO et al. 1993). The increase may also explain the controversial results in the effect of anti-CD3 on thymocytes in vivo and in vitro. SHI et al. (1989) have shown that the anti-CD3-induced apoptosis in murine thymocytes in vivo was inhibited by an injection of cyclosporin A (CsA), while McCARTHY et al. (1992) found that CsA and FK506 failed to inhibit anti-CD3-induced DNA fragmentation in thymocytes in vitro. CsA and FK506 are known to inhibit the TCR/CD3-mediated activation of mature T cells including the production of lymphokines.

In T cell hybridomas, it has been shown that activation-induced death is inhibited by glucocorticoids, but we do not have any direct evidence that glucocorticoids also inhibit activation-induced apoptosis in thymocytes or negative selection.

It has been shown that cAMP analogs or agents that elevate cAMP potentiate the apoptotic response to glucocorticoids (DURANT 1986; McCONKEY et al. 1993). Endogenous glucocorticoids in concert with cAMP-elevating stimuli may modulate thymic selection.

4 Effect of Glucocorticoids on Mature T Cells

The major population of thymocytes is sensitive to the apoptosis-inducing activity of glucocorticoids, while the major population of mature peripheral T cells is resistant to it. Glucocorticoid-sensitive and resistant thymocytes and mature I cells have almost the same number of GR per cell, with similar binding properties (HOMO et al. 1980). Recently, PERANDONES et al. (1993) have reported that a significant population of murine splenic T cells succumbs to apoptotic death by DEX at the same concentration range that induces apoptosis in thymocytes. They have shown that cycloheximide failed to inhibit glucocorticoid-induced death; thus, the mechanism of glucocorticoid-induced apoptosis in mature T cells is different from that in immature T cells. Concordantly, GONZALO et al. (1994) have reported that linomide (quinoline-3-carboxamide), a immunomodulator with predominantly stimulatory properties, inhibited the depletion of splenic CD4[+] and

CD8[+] cells induced by an in vivo treatment with DEX, but that linomide failed to inhibit glucocorticoid-induced apoptosis in double-positive thymocytes.

In concert with naturally produced glueocorticoids in vivo, T cell growth factors may play a role in modulation of the immune response. ZUBIAGA et al. (1992) have shown that IL-4 specifically rescues Th2 cells from glucocorticoid-induced apoptosis, whereas IL-2 and IL-1 are ineffective in these cells. However, IL-2 is the relevant rescue factor of glucocorticoid-treated Th1 cells. PKC activation appears to be involved in the IL-4- or IL-2-dependent protection of Th cells, as a PKC inhibitor blocks the protective effect of the lymphokines (ZUBIAGA et al. 1992).

Glucocorticoids may also be essential cofactors for the superantigen-driven deletion of T cells in vivo. GONZALO et al. (1993) have shown that an injection of the Vβ8-specific superantigen SEB into a mouse induces an increase in circulating corticosterone levels, and that an administration of RU 38486 abolishes the early deletion of Vβ8-expressing spleen cells detectable 12 h after the injection of SEB.

5 Inhibition of Apoptosis by Glucocorticoids

Under certain conditions, glucocorticoids inhibit apoptosis, as is observed in T cell hybridomas. Glucocorticoid hormones appear to be involved directly or indirectly in the survival of the granule cells in the rat hippocampal dentate gyrus, Adrenalectomy induces apoptosis in these cells, while apoptosis is not induced in rats that are maintained on corticosterone in saline (SLOVITER et al. 1993).

Glucocorticoids administered to various mammals at midgestation stage inhibit complete formation of the secondary palate in the fetus. It is partly because glucocorticoids inhibit the programmed cell death in the medial epithelial cells in the fetal shelves. In normal development, the palatal shelves grow from the maxillary process in both sides until the apposing epithelia into contact at the midline. The medial epithelial cells then undergo a programmed cell death, and the two shelves fuse into a single tissue, the secondary palate, which separates the oral and nasal cavities (PRATT et al. 1984). GUPTA et al. (1984) have shown that DEX itself or phospholipase A_2 inhibitory proteins obtained from DEX-treated thymocytes or embryonic palates inhibited programmed cell death in the medial edge epithelium of single mouse embryonic palatal shelves in culture. The capacity of glucocorticoids to induce cleft palate and thymocyte apoptosis is correlated with their anti-inflammatory potency (GOLDMAN 1984) (Fig. 2). It is known that at least some of the anti-inflammatory effects of glucocorticoids can be explained by the inhibition of oxygenated arachidonate metabolite release. Glucocorticoid-induced inhibition of arachidonate release may be related to glucocorticoid-induced cleft palate (GOLDMAN 1984), whereas PRATT et al. (1984) concluded that the regulation of arachidonate release is not directly involved.

Susceptibility to both glucocorticoid-induced cleft palate and glucocorticoid-induced thymolytic activity appears to be controlled by genes that map within the

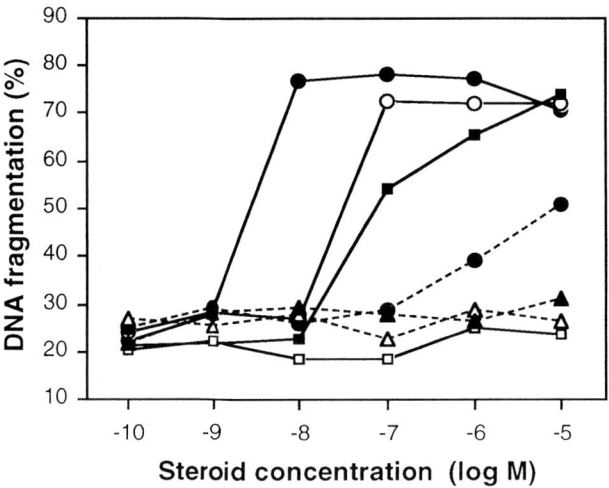

Fig. 2. Effects of various steroids on DNA fragmentation in murine thymocytes. BALB/c thymocytes were cultured with graded concentrations of various steroids for 18 h. Cortisone, which exerts anti-inflammatory activity after it is converted to active forms such as hydrocortisone in vivo, failed to affect thymocyte apoptosis in vitro. The sex steroids, except high concentrations of progesterone, hardly induced DNA fragmentation. *Closed circles*, dexamethasone; *open circles*, hydrocortisone; *closed squares*, corticosterone; *open squares*, cortisone; *closed circles, dotted line*, progesterone; *open triangles, dotted line*, β-estradiol; *closed triangle, dotted line*, testosterone

I region of the H-2 complex and involve genetic complementation (PLA et al. 1976; TYAN 1979; BONNER 1984; GOLDMAN 1984).

6 Effect of Gonadal Steroids

Testosterone and β-estradiol hardly affect apoptosis in thymocytes and T cell hybridomas in vitro (Fig. 2) (IWATA et al. 1991). BARR et al. (1982) found, however, that dihydrotestosterone and estradiol delete the same cortical population of thymocytes as glucocorticoids do in mice. As they could not find receptors for estradiol of dehydrotestosterone in this cortical population, they suggested that the sex steroids bind to other thymic elements, possibly thymic reticular epithelial cells, which may in turn act secondarily on cortical thymocytes within the thymus.

The survival of some types of cells is dependent on sex steroids. Deprivation of testosterone induces apoptosis in epithelial cells in the ventral prostate in rats, and the apoptosis requires RNA and protein synthesis (LEGER et al. 1987; SALTZMAN et al. 1987). The regression of human mammary cancers following estrogen ablation is partly due to the induction of apoptosis (KYPRIANOU et al. 1991).

In mammals, the embryogenesis of the urogenital tract is identical in males and females during the first phase of gestation (WILSON et al. 1981). Only after the onset of endocrine function of the testis do anatomic and physiologic development of male and female embryos diverge. Testosterone and Müllerian-inhibiting substance appear to play critical roles in these processes partly through the control of apoptotic cell death.

Sexual dimorphism can be seen also in the nervous system. In the absence of gonadal secretions, the nervous system also develops in a primarily female fashion (BREEDLOVE 1992). The spinal nucleus of the bulbocavernosus contains many more motoneurons in adult male rats than in females. NORDEEN et al. (1985) have suggested that androgens attenuate normally occurring cell death in the motoneurons during a critical period of the development. By contrast, the anteroventral nucleus of the preoptic area is larger and more densely cellular in females than males because androgen induces apoptosis in neurons in the nucleus (MURAKAMI and ARAI 1989). In many songbird species, the male sings and the female does not. The vocal control regions of the male brain are five to six times larger in volume than those of the female brain. KONISHI and AKUTAGAWA (1985) reported that one of the forebrain vocal control regions, the robust nucleus of the archistriatum, undergoes ontogenetic cell death that is more pronounced in females than males. Testosterone, which can be converted to estradiol in the brain or, surprisingly, estrogen itself apparently prevents cell death.

7 Effect of Retinoic Acids on Apoptosis

Ligands of other members of the steroid receptor superfamily are also involved in the regulation of apoptosis. It is well-known that thyroid hormones induce massive programmed cell death and cell transformation in the tadpole tail at metamorphosis.

All-trans retinoic acid (RA), a metabolite of vitamin A, is known to play an essential role in embryonic development. RA also affects apoptosis in thymocytes and T cell hybridomas (IWATA et al. 1992). RA at near physiological concentrations (0.01–1 µM) significantly inhibits the induction of thymocyte apoptosis by coimmobilized antibodies to CD3 and LFA-1 molecules, but enhances glucocorticoid-induced apoptosis. The inhibitory effect of RA might be correlated to the finding that acquired immunological tolerance of foreign cells is impaired by a vitamin A acetate-supplemented diet (MALKOVSKY et al. 1985). Apoptosis induced in T cell hybridomas by TCR/CD3-mediated stimulation or by the combination of ionomycin and PMA is also inhibited by RA at 0.1–10 µM. RA appears to interfere with the apoptotic process at some point after its initiation stage (IWATA et al. 1992). YANG et al. (1993) raised the possibility that retinoid X receptors (RXRs) might take part in the RA effect, since they found that 9-*cis*-retinoic acid, which binds to RXRs with high affinity in addition to binding to RA receptors, was approximately tenfold more potent than RA.

There is a possibility that retinoid receptors may not be necessarily involved in the inhibitory activity of RA. BUTTKE and SANDSTROM (1994) have reported that antioxidants such as glutathione and N-acetylcysteine (NAC) can inhibit activation-induced death in T cell hybridomas. The antioxidant N-(2-mercaptoethyl)-1,3-propanediamine (WR-1065) protects thymocytes from apoptosis induced by glucocorticoids, γ-irradiation, and calcium ionophores (RAMAKRISHNAN and CATRAVAS 1992). As RA is also known to have antioxidant potential, RA might inhibit apoptosis by reducing the oxidative stress. HOCKENBERY et al. (1993) have reported that treatment of T hybridoma cells with DEX resulted in quantifiable lipid peroxidation and that overexpression of Bcl-2 suppressed the lipid peroxidation and the glucocorticoid-induced cell death. As Bcl-2 protected cells from H_2O_2-and menadione-induced oxidative deaths, they proposed a model in which Bcl-2 regulates an antioxidant pathway at sites of free radical generation. Interestingly, either A23187 or DEX enhances the expression of glutathione S-transferase (GST) gene and apoptosis in murine thymoma cells (FLOMERFELT et al. 1993). GST is an antioxidant defense enzyme. As GST gene expression was also elevated in the regressing prostate of androgen-ablated rats (SALTZMAN et al. 1987), FLOMERFELT et al. (1993) suggested that activation of GST gene expression is a likely indicator of oxidative stress, rather than a required step in the pathway. Considering that RA enhances glucocorticoid-induced apoptosis in thymocytes, RA- and glucocorticoid-dependent regulation of oxidant-redox metabolism in apoptosis still remain to be elucidated.

8 Conclusions

Induction or inhibition of apoptosis is one of the physiologic roles of steroids. For some types of cells, steroids are physiological survival factors or trophic factors at certain periods of ontogeny. Among steroid-dependent apoptosis, thymocyte death induced by glucocorticoids and epithelial cell death in the ventral prostate induced by androgen depletion have been most intensively studied with respect to their biochemistry and molecular biology. The mechanism of glucocorticoid-induced apoptosis in thymocytes and its inhibition may have some relation to the mechanism of thymic selection that molds the T cell repertoire. The blood glucocorticoid hormone level has a circadian rhythm. As the peak concentrations can induce apoptosis in immature thymocytes in vitro, especially in mouse or rat cells, the immature T clones that are positively selected in the thymus should receive protective signals through the TCR and accessory molecules at least against the glucocorticoid effect. Glucocorticoids appear to activate Ca^{2+}-independent PKC-ε, while the protective signals can be provided by a proper increase in $[Ca^{2+}]_i$ and proper activation of PKC, probably other than PKC-ε. The molecular mechanisms of these events are, however, still largely unknown. Depending on the glucocorticoid effect, in vitro experimental modeling may help to discover the essential signals for clonal selection in thymocytes.

References

Alegre M, Vandenabeele P, Flamand V, Moser M, Leo O, Abramowicz D, Urbain J, Fiers W, Goldman M (1990) Hypothermia and hypoglycemia induced by anti-CD3 monoclonal antibody in mice: role of tumor necrosis factor. Eur J Immunol 20: 707–710

Allbritton NL, Verret CR, Wolley RC, Eisen HN (1988) Calcium ion concentrations and DNA fragmentation in target cell destruction by murine cytotoxic T lymphocytes. J Exp Med 167: 514–527

Allen PM (1994) Peptides in positive and negative selection: a delicate balance. Cell 76: 593–596

Alnemri ES, Litwack G (1990) Activation of internucleosomal DNA cleavage in human CEM lymphocytes by glucocorticoid and novobiocin evidence for a non-Ca^{2+}-requiring mechanism(s). J Biol Chem 265: 17323–17333

Bansal N, Houle AG, Melnykovych G (1990) Dexamethasone-induced killing of neoplastic cells of lymphoid derivation: lack of early calcium involvement. J Cell Physiol 143: 105–109

Barr IG, Khalid BAK, Pearce P, Toh BH, Bartlett PF, Scollay RG, Funder JW (1982) Dihydrotestosterone and estradiol deplete corticosensitive thymocytes lacking in receptors for these hormones. J Immunol 128: 2825–2828

Bonner JJ (1984) The H-2 genetic complex, dexamethasone-induced cleft palate and other craniofacial anomalies. Curr Top Dev Biol 19: 193–215

Breedlove SM (1992) Sexual dimorphism in the vertebrate nervous system. J Neurosci 12: 4133–4142

Brown OG, Sun XM, Cohen GM (1993) Dexamethasone-induced apoptosis involves cleavage of DNA to large fragments prior to internucleosomal fragmentation. J Biol Chem 268: 3037–3039

Buttke TM, Sandstrom PA (1994) Oxidative stress as a mediator of apoptosis. Immunol Today 15: 7–10

Carrera AC, Baker C, Roberts TM, Pardoll DM (1992) Tyrosine kinase triggering in thymocytes undergoing positive selection. Eur J Immunol 22: 2289–2294

Claman HN (1972) Corticosteroids and lymphoid cells. N Engl J Med 287: 388–397

Cohen JJ, Duke RC (1984) Glucocorticoid activation of a calcium-dependent endonuclease in thymocyte nuclei leads to cell death. J Immunol 132: 38–42

Cohen GM, Sun XM, Snowden RT, Dinsdale D, Skilleter DN (1992) Key morphological features of apoptosis may occur in the absence of internucleosomal DNA fragmentation. Biochem J 286: 331–334

Deckers CLP, Lyons AB, Samuel K, Sanderson A, Maddy AH (1993) Alternative pathways of apoptosis induced by methylprednisolone and valinomycin analyzed by flow cytometry. Exp Cell Res 208: 362–370

Durant S (1986) In vivo effects of catecholamines and glucacorticoids on mouse thymic cAMP content and thymolysis. Cell Immunol 102: 136–143

Duval D, Durant S, Homo-Delarche F (1984) Effect of antiglucocorticoids on dexamethasone-induced inhibition of uridine incorporation and cell lysis in isolated mouse thymocytes. J Steroid Biochem 20: 283–287

Egerton M, Scollay R, Shortman K (1990) Kinetics of mature T-cell development in the thymus. Proc Natl Acad Sci USA 87: 2579–2582

Ferran C, Sheeha K, Dy M, Schreiber R, Merite S, Landais P, Noel LH, Grau G, Bluestone J, Bach JF, Chatenoud L (1990) Cytokine-related syndrome following injection of anti-CD3 monoclonal antibody further evidence for transient in vivo T cell activation. Eur J Immunol 20: 509–515

Fine JS, Kruisbeek AM (1991) The role of LFA-1/ICAM-1 interactions during murine lymphocyte development. J Immunol 147: 2852–2859

Flomerfelt FA, Briehl MM, Dowd DR, Dieken ES, Miesfeld RL (1993) Elevated glutathione S-transferase gene expression is an early event during steroid-induced lymphocyte apoptosis. J Cell Physiol 154: 573–581

Goldman AS (1984) Biochemical mechanism of glucocorticoid and phenytoin-induced cleft palate. Curr Top Dev Biol 19: 217–239

Gonzalo JA, Gonzalez-Garcia A, Martinez-A C, Kroemer G (1993) Glucocorticoid mediated control of the activation and clonal detetion of peripheral T cells in vivo. J Exp Med 177: 1239–1246

Gonzalo JA, Gonzalez-Garcia A, Kalland T, Hedlund G, Martinez-AC, Kroemer G (1994) Linomide inhibits programmed cell death of peripheral T cells in vivo. Eur J Immunol 24: 48–52

Gupta C, Katsumata M, Goldman AS, Herold R, Piddington R (1984) Glucocorticoid induced phospholipase A_2 inhibitory proteins mediate glucocorticoid teratogenicity in vitro. Proc Natl Acad USA 81: 1140–1143

Hockenbery DM, Oltval ZN, Yin XM, Milliman CL, Korsmeyer SJ (1993) Bcl-2 functions in an antioxidant pathway to prevent apoptosis. Cell 75: 241–251

Homo F, Duval D, Hatzfeld J, Evrard C (1980) Glueocorticoid sensitive and resistant cell populations in the mouse thymus. J Steroid Biochem 13: 135–143

Hoshino J, Beckmann G, Kroger H (1993) 3-Aminobenzamide protects the mouse thymocytes in vitro from dexamethasone-mediated apoptotic cell death and cytolysis without changing DNA strand breakage. J Steroid Biochem Mol Biol 44: 113–119

Hugo P, Boyd RL, Waanders GA, Scollay R (1991) CD4+CD8+CD3[high] thymocytes appear transiently during ontogeny: evidence from phenotypic and functional studies. Eur J Immunol 21: 2655–2660

Iseki R, Mukai M, Iwata M (1991) Regulation of T lymphocyte apoptosis: Signals for the antagonism between activation- and glucocorticoid-induced death. J Immunol 147: 4286–4292

Iseki R, Kudo Y, Iwata M (1993) Early mobilization of Ca^{2+} is not required for glucocorticoid-induced apoptosis in thymocytes. J Immunol 151: 5198–5207

Iwata M, Hanaoka S, Sato K (1991) Rescue of thymocytes and T cell hybridomas from glucocorticoid-induced apoptosis by stimulation via the T cell receptor/CD3 complex: a possible in vitro model for positive selection of the T cell repertoire. Eur J Immunol 21: 643–648

Iwata M, Mukai M, Nakai Y, Iseki R (1992) Retinoic acids inhibit activation-induced apoptosis in T cell hybridomas and thymocytes. J Immunol 149: 3302–3308

Iwata M, Iseki R. Kudo Y (1993) Regulation of thymocyte apoptosis: glucocorticoid-induced death and its inhibition by T-cell receptor/CD3 complex-mediated stimulation. In: Lavin ML, Watters D (ed) Programmed cell death: the cellular and molecular biology of apoptosis. Harwood Academic Publishers, Chur, Switzerland, pp 31–44

Iwata M, Iseki R, Sato K, Tozawa Y, Ohoka Y (1994) Involvement of protein kinase C-ε in glucocorticoid induced apoptosis in thymocytes. Int Immunol 6: 431–438

Jondal M, Okret S, McConkey D (1993) Killing of immature CD4+CD8+ thymocytes in vivo by anti-CD3 or 5'-(N-ethyl)-carboxamido adenosine is blocked by glucocorticoid receptor antagonist RU-486. Eur J Immunol 23: 1246–1250

Jones DP, McConkey DJ, Nicotera P, Orrenius S (1989) Calcium-activated DNA fragmentation in rat liver nuclei. J Biol Chem 264: 6398–6403

Kaiser N, Edelman IS (1977) Calcium dependence of glucocorticoid-induced lymphocytolysis. Proc Natl Acad Sci USA 74: 638–642

Konishi M, Akutagawa E (1985) Neuronal growth, atrophy and death in a sexually dimorphic song nucleus in the zebra finch brain. Nature 315: 145–147

Kyprianou N, English HF, Davidson NE, Isaacs JT (1991) Programmed cell death during regression of the MCF-7 human breast cancer following estrogen ablation. Cancer Res 51: 162–166

Leger JG, Montpetit ML, Tenniswood MP (1987) Characterization and cloning of androgen-repressed mRNAs from rat ventral prostate. Biochem Biophys Res Commun 147: 196–203

MacDonald HR, Hengartner H, Pedrezzini T (1988) Intrathymic deletion of self-reactive cells prevented by neonatal anti-CD4 antibody treatment. Nature 335: 174–176

Malkovsky M, Medawar PB, Thatcher DR, Toy J, Hunt R, Rayfield LS, Dore C (1985) Acquired immunological tolerance of foreign cells is impaired by recombinant interleukin 2 or vitamin A acetate. Proc Natl Acad Sci USA 82: 536–538

McCarthy SA, Cacchione RN, Mainwaring MS, Cairns JS (1992) The effects of immunosuppressive drugs on the regulation of activation-induced apoptotic cell death in thymocytes. Transplantation 54: 543–547

McConkey DJ, Nicotera P, Hartzell P, Bellomo G, Wyllie AH, Orrenius S (1989) Glucocorticoids activate a suicide process in thymocytes through an elevation of cytosolic Ca^{2+} concentration. Arch Biochem Biophys 269: 365–370

McConkey DJ, Orrenius S, Jondal M (1990) Agents that elevates cAMP stimulate DNA fragmentation in thymocytes. J Immunol 145: 1227–1230

McConkey DJ, Orrenius S, Okret S, Jondal M (1993) Cycic AMP potentiates glueocorticoid-induced endogenous endonuclease activation in thymocytes. FASEB J 7: 580–585

Murakami S, Arai Y (1989) Neuronal death in the developing sexually dimorphic periventricular nucleus of the preoptic area in the female rat: effect of neonatal androgen treatment. Neurosci Lett 102: 185–190

Nicholson ML, Young DA (1979) Independence of the lethal actions of glucocorticoids on lymphoid cells from possible hormone effects on calcium uptake. J Supramol Struct 10: 165–174

Nicotera P, Bellomo G, Orrenius S (1992) Calcium-mediated mechanisms in chemically induced cell death. Annu Rev Pharmacol Toxicol 32: 449–470

Nordeen EJ, Nordeen KW, Sengelaub DR, Arnold AP (1985) Androgens prevent normally occurring cell death in a sexually dimorphic spinal nucleus. Science 229: 671–673

Ohoka Y, Nakai Y, Mukai M, Iwata M (1993) Okadaic acid inhibits glucocorticoid-induced apoptosis in T cell hybridomas at its late stage. Biochem Biophys Res Commun 197: 916–921

Ojeda F, Guarda MI, Maldonado C, Foich H (1990) Protein kinase C involvement in thymocyte apoptosis induced by hydrocortisone. Cell Immunol 125: 535–539

Owens GP, Hahn WE, Cohen JJ (1991) Identification of mRNAs associated with programmed cell death in immature thymocytes. Mol Cell Biol 11: 4177–4188

Perandones CE, Illera VA, Peckham D, Stunz LL, Ashman RF (1993) Regulation of apoptosis in vitro in mature murine spleen T cells. J Immunol 151: 3521–3529

Pla M, Zakany J, Fachet J (1976) H-2 influence on corticosteroid effects on thymus cells. Folia Biol (Praha) 22: 49–50

Pratt RM, Kim CS, Grove RI (1984) Role of glucocorticoids and epidermal growth factor in normal and abnormal palatal devetopment. Curr Top Dev Biol 19: 81–101

Pratt WB, Sanchez ER, Bresnick EH, Meshinchi S, Scherrer LC, Dalman FC, Welsh MJ (1989) Interaction of the glucocorticoid receptor with Mr 90,000 heat shock protein: An evolving model of ligand-mediated receptor transformation and translocation. Cancer Res [Suppl] 49: 2222s–2229s

Preston RR (1990) A magnesium current in paramecium. Science 250: 285–288

Punt JA, Hosono M, Hashimoto Y (1993) CD4+/CD8- thymocytes dominate the fetal thymus treated with a combination of anti-T cell receptor-β and anti-CD4 antibodies. J Immunol 151: 1290–1302

Ramakrishnan N, Catravas GN (1992) N-(2-mercaptoethyl)-1,3-propanediamine (WR-1065) protects thymocytes from programed cell death. J Immunol 148: 1817–1821

Ramsdell F, Fowlkes BJ (1989) Engagement of CD4 and CD8 accessory molecules is required for T-cell maturation. J Immunol 143: 1467–1471

Saltzman AG, Hiipakka RA, Chang C, Liao S (1987) Androgen repression of the production of a 29-Kilodalton protein and its mRNA in the rat ventral prostate. J Biol Chem 262: 432–437

Schraufstatter IU, Hyslop PA, Hinshaw DB, Spragg RG, Sklar LA, Cochrane CG (1986) Hydrogen peroxide-induced injury of cells and its prevention by inhibitors of poly(ADP-ribose) polymerase. Proc Natl Acad Sci USA 83: 4908–4912

Schwartz LM, Osborne BA (1993) Programmed cell death, apoptosis and killer genes. Immunol Today 14: 582–590

Schwartzman RA, Cidlowski JA (1993) Mechanism of tissue-specific induction of internucleosomal deoxyribonucleic acid cleavage activity and apoptosis by glucocorticoids. Endocrinology 133: 591–599

Shi Y, Sahai BM, Green DR (1989) Cyclosporin A inhibits activation-induced death in T cell hybridomas and in thymocytes. Nature 339: 625–626

Shi Y, Glynn JM, Guilbert LJ, Cotter TG, Bissonnette RP, Green DR (1992) Role for c-myc in activation-induced apoptotic cell death in T cell hybridomas. Science 257: 212–214

Shortman K, Jackson H (1974) The differentiation of T lymphocytes. 1. Proliferation kinetics and interrelationships of subpopulations of mouse thymus cells. Cell Immunol 12: 230–246

Sloviter RS, Sollas AL, Dean E, Noubort S (1993) Adrenalectomy-induced granule cell degeneration in the rat hippocampal dentate gyrus: characterization of an in vivo model of controlled neuronal death. J Comparative Neurol 330: 324–336

Sun DY, Jiang S, Zheng LM, Ojcius DM, Young JDE (1994) Separate metabolic pathways teading to DNA fragmentation and apoptotic chromatin condensation. J Exp Med 179: 559–568

Suzuki K, Kizaki H, Tadakuma T, Ishimura Y (1990) 12-O-tetradecanoylphorbol 13-acetate potentiates the action of cAMP in inducing DNA cleavage in thymocytes. Biochem Biophys Res Commun 171: 827–831

Tyan ML (1979) Genetic control of hydrocortisone induced thymus atrophy. Immunogenetics 8: 177–181

Wilson JD, George FW, Griffen JE (1981) The harmonal control of sexual development. Science 211: 1278–1284

Wyllie AH (1980) Glucocorticoid-induced thymocyte apoptosis is associated with endogenous ondonuclease activation. Nature 284: 555–556

Wyllie AH, Kerr JFR, Currie AR (1980) Cell death: the significance of apoptosis. Int Rev Cytol 68: 251–306

Wyllie AH, Morris RG, Smith AL, Dunlop D (1984) Chromatin cleavage in apoptosis: association with condensed chromatin morphology and dependence on macromolecular synthesis. J Pathol 142: 67–77

Yang Y, Vacchio MS, Ashwell JD (1993) 9-cis-Retinoic acid inhibits activation-driven T-cell apoptosis: implication for retinoid X receptor involvement in thymocyte development. Proc Natl Acad Sci USA 90: 6170–6174

Zacharchuk CM, Mercep M, Chakraborti PK, Simons SS, Ashwell JD (1990) Programmed T lymphocyte death: cell activation- and steroid-induced pathways are mutually antagonistic. J Immunol 145: 4037–4045

Zubiaga AM, Munoz E, Huber BT (1992) IL-4 and IL-2 selectively rescue Th cell subsets from glucocorticoid-induced apoptosis. J Immunol 149: 107–112

Calcium and Cyclosporin A in the Regulation of Apoptosis

D.J. McConkey[1] and S. Orrenius[2]

1 Introduction

Calcium concentration control is critical to cell viability and function. Energy-dependent Ca^{2+} transport systems located in the plasma membrane, endoplasmic reticulum, mitochondria, and nucleus maintain a cytosolic Ca^{2+} concentration (about 100 nM) that is roughly three orders of magnitude lower than that present in the extracellular millieu (about 1 mM). During cellular activation the presence of this gradient is exploited, and controlled elevations in the cytosolic Ca^{2+} level mediate the effects of hormones and other growth stimuli. However, damage to the cell can impair the proper function of Ca^{2+} homeostatic mechanisms and lead to uncontrolled, sustained Ca^{2+} increases that mediate cell killing in many pathological situations (ORRENIUS et al. 1989). Thus, Ca^{2+} can promote proliferation or death depending upon the cellular context.

The implication of apoptosis as the mechanism of cell deletion in both physiological and pathological circumstances has led to growing interest in the biochemical and molecular control of the process. Accumulating evidence indicates that Ca^{2+} plays a central role in regulating apoptosis in many tissues. This chapter will summarize these findings.

[1] Department of Cell Biology, MD Anderson Cancer Center, 1515 Holcombe Boulevard, Houston, TX 77030, USA
[2] Institute of Environmental Medicine, Division of Toxicology, Karolinska Institute, Box 210, 17177 Stockholm, Sweden

2 Calcium-Dependent Endonuclease Activation

Chromatin condensation is one of the most characteristic morphological features of apoptosis (WYLLIE et al. 1980). This change can be induced by Ca^{2+} in isolated nuclei (SUN et al. 1994) and has been linked to the activation of an endogenous endonuclease that cleaves host chromatin into oligonucleosome length DNA fragments (WYLLIE 1980; WYLLIE et al. 1984; ARENDS et al. 1990). Such DNA fragmentation gives rise to a ladder pattern on agarose gels (WYLLIE 1980) and serves as the most characteristic biochemical feature of the process.

Prior to the implication of oligonucleosomal DNA fragmentation in apoptosis, several investigators noted the presence of an enzyme activity capable of generating such fragments in preparations of isolated nuclei. HEWISH and BURGOYNE (1973), studying the Ca^{2+} requirement for activation of transcription in vitro, demonstrated that concentrations optimal for RNA synthesis also stimulated oligonucleosomal DNA cleavage. The activity was found in a variety of tissues. Later work by VANDERBILT et al. (1982) with isolated liver nuclei demonstrated that the enzyme possessed a strict Ca^{2+}/Mg^{2+} requirement and could be inhibited by polyamines (spermine, spermidine). Following Wyllie's demonstration of oligonucleosomal DNA fragmentation in apoptotic thymocytes, COHEN and DUKE (1984) and WYLLIE et al. (1984) presented evidence that the Ca^{2+}-dependent endonuclease was responsible for DNA cleavage in intact apoptotic cells. More recently, KYPRIANOU et al. (1988) implicated a Ca^{2+}-dependent endonuclease in DNA fragmentation during apoptosis in the prostate induced by androgen withdrawal. Together, these results clearly establish the presence of Ca^{2+}-dependent endonuclease activity in nuclei from various tissue sources and strongly suggest that its activation mediates DNA fragmentation in many examples of apoptosis.

Several candidate Ca^{2+}-dependent endonucleases have been purified. GAIDO and CIDLOWSKI (1990) reported the purification of an 18 kDa enzyme from rat thymocyte nuclei whose biochemical characteristics are consistent with those exhibited by the apoptosis endonuclease in intact cells. Microsequence analysis has revealed its identity with the 18 kDa cyclosporin A binding protein cyclophilin, and experiments with recombinant cyclophilin confirm that it possesses intrinsic endonuclease activity (personal communication). ARENDS et al. (1990) have isolated another, higher molecular weight protein from apoptotic thymocytes that they have linked to the chromatin condensation observed in whole apoptotic cells. Finally, PEITSCH et al. (1993) presented evidence that deoxyribonuclease I (DNase I), an enzyme that is also found in serum and several other tissues, is involved in DNA cleavage in rat thymocytes. Thus, at least three different Ca^{2+}-dependent enzymes have been isolated that could mediate DNA fragmentation in apoptotic cells. Whether redundant Ca^{2+}-dependent pathways of DNA cleavage actually exist in cells requires further study.

We have shown that addition of adenosine triphosphate (ATP) and nicotine adenine dinucleotide (NAD^+) to preparations of isolated nuclei allows DNA fragmentation to occur at submicromolar Ca^{2+} concentrations (JONES et al. 1989). The ATP requirement is linked to the function of a nuclear Ca^{2+} uptake system capable of raising intranuclear free Ca^{2+} levels and maintaining a concentration gradient between the nuclear matrix and the extranuclear millieu (the cytoplasm in intact cells) (NICOTERA et al. 1989). The pump is also dependent on the Ca^{2+} binding protein calmodulin. Preliminary evidence suggests that NAD^+ is required for activation of poly(ADP-ribose) polymerase (JONES et al. 1989), an enzyme that has been implicated in the response to DNA damage. Supporting its involvement in the regulation of apoptosis, previous work has suggested that the Ca^{2+}-dependent endonuclease is a substrate for this enzyme (YOSHIHARA et al. 1975), although how ADP-ribosylation modulates its activity is unclear.

Although Ca^{2+} appears critical in the regulation of chromatin degradation in many systems, calcium-independent endonuclease(s) may mediate apoptotic DNA fragmentation in other models. For example, although glucocorticoid-induced apoptosis in the human CEM (T cell acute lymphoblastic leukemia) cell line clearly involves oligonucleosomal DNA fragmentation, endonuclease activation does not occur in isolated CEM nuclei incubated in the presence of Ca^{2+} and Mg^{2+} (ALNEMRI and LITWACK 1990). This could be due to localization of the enzyme in another subcellular compartment or to the involvement of a different enzymatic activity. Supporting the latter, BARRY and EASTMAN (1993) have implicated deoxyribonuclease II (DNase II) in the apoptotic DNA fragmentation induced in Chinese hamster ovary (CHO) cells by various chemotherapeutic agents. This 40 kDa enzyme is dependent upon Mg^{2+} but not Ca^{2+} and is active at low (acidic) pH. At least a fraction of DNase II may be localized to nuclei, consistent with a role in DNA fragmentation. Ongoing work is aimed at determining whether the pH levels required for DNase II activation are reached in apoptotic cells.

3 High Molecular Weight DNA Fragmentation in Apoptosis

Scattered throughout the literature are notable exceptions to the idea that chromatin condensation is invariably linked to oligonucleosomal DNA fragmentation. For example, OBERHAMMER et al. (1993) have shown that, although typical apoptotic chromatin condensation can be demonstrated in hepatocytes treated with TGF-β and in DU-145 prostatic carcinoma cells treated with etoposide, oligonucleosomal DNA fragmentation cannot be detected. Similarly, COHEN et al. (1992) have reported that zinc is capable of blocking oligonucleosomal DNA fragmentation but not chromatin condensation in glucocorticoid-treated rat thymocytes. Instead, both groups have demonstrated that apoptosis in these models is associated with cleavage of chromatin into large (50–300 kilobase) DNA fragments (OBERHAMMER et al. 1993; BROWN et al. 1993). Since these observations were published a number of other groups have reported similar findings in other cell types. Thus, the formation of large DNA fragments may be another characteristic biochemical marker for apoptosis in certain cell types.

Although the mechanisms underlying the formation of the large DNA fragments remain unclear, it has been suggested that they may result from sequential disorganization of chromatin structure occurring as a consequence of apoptosis, leading to the release of loop domains that are subsequently vulnerable to endonuclease attack (FILIPSKI et al. 1990; ZHIVOTOVSKY et al. 1994). FILIPSKI and coworkers have proposed that higher order folding of chromatin involves the formation of 50 kb loops that are wound in groups of six into structures termed rosettes; the sizes of these structures correspond well with the sizes of the large fragments produced within apoptotic cells. It is therefore possible that changes in chromatin structure regulate endonuclease activation and apoptosis by altering substrate availability, an idea that is under investigation at present. In addition, it will be interesting to determine whether formation of the large fragments plays a direct role in promoting chromatin condensation.

The biochemical characteristics of the enzyme(s) responsible for forming the large DNA fragments are also still poorly defined, but preliminary results support a central role for Ca^{2+} in this process when studied in human thymocytes (ZHIVOTOVSKY et al. 1994). Thus, the large fragments can be induced by Ca^{2+} ionophore or thapsigargin treatment in whole cells. Moreover, incubation of isolated nuclei in the presence of Ca^{2+} also leads to their formation (FILIPSKI et al. 1990). Therefore, it appears that the large DNA fragments may be produced by the same enzymatic activity responsible for the subsequent oligonucleosomal DNA fragmentation and that sequential changes in chromatin structure alone are what dictate its substrate specificity. Support for this notion comes from the observation that zinc, calcium chelators, protease antagonists, and the endonuclease inhibitor aurintricarboxylic acid can all block the formation of large DNA fragments in human thymocytes and isolated human thymocyte nuclei (FILIPSKI et al. 1990; ZHIVOTOVSKY et al. 1994), although Cohen and coworkers have suggested that the activities responsible for production of the large DNA fragments and the subsequent oligonucleosomal DNA fragmentation may be distinguished on the basis of differential sensitivity to inhibition by zinc in rat thymocytes (BROWN et al. 1993). Nonetheless, the observation that all of the morphological (chromatin condensation, reduction in cell volume) and biochemical (production of large chromatin fragments, oligonucleosomal DNA fragmentation) features of apoptosis can be induced in human and rat thymocytes by thapsigargin suggests that the increase in intracellular Ca^{2+} can account for all of the events observed.

4 Calcium Signaling in Apoptosis

Early work by Kaiser and Edelman provided the first evidence that increases in intracellular Ca^{2+} might be involved in triggering apoptosis. Working with immature thymocytes, the authors showed that glucocorticoid-stimulated apoptosis is associated with Ca^{2+} influx (KAISER and EDELMAN 1977) and that the cytolytic process can be mimicked by treating the cells with Ca^{2+} ionophores (KAISER and EDELMAN 1978). We have since confirmed that glucocorticoids induce cytosolic Ca^{2+} increases in thymocytes (MCCONKEY et al. 1989a). WYLLIE and coworkers showed that Ca^{2+} ionophores induce many of the morphological changes and endogenous endonuclease activation in thymocytes that are typical of apoptosis (Wyllie et al. 1984). Rapid, sustained Ca^{2+} increases precede the cytolysis of the targets of cytotoxic T lymphocytes (ALLBRITTON et al. 1988) and natural killer (NK) cells (MCCONKEY et al. 1990). In developing T lymphocytes high affinity engagement of the T cell receptor induces apoptosis (SMITH et al. 1989; SHI et al. 1989; MCCONKEY et al. 1989b; MURPHY et al. 1990) that is preceded by Ca^{2+} increases (MCCONKEY et al. 1989b; NAKAGAMA et al. 1992). Calcium increases have since been observed in many other examples of apoptosis as well (PEROTTI et al. 1990; ZHENG et al. 1991; MCCONKEY et al. 1991; BELLOMO et al. 1992).

Direct evidence that Ca^{2+} increases are necessary for apoptotic endonuclease activation and cell death has been obtained from experiments with intracellular Ca^{2+} buffering agents and extracellular Ca^{2+} chelators. We (MCCONKEY et al. 1989a, b, 1990, 1991; Aw et al. 1990; PEROTTI et al. 1990; BELLOMO et al. 1992) and others (STORY et al. 1992; ROBERTSON et al. 1993) have shown that intracellular Ca^{2+} buffers and extracellular EGTA can inhibit both DNA fragmentation and death in apoptotic cells, suggesting that sustained Ca^{2+} elevations are required for both responses. Furthermore, increases in calmodulin expression are linked to apoptosis in glucocorticoid-treated thymoma cells (DOWD et al. 1991) and in the prostate following withdrawal of androgen (FURUYA and ISAACS 1993), and we and others have shown that calmodulin antagonists can interfere with apoptosis in some of these systems (MCCONKEY et al. 1989a; DOWD et al. 1991). Independent evidence for the involvement of Ca^{2+} influx has come from studies with specific Ca^{2+} channel blockers, which abrogate apoptosis in the regressing prostate following testosterone withdrawal (MARTIKAINEN and ISAACS 1990) and in pancreatic β-cells treated with serum from patients with type I diabetes (JUNTTI-BERGGREN et al. 1993). Thus, elevations of the cytosolic Ca^{2+} level appear to represent a relatively common trigger for apoptosis in cells of diverse tissue origins.

Calcium-dependent mechanisms also appear to play important roles in promoting apoptosis in the brain. Apoptosis has been proposed as the mechanism underlying neuronal death in Huntington's and Alzheimer's diseases (FORLONI et al. 1993), ischemia, and glutamate toxicity (CHOI 1992). The latter is somewhat similar to activation-induced cell death in T lymphocytes, as it is triggered by a surface receptor (the NMDA receptor) and is mediated by elevations in the cytosolic Ca^{2+} concentration. Cell-permeant Ca^{2+} chelators (TYMIANSKI et al. 1993) or overexpression of the Ca^{2+} binding protein calbindin (MATTSON et al. 1991) block NMDA receptor-mediated cell killing, as is true in T cells treated with T cell receptor agonists, glucocorticoid hormones, or calcium ionophores (DOWD et al. 1992). Calcium may act directly by promoting endonuclease activation in neurons. In addition, Ca^{2+} may also promote the function of nitric oxide synthase (DAWSON et al. 1991). This in turn can lead to the accumulation of nitric oxide, a second messenger that has also been implicated in triggering apoptosis in several different experimental systems (CHOI 1992; ALBINA et al. 1993; XIE et al. 1993).

In some model systems elevations of the cytosolic Ca^{2+} level have been shown to block apoptosis. Hematopoietic cells dependent upon interleukin 3 (IL-3) for their growth and survival die by apoptosis when the cytokine is removed. RODRIGUEZ-TARDUCHY and colleagues have shown that the calcium ionophore A23187, which triggers apoptosis in thymocytes (WYLLIE et al. 1984), suppresses apoptosis at similar concentrations in IL-3-dependent hematopoietic progenitor cells (RODRIGUEZ-TARDUCHY et al. 1990,1992). The effect of ionophore is dependent upon the production of IL-4 which in turn promotes survival. Interestingly, isolated nuclei from these cells lack Ca^{2+}-dependent endonuclease activity (RODRIGUEZ-TARDUCHY et al. 1992). This observation supports the idea that multiple chromatin cleavage mechanisms exist within different types of mammalian cells and that the response to elevated Ca^{2+} may be dictated by tissue-specific genetic programming. The observation that nerve growth factor (NGF)-deprived neurons can be saved from apoptosis by depolarization (EDWARDS et al. 1991; MARTIN et al. 1992), which induces Ca^{2+} increases in the cells, strengthens this conclusion. Whether an acidic nuclease mediates DNA fragmentation in IL-3-dependent cells has not been determined, although the observation that phorbol esters protect the cells via a mechanism that involves intracellular alkalinization suggests that this may be the case (RAJOTTE et al. 1992).

5 Role of Cyclosporin A

An important aim of ongoing research is to define the targets of Ca^{2+} in apoptotic cells. It is conceivable that elevations in the cytosolic Ca^{2+} level might promote apoptosis by directly stimulating the enzymatic activities of proteases, phospholipases, and/or endonucleases responsible for mediating cellular demise in apoptosis. Alternatively, Ca^{2+} rises may activate intracellular signal transduction pathways involving protein kinases and/or phosphatases that could regulate downstream effectors of apoptosis via posttranslational modification. The observation that calmodulin antagonists can interfere with apoptosis supports this interpretation, as calmodulin is a well-known mediator of Ca^{2+} signal transduction pathways.

The immunosuppressive drugs cyclosporin A (CsA) and FK-506 are known to interfere with a Ca^{2+}-sensitive signal transduction pathway in B and T lymphocytes. In the latter, the effects are due to interference with critical elements within the IL-2 promoter, resulting in an inhibition of IL-2 production. The drugs act by binding to small polypeptides, cyclophilin A and FK506-binding protein (FKBP), known as immunophilins, which possess peptidyl-prolyl isomerase activities. The three-dimensional structures of CsA-cyclophilin and FK506-FKBP complexes have recently been solved (RINGE 1991 and references therein), and although they are in general distinct, certain aspects of their structures are quite similar, particularly a hybrid surface created in each case by the drug-immunophilin interaction. Interaction of the drug with its receptor is absolutely required for immunosuppression. Evidence that inhibition of immunophilin peptidyl-prolyl isomerase activity is not involved in the drugs' actions has been obtained from experiments with 506BD and rapamycin, inactive analogs of FK506 which inhibit peptidyl-prolyl isomerase activity without suppressing T cell activation (BIERER et al. 1990).

Recent work has demonstrated that CsA-cyclophilin and FK506-FKBP complexes block T cell activation by sequestering the Ca^{2+}/calmodulin-dependent protein phosphatase calcineurin (LIU et al. 1991). This is accomplished by direct binding to the phosphatase, resulting in an inhibition of its enzymatic activity. It has been suggested that the conserved surfaces formed by the immunosuppressant/immunophilin interaction may represent the complex contact sites for calcineurin. Calcineurin function appears to be required for the translocation of a component of the transcription factor, nuclear factor of activated T cells (NFAT), from the cytoplasm to the nucleus (FLANAGAN et al. 1991), an effect that appears sufficient to explain the inhibitory actions of CsA and FK506 on the IL-2 promoter. Further evidence that inhibition of calcineurin function is solely responsible for mediating the inhibitory effects of CsA and FK506 comes from the observation that overexpression of calcineurin makes Jurkat T cells resistant to CsA action (CLIPSTONE and CRABTREE 1992).

Studies with CsA and FK506 suggest that calcineurin is one target for Ca^{2+} in apoptosis. SHI et al. (1989) demonstrated that CsA could block activation-induced cell death in T cell hybridomas in vitro and thymocyte apoptosis in response to anti-CD3 antibody treatment in vivo. However, not all pathways of apoptosis are affected by the drugs: Although CsA and FK506 can almost completely inhibit DNA fragmentation induced by anti-CD3 antibodies, Ca^{2+} ionophores, or thapsigargin in T cell hybridomas, they have no effect on endonuclease activation induced by the zinc chelator TPEN or the synthetic glucocorticoid methylprednisolone (S. Jiang, S.C. Chow, S. Orrenius, unpublished observation). Significantly, the FKBP ligand rapamycin, which interferes with a different signal transduction pathway but does not inhibit calcineurin, has no effect on DNA fragmentation observed in response to any of these treatments. Thus, CsA and FK506 may specifically and selectively inhibit some of the important Ca^{2+}-dependent pathways of apoptosis in certain T cell model systems. Together with the observations discussed above, these results indicate that calcineurin and perhaps NFAT are involved in at least some of the apoptotic pathways operative in thymocytes. Intriguingly, CsA and FK506 may also interfere with certain apoptotic pathways in neurons, where their mechanism of action appears to involve inhibition of calcineurin-dependent dephosphorylation and activation of nitric oxide synthase (DAWSON et al. 1993). Whether CsA and FK506 have similar effects in other tissues requires further investigation.

6 Summary and Future Directions

A large body of evidence supports the idea that Ca^{2+} plays an important role in regulating apoptosis. A schematic model illustrating potential targets for Ca^{2+} in apoptosis is presented in Fig. 1. Studies on the endonuclease(s) involved suggest that many cell types constitutively express a Ca^{2+}-dependent activity that by all available biochemical criteria is a strong candidate for the enzyme that mediates oligonucleosomal DNA cleavage in intact apoptotic cells. Sustained elevations in the cytosolic Ca^{2+} level are observed in diverse cell types undergoing apoptosis, and it is possible that they directly trigger endogenous endonuclease activation, perhaps via a mechanism that involves the function of a nuclear Ca^{2+} pump. Alternatively, results from experiments with CsA suggest that Ca^{2+} may also exert its effects in some model systems via activation of the Ca^{2+}/calmodulin-dependent protein phosphatase calcineurin. Results obtained with Ca^{2+} buffering agents, Ca^{2+} chelators, and Ca^{2+} channel blockers confirm that these increases are required for both endonuclease activation and subsequent cell death. In some model systems Ca^{2+} appears either not to be involved in apoptosis regulation or it inhibits the process, and the involvement of alternative endonuclease(s) has been proposed to explain these differences.

Further efforts are required to identify and clone candidate apoptosis endonuclease(s) to directly test whether they are necessary for DNA fragmentation in apoptotic cells by gene targeting. In addition, investigation into the potential involvement of selective regulation of subcellular Ca^{2+} localization in apoptosis, including the potential involvement of Bcl-2 in these processes (BAFFY et al. 1993; ANDJELIC et al. 1993), deserves additional attention. The possible involvement of other Ca^{2+}-dependent processes, particularly protease activation, should reveal additional targets of Ca^{2+} action in apoptosis. Finally, identification of apoptosis-regulating genes whose expression is controlled by Ca^{2+} may help to reveal a molecular basis for the Ca^{2+} dependency.

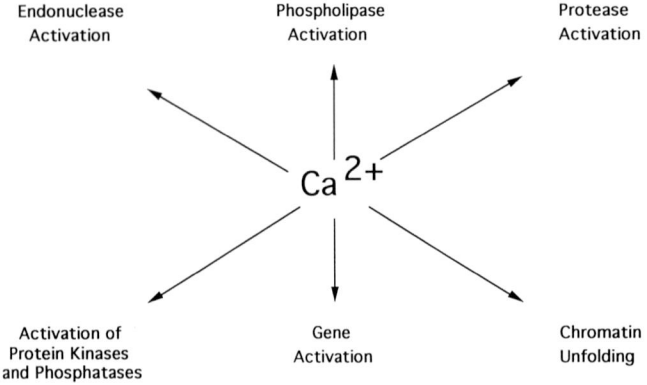

Fig. 1. Summary of the most likely targets for Ca^{2+} in apoptotic cells, based on experimental data available at present. For detailed explanation of these processes, see text

References

Albina JE, Cui S, Mateo RB, Reichner JS (1993) Nitric oxide-mediated apoptosis in murine peritoneal macrophages. J Immunol 150: 5080–5085

Allbritton NL, Verret CR, Wolley RC, Eisen HN (1988) Calcium ion concentrations and DNA fragmentation in target cell destruction by murine cloned cytotoxic T lymphocytes. J Exp Med 167: 514–527

Alnemri ES, Litwack G (1990) Activation of internucleosomal DNA cleavage in human CEM lymphocytes by glucocorticoid and novobiocin. Evidence for a non-Ca^{2+}-requiring mechanism(s). J Biol Chem 265: 17323–17333

Andelic S, Jain N, Nikolic-Zugic J (1993) Immature thymocytes become sensitive to calcium-mediated apoptosis with the onset of CD8, CD4, and the T cell receptor expression: a role for bcl-2? J Exp Med 178: 1745–1751

Arends MJ, Morris RG, Wyllie AH (1990) Apoptosis. The role of the endonuclease. Am J Pathol 136: 593–608

Aw TY, Nicotera P, Manzo L, Orrenius S (1990) Tributyltin stimulates apoptosis in rat thymoctyes. Arch Biochem Biophys 283: 46–50

Baffy G, Miyashita T, Williamson JR, Reed JC (1993) Apoptosis induced by withdrawal of interleukin 3 (IL-3) from an IL-3-dependent hematopoietic cell line is associated with repartitioning of intracellular calcium and is blocked by enforced Bcl-2 oncoprotein production. J Biol Chem 268: 6511–6519

Barry MA, Eastman A (1993) Identification of deoxyribonuclease II as an endonuclease involved in apoptosis. Arch Biochem Biophys 300: 440–450

Bellomo G, Perotti M, Taddei F, Mirabelli F, Finardi G, Nicotera P, Orrenius S (1992) Tumor necrosis factor alpha induces apoptosis in mammary adeoncarcinoma cells by an increase in intranuclear free Ca^{2+} concentration and DNA fragmentation. Cancer Res 52: 1342–1346

Bierer BE, Mattila PS, Standaert RF, Herzenberg LA, Burakoff SJ, Crabtree GR, Schreiber SL (1990) Two distinct signal transmission pathways in T lymphocytes are inhibited by complexes formed between an immunophilin and either FK506 or rapamycin. Proc Natl Acad Sci USA 87: 9231–9235

Brown DG, Sun X-M, Cohen GM (1993) Dexamethasone-induced apoptosis involves cleavage of DNA to large fragments prior to internucleosomal fragmentation. J Biol Chem 268: 3037–3039

Choi DW (1992) Bench to bedside: the glutamate connection. Science 258: 241–243

Clipstone NA, Crabtree GR (1992) Identification of calcineurin as a key signalling enzyme in T-lymphocyte activation. Nature 357: 695–697

Cohen GM, Sun X-M, Snowden RT, Dinsdale D, Skilleter DN (1992) Key morphological features of apoptosis may occur in the absence of internucleosomal DNA fragmentation. Biochem J 286: 331–334

Cohen JJ, Duke RC (1984) Glucocorticoid-induced activation of a calcium dependent endonuclease in thymocyte nuclei leads to cell death. J Immunol 132: 38–42

Dawson VL, Dawson TM, London GD, Bredt DS, Snyder SH (1991) Nitric oxide mediates glutamate neurotoxicity in primary cortical cultures. Proc Natl Acad Sci USA 88: 6368–6371

Dawson TM, Stiner JP, Dawson VL, Dinerman JL, Uhl GR, Snyder SH (1993) Immunosuppressant FK506 enhances phosphorylation of nitric oxide synthase and protects against glutamate neurotoxicity. Proc Natl Acad Sci USA 90: 9808–9812

Dowd DR, MacDonald PN, Komm BS, Haussler MR, Miesfeld R (1991) Evidence for early induction of calmodulin gene expression in lymphocytes undergoing glucocorticoid-mediated apoptosis. J Biol Chem 266: 18423–18426

Dowd DR, MacDonald PN, Komm BS, Haussler MR, Miesfeld RL (1992) Stable expression of the calbindin-D28K complementary DNA interferes with the apoptotic pathway in lymphocytes. Mol Endocrinol 6: 1843–1648

Edwards SN, Buckmaster AE, Tolkovsky AM (1991) The death programme in cultured sympathetic neurones can be suppressed at the posttranslational level by nerve growth factor, cyclic AMP, and depolarization. J Neurochem 57: 2140–2143

Filipski J, Leblanc J, Youdale T, Sikorska M, Walker PR (1990) Periodicity of DNA folding in higher order chromatin structures. EMBO J 9: 1319–1327

Flanagan WM, Corthesy B, Bram RJ, Crabtree GR (1991) Nuclear association of a T-cell transcription factor blocked by FK-506 and cyclosporin A. Nature 352: 803–807

Forloni G, Chiesa R, Smiroldo S, Verga L, Salmona M, Tagliavini F, Angeretti N (1993) Apoptosis mediated neurotoxicity induced chronic application of beta amyloid fragment 25–35. Neuro Report 4: 523–526

Furuya Y, Isaacs JT (1993) Differential gene regulation during programmed death (apoptosis) versus proliferation of prostatic glandular cells induced by androgen manipulation. Endocrinology 133: 2660–2666

Gaido ML, Cidlowski JA (1990) Identification, purification, and characterization of a calcium-dependent endonuclease (NUC18) from apoptotic rat thymoytes. J Biol Chem 266: 18580–18585

Hewish DR, Burgoyne LA (1973) Chromatin sub-structure. The digestion of chromatin DNA at regularly spaced sites by a nuclear deoxyribonuclease. Biochem Biophys Res Commun 52: 504–510

Jones DP, McConkey DJ, Nicotera P, Orrenius S (1989) Calcium-activated DNA fragmentation in rat liver nuclei. J Biol Chem 264: 6398–6403

Juntti-Berggren L, Larsson O, Rorsman P, Ammala C, Bokvist K, Wahlander K, Nicotera P, Dypbukt J, Orrenius S, Hallberg A, Berggren PO (1993) Increased activity of L-type Ca^{2+} channels in cells exposed to serum from patients with type I diabetes. Science 261: 86–90

Kaiser N, Edelman IS (1977) Calcium dependence of glucocorticoid-induced lymphocytolysis. Proc Natl Acad Sci USA 74: 638–642

Kaiser N, Edelman IS (1978) Further studies on the role of calcium in glucocorticoid-induced lymphocytolysis. Endocrinology 103: 936–942

Kyprianou N, English HF, Isaacs JT (1988) Activation of a Ca^{2+}-Mg^{2+}-dependent endonuclease as an early event in castration-induced prostatic cell death. Prostate 13: 103–117

Liu J, Farmer JD Jr, Lane WS, Friedman J, Weissman I, Schreiber SL (1991) Calcineurin is a common target of cyclophilin-cyclosporin A and FKBP-FK506 complexes. Cell 66: 807–815

Martikainen P, Isaacs J (1990) Role of calcium in the programmed death of rat prostatic glandular cells. Prostate 17: 175–187

Martin DP, Ito A, Harigame K, Lampe PA, Johnson EM Jr (1992) Biochemical characterization of programmed cell death in NGF-deprived sympathetic neurons. J Neurobiol 23: 1205–1220

Mattson MP, Rychlik B, Chu C, Shristakos S (1991) Evidence for calcium-reducing and excito-protective roles for calcium-binding protein calbindin-D28K in cultured hippocampal neurons. Neuron 6: 41–51

McConkey DJ, Nicotera P, Hartzell P, Bellomo G, Wyllie AH, Orrenius S (1989a) Glucocorticoids activate a suicide process in thymocytes through an elevation of cytosolic Ca^{2+} concentration. Arch Biochem Biophys 269: 365–370

McConkey DJ, Hartzell P, Perez JF, Orrenius S, Jondal M (1989b) Calcium-dependent killing of immature thymocytes by stimulation via the CD3/T cell receptor complex. J Immunol 143: 1801–1806

McConkey DJ, Chow SC, Orrenius S, Jondal M (1990) NK cell-induced cytotoxicity is dependent on a Ca^{2+} increase in the target. FASEB J 4: 2661–2664

McConkey DJ, Aguilar-Santelises A, Hartzell P, Eriksson I, Mellstedt H, Orrenius S, Jondal M (1991) Induction of DNA fragmentation in chronic B-lymphocytic leukemia cells. J Immunol 146: 1072–1076

Murphy KM, Heimberger KB, Loh DY (1990) Induction by antigen of intrathymic apoptosis of $CD4^+CD8^+$ TCR^{lo} thymocytes in vivo. Science 250: 1720–1723

Nakagama T, Ueda Y, Yamada H, Shores EW, Singer A, June CH (1992) In vivo calcium elevations in thymocytes with T cell receptors that are specific for self ligands. Science 257: 96–99

Nicotera P, McConkey DJ, Jones DP, Orrenius S (1989) ATP stimulates Ca^{2+} uptake and increases the free Ca^{2+} concentration in isolated rat liver nuclei. Proc Natl Acad Sci USA 86: 453–457

Oberhammer F, Wilson JW, Dive C, Morris ID, Hickman JD, Wakeling AE, Walker PR, Sikorska M (1993) Apoptotic death in epithelial cells: cleavage of DNA to 300 and/or 50 kb fragments prior to or in the absence of internucleosomal fragmentation. EMBO J 12: 3679–3684

Orrenius S, McConkey DJ, Bellomo G, Nicotera P (1989) Role of calcium in toxic cell killing. Trends Pharmacol Sci 7: 281–285

Peitsch M, Polzar B, Stephan H, Crompton T, MacDonald HR, Mannherg HG, Tschopp J (1993) Characterization of the endogenous deoxyribonuclease involved in nuclear DNA degradation during apoptosis (programmed cell death). EMBO J 12: 371–377

Perotti M, Toddei F, Mirabelli F, Vairetti M, Bellomo G, McConkey DJ, Orrenius S (1990) Calcium-dependent DNA fragmentation in human synovial cells exposed to cold shock. FEBS Lett 259: 331–334

Rajotte D, Haddad P, Haman A, Cragoe EJ Jr, Hoang T (1992) Role of protein kinase C and the Na^+/H^+ antiporter in suppression of apoptosis by granulocyte macrophage colony-stimulating factor and interleukin-3. J Biol Chem 267: 9980–9987

Ringe D (1991) Binding by design. Nature 351: 185–186

Robertson LE, Chubb S, Meyn RE, Story M, Ford R, Hittelman WN, Plunkett W (1993) Induction of apoptotic cell death in chronic lymphcytic leukemia by 2-chloro-2'-deoxyadenosine and 9-beta-D-arabinosyl-2'-fluroadenine. Blood 81: 143–150

Rodriguez-Tarduchy G, Collins M, Lopez-Rivas A (1990) Regulation of apoptosis in interleukin-3-dependent hemapoietic cells by interleukin-3 and calcium ionophores. EMBO J 9: 2997–3002

Rodriguez-Tarduchy G, Malde P, Lopez-Rivas A, Collins MKL (1992) Inhibition of apoptosis by calcium ionophores in IL-3-dependent bone marrow cells is dependent upon production of IL-4. J Immunol 148: 1416–1422

Shi Y, Sahai BM, Green DR (1989) Cyclosporin A inhibits activation-induced cell death in T cell hybridomas and thymocytes. Nature 339: 615–616

Smith CA, Williams GT, Kingston R, Jenkinson EJ, Owen JJT (1989) Antibodies to CD3/T cell receptor complex induce death by apoptosis in immature T cells in thymic cultures. Nature 337: 181–183

Story MD, Stephens LC, Tomasovic SP, Meyn RE (1992) A role for calcium in regulating apoptosis in rat thymocytes irradiated in vitro. Int J Rad Biol 61: 243–251

Sun DY, Jiang S, Zheng LM, Ojcius DM, Young JDE (1994) Separate metabolic pathways leading to DNA fragmentation and apoptotic chromatin condensation. J Exp Med 179: 559–568

Tymianski M, Wallace MC, Spigelman I, Uno M, Carlen PL, Tator CH, Charlton MP (1993) Cell permeant Ca^{2+} chelators reduce early excitotoxicity and ischemic neuronal injury in vivo and in vitro. Neuron 11: 221–235

Vanderbilt JN, Bloom KS, Anderson JN (1982) Endogenous nuclease: Properties and effects on transcribed genes in chromatin. J Biol Chem 257: 13009–13017

Wyllie AH (1980) Glucocorticoid-induced thymocyte apoptosis is associated with endogenous endonuclease activation. Nature 284: 555–556

Wyllie AH, Kerr JFR, Currie AR (1980) Cell death: the significance of apoptosis. Int Rev Cytol 68: 251–305

Wyllie AH, Morris RH, Smith AL, Dunlop D (1984) Chromatin cleavage in apoptosis: association with condensed chromatin morphology and dependence on macromolecular synthesis. J Pathol 142: 67–77

Xie K, Huang S, Dong Z, Fidler IJ (1993) Cytokine-induced apoptosis in transformed murine fibroblasts involves synthesis of endogenous nitric oxide. Int J Oncol 3: 1043–1048

Yoshihara K, Tanigawa Y, Burzio L, Koide SS (1975) Evidence for adenosine diphosphate ribosylation of Ca^{2+}, Mg^{2+}-dependent endonuclease. Proc Natl Acad Sci USA 72: 289–293

Zheng LM, Zychlinsky A, Liu CC, Ojeius DM, Young JDE (1991) Extracellular ATP as a trigger for apoptosis or programmed cell death. J Cell Biol 112: 279–288

Zhivotovsky B, Cedervall B, Jiang S, Nicotera P, Orrenius S (1994) Involvement of Ca^{2+} in the formation of high molecular weight DNA fragments in thymocyte apoptosis. Biochem Biophys Res Commun 202: 120–127

Bcl-2 and Bcl-2-Related Proteins in Apoptosis Regulation

L.H. Boise[1], A.R. Gottschalk[2], J. Quintáns[2], and C.B. Thompson[1]

1 Apoptosis

Apoptosis is a form of cell death critical for the normal development of multi-cellular organisms (KERR et al. 1972; WYLLIE et al. 1980). It is characterized by morphological and biochemical criteria consisting of: nuclear shrinkage, chroma-tin condensation, cytoplasmic blebbing, and internucleosomal DNA fragmenta-tion (KERR et al. 1972; WYLLIE et al. 1980; COHEN and DUKE 1992). As opposed to other forms of cell death, apoptosis does not induce an inflammatory response. There are a number of ways by which cell death by apoptosis can be induced, including growth factor deprivation, cytokine treatment, antigen-receptor en-gagement, cell–cell interactions, irradiation, and glucocorticoids (COHEN and DUKE 1992). Within the immune system, the regulation of cell death appears to be crucial for the prevention of autoimmune disease. Immature lymphocytes are particularly susceptible to apoptosis, as 95% of thymocytes die in situ during development. Self-reactive lymphocytes are eliminated from the immune repertoire following engagement of their antigen-specific receptors. Thus, the

[1] Gwen Knapp Center for Lupus and Immunology Research, University of Chicago, 924 E, 57th Street, Chicago, IL 60637, USA
[2] Department of Pathology, University of Chicago, 5841 S. Maryland Ave, MC1089, Chicago, IL 60637, USA

process of clonal deletion by apoptosis allows for the elimination of self-reactive lymphocytes without initiating an inflammatory response.

The process of apoptosis is conserved in virtually all complex organisms. In recent years, much attention has been focused on the molecular mechanism of apoptosis and a number of genes have been identified that regulate programmed cell death (PCD) in both mammalian and nematode systems. The genetics of apoptotic cell death are best worked out in the nematode *C. elegans*, where the developmental fate of every cell is known. The *C. elegans* genes *ced*-3 and *ced*-4 are required for apoptosis to occur (ELLIS and HORVITZ 1986; YUAN and HORVITZ 1990), while *ced*-9 functions as a suppressor (HENGARTNER et al. 1992). Although a mammalian homologue for *ced*-4 has not yet been identified, the interleukin (IL)-1β converting enzyme (ICE) has structural homology to *ced*-3 (YUAN et al. 1993). Overexpression of either *ced*-3 or the murine ICE gene induces apoptosis in rat fibroblasts (MIURA et al. 1993), suggesting that ICE may function during mammalian development to cause apoptosis. Interestingly, the activity of ICE is inhibited by a cowpox protein encoded by the *crm*A gene (RAY et al. 1992; GAGLIARDINI et al. 1994), indicating one mechanism viruses may have evolved to regulate death in infected cells. *Ced*-9 is most homologous to the mammalian gene *bcl*-2, but also shows homology to other members of the Bcl-2 family (HENGARTNER and HORVITZ 1994). In *ced*-9 mutants it has been shown that human *bcl*-2 can reduce the number of PCDs in *C. elegans* (VAUX et al. 1992b; HENGARTNER and HORVITZ et al. 1994), suggesting that the mechanism of apoptosis controlled by *bcl*-2 in human is the same as that in nematodes.

2 *bcl-2*

bcl-2 was initially described as the oncogene that was present in the immunoglobulin locus as a result of a translocation [t(14;18)] that is seen in human B cell leukemias and lymphomas (TSUJIMOTO et al. 1984; BAKHASHI et al. 1985; CLEARY et al. 1986; TSUJIMOTO and CROCE 1986). Upon cloning of the normal *bcl*-2 gene it was determined that the oncogenic form of the protein was identical to the normal gene product (CHEN-LEVY et al. 1989). Thus inappropriate expression of the normal protein was the cause of tumor formation. The *bcl*-2 gene product is a 26 kDa membrane-associated protein that localizes to the mitochondrial membrane and to the perinuclear membrane (CHEN-LEVY et al. 1989; Hockenbery et al. 1990; CHEN-LEVY and CLEARY 1990; DE JONG et al. 1992). In vitro analysis revealed that *bcl*-2 was distinct from previously described oncogenes in that it did not enhance the growth or proliferation of transfected cell lines (NUÑEZ et al. 1990). Upon closer inspection it was discovered that *bcl*-2 expression could enhance cell survival in the absence of growth factors in some cells (VAUX et al. 1988; NUÑEZ et al. 1990). Normally, growth factor-dependent cell lines die rapidly in the absence of growth factor through the process of apoptosis. In contrast, *bcl*-2 transfected cells remained in a quiescent state following growth factor withdrawal but could be induced to proliferate by reintroduction of growth factor. The ability of *bcl*-2 to block this form of PCD sets it apart from other oncogenes in that *bcl*-2 expression can affect normal homeostasis by allowing cells destined to die to survive instead of affecting the proliferation rate of the cell.

PCD represents the mechanism by which cell death occurs in response to a wide range of conditions. The ability of *bcl*-2 to prevent a variety of dissimilar causes of cell death has been tested. *bcl*-2 has been shown to function in many systems, including cell death induced by nerve growth factor (NGF) withdrawal (GARCIA et al. 1992; ALLSOPP et al. 1993; KANE et al. 1993), γ-irradiation (SENTMAN et al. 1991; STRASSER et al. 1991a), and cancer chemotherapeutics (MIYASHITA and REED 1992; OHMORI et al. 1993; WALTON et al. 1993). Recently, genes such as the oncogene c-*myc*, the tumor suppressor gene *p53*, and the Fas antigen have all been shown to cause cell death. When c-*myc* is ectopically expressed in serum-starved fibroblasts it causes these cells to undergo apoptosis (BISSONNETTE et al. 1992; FANIDI et al. 1992; WAGNER et al. 1993). *bcl*-2 is capable of preventing the c-*myc*-induced cell death. *p53* has been demonstrated to be necessary for irradiation-induced cell death in thymocytes (CLARKE et al. 1993; LOWE et al. 1993). The wild-type homologue of *p53* is necessary for induction of PCD as demonstrated by the transfection of cell lines with a temperature-sensitive form of *p53* (YONISH-ROUACH et al. 1991). When cells are incubated at the permissive temperature (37° C), the protein takes on the wild-type confirmation and the cells die. Conversely, if cells are placed at the nonpermissive temperature (32.5° C) the p53 protein reverts to a mutant confirmation and the cells can survive. As seen with c-*myc*-induced cell death, *bcl*-2 is capable of preventing p53-induced cell death (CHIOU et al. 1994). Wild-type p53-mediated apoptosis is also inhibited by the adenovirus E1B gene which may have some structural similarities with *bcl*-2 (DEBBAS and WHITE 1993). It has also been shown that engagement of the Fas antigen induces apoptosis in a variety of murine cell lines transfected with the human Fas gene (ITOH et al. 1991). In addition, the lymphoproliferative disorder associated with *lpr* mice has been attributed to a defect in the Fas antigen (WATANABE-FUKUNAGA et al. 1992). Cotransfecting murine cell lines with human *Fas* and *bcl*-2 partially protected these cells from Fas-induced PCD, as more than 50% of the double transfectants were still viable following incubation with anti-Fas antibody (ITOH et al. 1993).

How the Bcl-2 protein functions to inhibit PCD is uncertain. Bcl-2 may block the generation of reactive oxygen species which are produced during cell death as a result of growth factor withdrawal (HOCKENBERY et al. 1993; KANE et al. 1993). The actual mechanism of antioxidant activity and whether this is the only mechanism of action of Bcl-2 remains to be determined. That Bcl-2 can block PCD induced by a variety of signals suggests that Bcl-2 plays a central role in the regulation of cell death from diverse signaling mechanisms.

2.1 Bcl-2: A Role in B Cell Development

Since *bcl*-2 was initially described as an oncogene active in B cells, its expression and function has been widely studied in lymphocytes. Bcl-2 expression is regulated in the developing B cells of the bone marrow in a biphasic fashion, with expression present in pro-B cells and mature B cells, but not in pre-B or immature B cells (MERINO et al. 1994). In splenic B cells, Bcl-2 is highly expressed in the IgM⁺/IgD⁺ cells of the follicular mantle (KORSMEYER et al. 1990). This is in contrast to the lack of expression seen in the proliferating centroblasts of the dark zone and the centrocytes of the basal light zone of the germinal center (KORSMEYER et al. 1990). Germinal center B cells spontaneously undergo apoptosis unless they are rescued by antigen receptor cross-linking or CD40 ligation, both of which induce expression of Bcl-2 protein (LIU et al. 1991). Expression of Bcl-2 reappears in the B cell blasts of the apical light zone (KORSMEYER et al. 1990).

Mice that carry a *bcl*-2 transgene with expression directed to lymphoid cells have increased numbers of resting B cells in the spleen and bone marrow (McDONNELL et al. 1989, 1990; STRASSER et al. 1991b; KATSUMATA et al. 1992). These cells had enhanced survival in culture and prolonged immune responses. In adoptive transfer experiments it was determined that as a consequence of prolonging cell survival, Bcl-2 can maintain B cell memory (NUÑEZ et al. 1991). The importance of normal Bcl-2 in B cell development is unclear since mice that are deficient in Bcl-2 or have reconstituted lymphoid systems that are lacking Bcl-2, can develop normal lymphocytes (NAKAYAMA et al. 1993; VEIS et al. 1993b). This suggests that Bcl-2 is not necessary for B cell development but, interestingly, these mice eventually show a loss of B and T cells suggesting a role for Bcl-2 in the maintenance of lymphoid cells. STRASSER et al. (1991b) found that Bcl-2 transgenic mice suffered from renal failure due to autoimmune disease, suggesting that Bcl-2 could override selective processes which deleted autoreactive B cells. Overexpression of Bcl-2 has also been shown to allow B cell development in the absence of immunoglobulin expression in SCID mice (STRASSER et al. 1994). These SCID/Bcl-2 mice have B220⁺ cells present in their periphery which, although lacking surface immunoglobulin, express other markers consistent with mature B cells. Thus, Bcl-2 allowed B cells to survive through developmental selection events and permitted B cell development in SCID mice to occur. These patterns of expression suggest that Bcl-2 may play a role in the development (including selection) of B cells and the maintenance of B cell memory.

2.2 Bcl-2 Is Regulated During T Cell Development

Bcl-2 protein expression in thymocytes parallels that of developing B cells in that the expression pattern is biphasic through the developmental pathway (GRATIOT-DEANS et al. 1993; VEIS et al. 1993a). The progenitor cells (CD4⁻, CD8⁻) express moderate levels of protein. This is down-regulated in double positive cells (CD4⁺, CD8⁺), the population which will undergo positive and negative selection.

Following selection, single positive, mature thymocytes express high levels of Bcl-2 protein. This pattern of expression predicts a role for Bcl-2 in the selection of the T cell repertoire, yet genetically manipulated mice have yielded equivocal results. Mice which are transgenic for *bcl*-2 have higher numbers of thymocytes (SENTMAN et al. 1991; STRASSER et al. 1991b; SIEGAL et al. 1992). These cells have enhanced survival in culture and in the presence of glucocorticoids, ionomycin, and irradiation, suggesting that Bcl-2 can allow survival of normally PCD-susceptible cells. When the *bcl*-2 transgenic mice were crossed with a mouse transgenic for the H-Y T cell receptor, it was shown that Bcl-2 could alter positive selection by skewing the repertoire of mature thymocytes to the CD4⁻ CD8⁺ lineage (TAO et al. 1994). However, *bcl*-2 transgenic animals had normal numbers of peripheral T cells. Thus, *bcl*-2 transgenes appear to have almost no effect on the processes of negative selection.

The expression of *bcl*-2 is maintained in the periphery, as T cells isolated from lymph nodes, spleen and blood express Bcl-2 protein (KONDO et al. 1992; CLEQ-DESHAMPS et al. 1993). The regulation of this pattern of expression is unique in that mRNA and protein levels can show an inverse correlation. When T cells are activated, *bcl*-2 mRNA levels increase, with little to no effect on protein levels (REED et al. 1987, 1992; BOISE et al. 1993, submitted). This regulation may be related to the long half-life of the Bcl-2 protein (MERINO et al. 1994).

2.3 Bcl-2 Is Ineffective in Several Systems

Although Bcl-2 can prevent apoptosis in a variety systems, there are several examples where the role of Bcl-2 is not clear. For example, while Bcl-2 could promote survival of an IL-3-dependent cell line, the effect of Bcl-2 on growth factor deprivation was not universal. Both an IL-2-dependent T cell line and an IL-6-dependent myeloma line that were infected with *bcl*-2 retroviral vector demonstrated no enhanced survival upon growth factor withdrawal (NUÑEZ et al. 1990). There are conflicting reports on the ability of Bcl-2 to block PCD in the IL-2-dependent T cell line CTLL2. NUÑEZ et al. (1990) show that CTLL2 cells overexpressing Bcl-2 do not survive following the removal of IL-2. In contrast, DENG and PODACK (1993) demonstrate that CTLL2 cells expressing even higher levels of Bcl-2, by the use of a high copy number episomal plasmid, can survive in the absence of IL-2. In the latter system, survival of the *bcl*-2 transfectants allowed for analysis of endogenous *bcl*-2 mRNA levels. Following the withdrawal of IL-2, expression of the endogenous *bcl*-2 gene was down-regulated within 8 h and was not detected after 3 days. Addition of IL-2 to growth factor-deprived cells induced endogenous *bcl*-2 expression within 8 h. Although these findings suggest that apoptosis is prevented in CTLL2 cells by the induction of Bcl-2 expression by IL-2, further investigation of Bcl-2 protein levels in the absence of a transgene is necessary.

There are also conflicting reports on the effect of Bcl-2 in tumor necrosis factor-α (TNF-α)-mediated cytotoxicity. Data by HENNET et al. (1993) demonstrated that the highly TNF-α-sensitive L929 mouse fibrosarcoma, cell line transfected with *bcl*-2 exhibited increased survival compared to wild-type cells. In contrast, VANHAESEBROECK et al. (1993) reported that overexpression of *bcl*-2 in the same L929 cells did not alter TNF sensitivity. In addition, transfection of the *bcl*-2 gene in the human MCF7 breast carcinoma, HL-60 promyelocytic leukemia and U937 histiocytic lymphoma cell lines did not reduce TNF sensitivity. Although the reason for this discrepancy is unclear, it may reflect the relative amount of Bcl-2 protein expressed in each system.

There are a number of other systems in which Bcl-2 fails to promote cell survival including negative selection in the thymus (SENTMAN et al. 1991), apoptosis induced by cytotoxic T lymphocytes (VAUX et al. 1992a), and anti-immunoglobulin-induced cell death in WEHI-231 cells (CUENDE et al. 1993). Although in transgenic mice, Bcl-2 protected immature CD4+ CD8+ thymocytes from glucocorticoid, irradiation and anti-CD3-induced apoptosis, clonal deletion of T cells that recognized endogenous superantigens still occurred (SENTMAN et al. 1991). In contrast to B cells, T cell development in SCID mice was not affected by the overexpression of *bcl*-2 (STRASSER et al. 1994). In addition, PCD mediated by cytotoxic T lymphocytes (CTLs) is not prevented by *bcl*-2 expression (VAUX et al. 1992a). It has been shown that CTL-induced killing occurs by apoptosis (RUSSELL et al. 1982; MARTZ and HOWELL 1989), but overexpression of *bcl*-2 in the target cells does not prevent cell death (VAUX et al. 1992a). In avian CNS, Bcl-2 can protect NGF-, brain derived neurotrophic factor-, and neurotrophin 3-dependent neurons, but not ciliary neurotrophic factor-dependent neurons from growth factor withdrawal (ALLSOPP et al. 1993). These differential effects of Bcl-2 on cell death could result from cell death inducing pathways that bypass a Bcl-2-dependent step or by the presence of additional factors that regulate the Bcl-2-dependent step of cell death. Consistant with this possibility, several Bcl-2-related proteins have recently been discovered in vertebrates.

3 *bcl*-2-Related Genes

3.1 *bcl*-x

The *bcl*-x gene was originally identified in chickens by low stringency hybridization with the murine *bcl*-2 cDNA in an attempt to clone the avian *bcl*-2 homologue (BOISE et al. 1993). A 2.7 kb mRNA was detected by northern analysis with the *bcl*-x-specific probe. Messenger RNA levels were highest in the thymus, bursa and CNS. The *bcl*-x clone hybridized efficiently to chicken, mouse and human DNA, implicating conservation during vertebrate evolution. In addition, the *bcl*-x and *bcl*-2 probes bound distinct genomic fragments in all three species, suggesting that the probes recognize separate genetic loci. By screening human libraries two types of *bcl*-x cDNAs were identified. These sequences contained different open

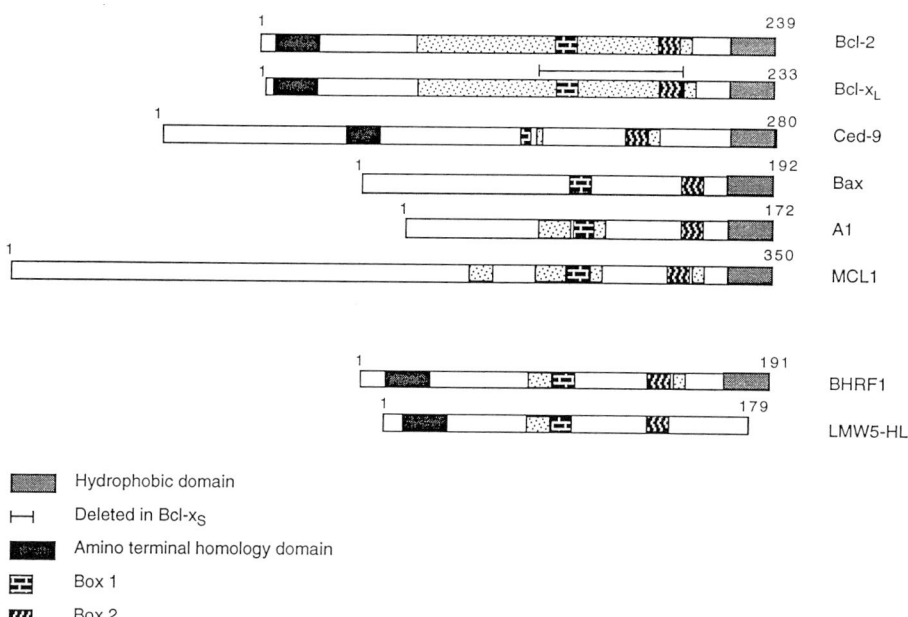

Fig. 1. The Bcl-2 family. The protein structure of the known Bcl-2 family members have been aligned, and regions of homology have been *highlighted*. The *numbers* above each protein represent the amino acid residues. The *gray shaded area* is the hydrophobic domain that has been shown to be important in Bcl-2 membrane association. The homology in this domain is based on predictions of hydrophobicity (CHEN-LEVY and CLEARY 1990). *Box 1* and *Box 2* have been previously defined as homology regions between various members of the Bcl-2 family (OLTVAI et al. 1993; WILLIAMS and SMITH 1993). The NH_2-terminal homology domain was initially defined as a region of high (50%) identity between Bcl-x and Bcl-2 (BOISE et al. 1993). More recently this region has been found in the NH_2-terminals of Ced-9 (HENGARTNER and HORVITZ et al. 1994). Upon inspection of other family members, the region is also found in the viral homologue BHRF1 and LMW5-HL. Other *shaded areas* represent other regions of homology between Bcl-2 and related proteins

reading frames, but identical 5' and 3' untranslated regions. The longer cDNA clone, *bcl* x_L contains an open reading frame with 233 amino acids. The other cDNA, *bcl-x_S*, encodes a 170 amino acid protein in which the area of highest homology to Bcl-2 has been deleted (Fig. 1). Additional regions of homology include the first 20 amino acids in the NH_2-terminal and the putative transmembrane domain in the COOH-terminal both of which are present in Bcl-2, Bcl-x_L and Bcl-x_S. The difference between the two *bcl-x* cDNAs arises from differential usage of two 5' splice sites within the first coding exon. Overexpression of *bcl-x_L* in an IL-3-dependent cell line prevented apoptosis upon growth factor withdrawal. In contrast, transfection of *bcl-x_S* into the same cell line neither caused PCD nor accelerated apoptotic cell death induced by IL-3 deprivation. Interestingly, *bcl-x_S* could prevent overexpression of *bcl-2* from inducing resistance to PCD.

3.2 *bax*

Bax (Bcl-2-associated X protein) is a 21 kDa protein originally identified by coimmunoprecipitation with human Bcl-2 (OLTVAI et al. 1993). The interaction of Bcl-2 and Bax is stable in 0.2% Nonidet P-40 (NP-40), but the association is interrupted by the addition of 0.1% SDS, arguing that Bax is noncovalently bound to Bcl-2. The 21 kDa Bax protein was partially microsequenced and two degenerate primers, corresponding to the amino acid regions of the sequenced peptide fragment, created a 71 bp PCR product that was used as a probe to screen both human and murine cDNA libraries. Both the murine and human open reading frames encoded a 192 amino acid protein that were 96% homologous to each other. Northern blot analysis of total RNA from a survey of organs revealed that *bax* was not lymphoid specific, but expressed in a wide variety of tissues. *bax* is alternatively spliced to form a 1.0 kb and 1.5 kb RNA transcript, but a function has only been attributed to the 1.0 kb RNA species. There is 20.8% identity and 43.2% similarity between human Bax and Bcl-2. The areas most highly conserved are box 1 on exon 4, box 2 on exon 5 and the putative COOH-terminal transmembrane domain on exon 6 (Fig. 1). Functionally, overexpression of Bax in FL5.12 cells accelerated cell death following the removal of IL-3. In addition, overexpression of Bax reversed protection conferred by Bcl-2. The ability of Bax to block Bcl-2-enhanced cell survival was critically dependent on the ratio of Bcl-2 to Bax. When Bcl-2 is in excess, Bax/Bcl-2 heterodimers are formed, and cells are protected. However, when Bax predominates, Bax homodimers are formed, and cells are susceptible to PCD.

3.3 MCL1/A1

Recently two *bcl*-2 homologues were cloned through screens designed to isolate myeloid-specific early response genes. *MCL1* was differentially cloned from ML-1 cells which were induced to differentiate with phorbol ester (KOZOPAS et al. 1993). *MCL1* expression is induced within the first few hours of differentiation and then gradually returns to baseline through the time course of differentiation (3 days). The MCL1 protein is homologous to Bcl-2 only in the COOH-terminal portion of the proteins including the boxes of high homology and the membrane-binding domain of the COOH-terminal (Fig. 1). The NH$_2$-terminal half of MCL1 contains PEST sequences which have been predicted to be important in protein–protein interactions.

The *A*1 gene was cloned by a similar strategy (LIN et al. 1993). In this case the search was for myeloid-specific genes which were induced by treatment of bone marrow cells with the growth factor GM-CSF. While *A*1 was screened for myeloid-specific expression, characterization of its expression at the mRNA level has revealed that *A*1 is also expressed in the T cell lineage but not the B cell or erythroid lineages. In addition, *A*1 is also induced in macrophages by lipopolysaccharide (LPS) and can be superinduced by cycloheximide. The homology between A1 and Bcl-2 is restricted to the homology boxes in the central portion of Bcl-2. To date, no function has been determined for either of these inducible members of the Bcl-2 family.

3.4 Viral Homologues

The Epstein-Barr virus (EBV) gene *BHRF-1* is expressed early in the lytic replication cycle and transiently in some latent infected cell lines and has been shown to be related to *bcl-2* (CLEARY et al. 1986). BHRF-1 is not required for B cell transformation or viral replication, but can function in a similar fashion as Bcl-2 (MARCHINI et al. 1991; LEE and YATES 1992). BHRF-1 can enhance the survival of serum-starved B cells in culture (HENDERSON et al. 1993). These characteristics of BHRF-1 predict that it may function to maintain a viable host cell for proper replication of the virus. Cells isolated from patients with Burkitt's lymphoma that express latent EBV proteins have also been shown to have high levels of *bcl-2* expression (HENDERSON et al. 1991; LIU et al. 1991). Thus, EBV may also utilize the host's own survival machinery to its benefit.

A second viral gene, the *LMW5*-HL open reading frame of the African swine fever virus, has also been shown to be homologous to *bcl-2*, but no function has yet been demonstrated for this gene (NEILAN et al. 1993). The role of cell survival genes in viral pathogenesis should be an area of focus in the study of PCD in the coming years.

4 WEHI-231: A Model of Bcl-2-Independent Cell Death

WEHI-231 is a murine B cell lymphoma commonly used to study immature B lymphocytes because it can readily undergo apoptosis (BENHAMOU et al. 1990; HASBOLD and KLAUS 1990). In contrast to the classical IgM$^+$/IgD$^-$ phenotype of immature B lymphocytes, WEHI-231 cells are IgM$^+$/IgD$^+$ (HAGGERTY et al. 1993; GOTTSCHALK et al. 1994a). However, as in immature B cells, cross-linking of surface IgM with anti-Ig reagents causes WEHI-231 cells to initially growth arrest in the G_0/G_1 phase of the cell cycle, followed by initiation of PCD 24–48 h later (GOTTSCHALK et al. 1994a). The anti-Ig-induced PCD in WEHI-231 cells has all the morphological and biochemical features of apoptosis (BENHAMOU et al. 1990; HASBOLD and KLAUS 1990; GOTTSCHALK et al. 1994a). Unlike *bcl-2* (CUENDE et al. 1993), overexpression of *bcl-x$_L$* in WEHI-231 cells enhanced cell survival (manuscript submitted), indicating that *bcl-2* and *bcl-x$_L$* can differentially regulate apoptosis in WEHI-231 cells. Following anti-IgM treatment, > 75% of the *bcl-x$_L$* transfectants remain viable, while less than 10% of the *bcl-2* transfectant survive. PCD in WEHI-231 cells induced by immunosuppressants, irradiation, and protein synthesis inhibitors is also blocked by overexpression of *bcl-x$_L$*, but not *bcl-2*

(GOTTSCHALK et al., 1994b). Cyclosporin A, FK-506 and rapamycin are immunosuppressants often used as pharmacological probes to study lymphocyte activation and PCD (SIGAL and DUMONT 1992). Cyclosporin A and FK-506 are known to prevent PCD in T cell hybridomas and thymocytes (SHI et al. 1989; BIERER et al. 1990; STARUCH et al. 1991). These reagents, as well as rapamycin, induced PCD only in WEHI-231 cells susceptible to anti-IgM-mediated apoptosis (GOTTSCHALK et al. 1994b). PCD was preceded by growth arrest and characterized by the DNA fragmentation pattern typical of apoptosis. In all the systems mentioned above, the degree of protection provided by bcl-x_L correlated with the level of Bcl-x_L protein expressed by the transfectants. These results suggest that the inability of bcl-2 to protect WEHI-231 cells from PCD is characteristic of the cell line and not specific for the mode in which cell death is induced.

There are several explanations for the differential abilities of Bcl-2 and Bcl-x_L to regulate apoptosis in WEHI-231 cells. One possibility is that Bcl-x_L regulates a pathway that is independent of Bcl-2 expression. An alternative interpretation is that Bcl-2 and Bcl-x_L regulate overlapping pathways to prevent apoptosis, but the function of Bcl-2 is actively inhibited in certain cellular systems. There are two recently identified antagonist of Bcl-2: Bcl-x_S and Bax. Transfection with either bcl-x_S or bax reversed the protection provided by the overexpression of bcl-2 in the IL-3-dependent FL5.12 cell line (BOISE et al. 1993; OLTVAI et al. 1993). Bax can form homodimers which is thought to accelerate cell death and heterodimers with Bcl-2 that neither enhance cell survival nor prevent it (OLTVAI et al. 1993). In contrast, in vitro translated Bcl-x_S does not bind Bcl-2, suggesting that Bax and Bcl-x_S regulate PCD by different mechanisms.

WEHI-231 cells express high levels of Bax protein and undetectable levels of Bcl-x_S protein, implicating Bax expression as a possible explanation for the inability of Bcl-2 to protect against anti-Ig-induced apoptosis in this cell line. Furthermore it has been found that although Bax can antagonize Bcl-2, overexpression of Bax in FL5.12 cells did not alter the ability of Bcl-x_L to enhance cell survival in the absence of IL-3 (manuscript in preparation). In addition, there may be differences in the ability of Bcl-x_L and Bcl-2 to physically associate with Bax in either FL5.12 or WEHI-231 cells. These differences suggest that Bcl-2 and Bcl-x_L probably regulate similar pathways to prevent apoptosis, and the interplay between these factors with Bax and potentially other members (known and unknown) of the Bcl-2 family may determine the susceptibility of a cell to undergo PCD.

5 Summary

In this review we have discussed the importance of Bcl-2 and related proteins in the regulation of apoptotic cell death in mammalian systems. It is clear that Bcl-2 plays a critical role in controlling many forms of PCD. Bcl-2 seems to have particular significance in lymphocyte development and the function of the immune system. We have also discussed the increasing size of the newly identified Bcl-2 family. There are a number of Bcl-2 homologues in human, murine, avian, nematode, and viral systems. The evolutionary conservation of the function of the Bcl-2 homologues, reinforces the importance of PCD in all complex organisms. Some of these *bcl*-2-like genes function as agonists and others as antagonists. Despite the seemingly universal importance of Bcl-2, it is unable to prevent PCD in all systems. In addition, we have described a role for other Bcl-2 family members in systems in which Bcl-2 is ineffective and supplied a potential rationale for the large number of genes involved in the regulation of PCD. Identification and functional analysis of the Bcl-2 family members reveals the complex nature of cell death regulation.

As we begin to appreciate the significance of PCD in the control of development and homeostasis, its regulation at the molecular level is becoming better understood. Bcl-2 has long been the only known intracellular regulator of the PCD pathway(s), although its ability to prevent apoptosis is not universal. We now know that *bcl*-2 is only one member of an evolutionary conserved family of genes which display different patterns of expression as well as function. At least two family members, Bcl-x_S and Bax, act in opposition to Bcl-2. The discovery of these new family members, including those with Bcl-2-like function and antagonists, should help clear up the discrepancies seen in Bcl-2's ability to protect cells from PCD. In doing so, we will be able to further define the pathways associated with cell death signaling. The study of these family members, as well as the non-related genes of the PCD pathways (*ced*-3, *ced*-4, *ice*) should lead us to understanding of how cells of multicellular organisms make decisions to die.

Acknowledgements.   We thank Jonathan Green for critical review of the manuscript. LHB is a fellow of the Leukemia Society of America. ARG is supported by the Medical Scientist Training Program. LHB and ARG contributed equally to this manuscript.

References

Allsopp TE, Wyatt S, Paterson HF, Davies AM (1993) The proto-oncogene bcl-2 can selectively rescue neurotrophic factor-dependent neurons from apoptosis. Cell 73: 295–307

Bakhshi A, Jensen JP, Goldman P, Wright JJ, McBride OW, Epstein AL, Korsmeyer SJ (1985) Cloning of the chromosomal breakpoint of t(14;18) human lymphomas: clustering around J_H on chromosome 14 and near a transcriptional unit on 18. Cell 41: 899–906

Benhamou LE, Cazenave PA, Sarthou P (1990) Anti-immunoglobulins induce death by apoptosis in WEHI-231 B lymphoma cells. Eur J Immunol 20: 1405–1407

Bierer BE, Mattila PS, Standaert RF, Herzenberg LA, Burakoff SJ, Crabtree G, Schreiber SL (1990) Two distinct signal transmission pathways in T lymphocytes are inhibited by complexes formed between an immunophilin and either FK506 or rapamycin. Proc Natl Acad Sci USA 87: 9231–9235

Bissonnette RP, Echeverri F, Mahboubi A, Green DR (1992) Apoptotic cell death induced by c-myc is inhibited by bcl-2. Nature 359: 552–554

Boise LH, González-Garcia M, Postema CE, Ding L, Lindsten T, Turka LA, Mao X, Nuñez G, Thompson CB (1993) bcl-x, a bcl-2-related gene that functions as a dominant regulator of apoptotic cell death. Cell 74: 597–608

Chen-Levy Z, Cleary ML (1990) Membrane topology of the Bcl-2 proto-oncogene protein demonstrated in vitro. J Biol Chem 265: 4929–4933

Chen-Levy Z, Nourse J, Cleary ML (1989) The bcl-2 candidate proto-oncogene product is a 24-kilodalton intergral membrane protein highly expressed in lymphoid cell lines and lymphomas carrying the t(14;18) translocation. Mol Cell Biol 9: 701–710

Chiou S K, Rao L, White E (1994) Bcl-2 blocks p53-dependent apoptosis. Mol Cell Biol 14: 2556–2563

Clarke AR, Purdie CA, Harrison DJ, Morris RG, Bird CC, Hooper ML, Wyllie AH (1993) Thymocyte apoptosis induced by p53-dependent and independent pathways. Nature 362: 849–852

Cleary ML, Smith SD, Sklar J (1986) Cloning and structural analysis of cDNAs for bcl 2 and the hybrid bcl-2/immunoglobulin transcript resulting from the t(14;18) translocation. Cell 47: 19–28

Cleq-Deschamps CM, LeBrun DP, Huie P, Besnier DP, Warnke RA, Sibley RK, Cleary ML (1993) Topographical dissociation of Bcl-2 messenger RNA and protein expression in human lymphoid tissues. Blood 81: 293–298

Cohen JJ, Duke RC (1992) Apoptosis and programmed cell death in immunity. Annu Rev Immunol 10: 2676–293

Cuende E, Ales-Martinez JE, Ding L, Gonzalez-Garcia M, Martinez-AC, Nuñez G (1993) Progammed cell death by bcl-2-dependent and independent mechanisms in B lymphoma cells. EMBO J 12: 1555–1560

Debbas M, White E (1993) Wild-type p53 mediates apoptosis by E1A, which is inhibited by E1B. Genes Dev 7: 546–554

de Jong D, Prins F, van Krieken HHJM, Mason DY, van Ommen GB, Kluin PM (1992) Subcellular localization of bcl-2 protein. In: Potter M, Melchers F (eds) Mechanisms in B-cell neoplasia. Springer, Berlin Heidelberg New York, pp 287–292 (Current topics in microbiology and immunology, vol 182)

Deng G, Podack ER (1993) Suppression of apoptosis in a cytotoxic T-cell line by interleukin 2-mediated gene transcription and deregulated expression of the protooncogene bcl-2. Proc Natl Acad Sci USA 90: 2189–2193

Ellis HM, Horvitz HR (1986) Genetic control of programmed cell death in the nematode C. elegans. Cell 44: 817–829

Fanidi A, Harrington EA, Evan GI (1992) Cooperative interaction between c-myc and bcl-2 proto-oncogenes. Nature 359: 554–556

Gagliardini V, Fernandez P-A, Lee RKK, Drexler HCA, Rotello RJ, Fishman MC (1994) Prevention of vertebrate neuronal death by the crmA gene. Science 263: 826–828

Garcia I, Martinou I, Tsujimoto Y, Martinou JC (1992) Prevention of programmed cell death of sympathetic neurons by the bcl-2 proto-oncogene. Science 258: 302–304

Gottschalk AR, McShan CL, Merino R, Nuñez G, Quintáns J (1994a) Physiological cell death in B lymphocytes: I. Differential susceptibility of WEHI-231 sublines to anti-Ig induced physiological cell death and lack of correlation with bcl-2 expression. Int Immunol 6: 121–130

Gottschalk AR, Boise LH, Thompson CB, Quintans J (1994b) Identification of immunosuppressant-induced apoptosis in murine B cell lines and its prevention by bcl-x but not bcl-2. Proc Natl Acad Sci USA 91: 7350–7354

Gratiot-Deans J, Ding L, Turka LA, Nuñez G (1993) bcl-2 proto-oncogene expression during human T cell development. J Immunol 151: 83–91

Haggerty HG, Wechsler RJ, Lentz VM, Monroe JG (1993) Endogenous expression of δ on the surface of WEHI-231. J Immunol 151: 4681–4693

Hasbold J, Klaus GGB (1990) Anti-immunoglobulin antibodies induce apoptosis in immature B cell lymphomas. Eur J Immunol 20: 1685–1690

Henderson S, Rowe M, Gregory CD, Croom-Carter D, Wang F, Kieff E, Rickinson AB (1991) Induction of bcl-2 expression by Epstein-Barr virus latent membrane protein 1 protects infected B cells from programmed cell death. Cell 65: 1107–1115

Henderson S, Huen D, Rowe M, Dawson C, Johnson G, Rickinson A (1993) Epstein-Barr virus-encoded BHRF1 protein, a viral homologue of Bcl-2, protects human B cells from programmed cell death. Proc Natl Acad Sci USA 90: 8479–8483

Hengartner MO, Ellis RE, Horvitz HR (1992) Caenorhabditis elegans gene ced-9 protects cells from programmed cell death. Nature 356: 494–499

Hengartner MO, Horvitz HR (1994) C. elegans cell survival gene ced-9 encodes a functional homolog of the mammalian proto-oncogene bcl-2. Cell 76: 665–676

Hennet T, Bertoni G, Richter C, Peterhans E (1993) Expression of Bcl-2 protein enhances the survival of mouse fibrosarcoid cells in tumor necrosis factor-mediated cytotoxcity. Cancer Res 53: 1456–1460

Hockenbery D, Nuñez G, Milliman C, Schreiber RD, Korsmeyer SJ (1990) Bcl-2 is an inner mitochondrial membrane protein that blocks programmed cell death. Nature 348: 334–336

Hockenbery DM, Oltvai ZN, Yin XM, Milliman CL, Korsmeyer SJ (1993) Bcl-2 functions in an antioxidant pathway to prevent apoptosis. Cell 75: 241–251

Itoh N, Yonehara S, Ishii A, Yonehara M, Mizushima S-I, Sameshima M, Hase A, Seto Y, Nagata S (1991) The polypeptide encoded by the cDNA for human cell surface antigen Fas can mediate apoptosis. Cell 66: 233–243

Itoh N, Tsujimoto Y, Nagata S (1993) Effect of bcl-2 on Fas antigen-mediated cell death. J Immunol 151: 621–627

Kane DJ, Sarafian TA, Anton R, Hahn H, Gralla EB, Valentine JS, Örd T, Bredesen, DE (1993) Bcl-2 inhibition of neural death: decreased generation of reactive oxygen species. Science 262: 1274–1277

Katsumata M, Seigel RM, Louie DC, Miyashita T, Tsujimoto Y, Nowell PC, Greene MI, Reed JC (1992) Differential effects of Bcl-2 on T and B cells in transgenic mice. Proc Natl Acad Sci USA 89: 11376–11380

Kerr JFR, Wyllie AH, Currie AR (1972) Apoptosis: a basic biological phenomenon with wide-ranging implications in tissue kinetics. Br J Cancer 26: 239–257

Kondo E, Nakamura S, Onoue H, Matsuo Y, Yoshino T, Aoki H, Hayashi K, Takahashi K, Minowada J, Nomura S, Akagi T (1992) Detection of bcl-2 protein and bcl-2 messenger RNA in normal and neoplastic lymphoid tissues by immunohistochemistry and in situ hybridization. Blood 80: 2044–2051

Korsmeyer SJ, McDonnell TJ, Nunez G, Hockenbery D, Young R (1990) Bcl-2: B cell life, death and neoplasia. In: Potter M, Melchers F (eds) Mechanisms in B-cell neoplasia 1990. Springer, Berlin Heidelberg New York, pp 203–207 (Current topics in Microbiology and immunology, vol 166)

Kozopas KM, Yang T, Buchan HL, Zhou P, Craig RW (1993) MCL1, a gene expressed in programmed myeloid cell differentiation, has sequence similarity to BCL2. Proc Natl Acad Sci USA 90: 3516–3520

Lee M-A, Yates JL (1992) BHRF1 of Epstein-Barr virus, which is homologous to human proto-oncogene bcl2, is not essential for transformation of B cells or for virus replication in vitro. J Virol 66: 1899–1906

Lin EY, Orlofsky A, Berger MS, Prystowsky (1993) Characterization of A1, a novel hemopoietic-specific early-response gene with sequence similarity to bcl-2. J Immunol 151: 1979–1988

Liu Y-J, Mason DY, Johnson GD, Abbot S, Gregory CD, Hardie DL, Gordon J, MacLennan ICM (1991) Germinal center B cells express bcl-2 protein after activation by signals which prevent their entry into apoptosis. Eur J Immunol 21: 1905–1910

Lowe SW, Schmitt EM, Smith SW, Osborne BA, Jacks T (1993) p53 is required for radiation-induced apoptosis in mouse thymoytes. Nature 362: 847–849

Marchini A, Tomkinson B, Cohen JI, Kieff E (1991) BHRF1, the Epstein-Barr virus gene with homology to Bcl2 is dispensable for B-lymphocyte transformation and virus replication. J Virol 65: 5991–6000

Martz E, Howell DM (1989) CTL: virus control cells first and cytolytic cells second? DNA fragmentation, apoptosis and the prelytic halt hypothesis. Immunol Today 10: 79–86

McDonnell TJ, Deane N, Platt FM, Nunez G, Jaeger U, McKearn JP, Korsmeyer SJ (1989) bcl-2-immunoglobulin transgenic mice demonstrate extended B cell survival and follicular lymphoproliferation. Cell 57: 79–88

McDonnell TJ, Nunez G, Platt FM, Hockenberry D, London L, McKearn JP, Korsmeyer SJ (1990) Deregulated Bcl-2-immunoglobulin transgene expands a resting but responsive immunoglobulin M and D-expressing B-cell population. Mol Cell Biol 10: 1901–1907

Merino R, Ding L, Veis DJ, Korsmeyer SJ, Nuñez G (1994) Developmental regulation of the Bcl-2 protein and susceptibility to cell death in B lymphocytes. EMBO J 13: 683–691

Miura M, Zhu H, Rotello R, Hartwieg EA, Yuan J (1993) Induction of apoptosis in fibroblasts by IL-1β-coverting enzyme, a mammalian homolog of the C. elegans cell death gene ced-3. Cell 75: 653 660

Miyashita T, Reed JC (1992) bcl-2 gene transfer increases relative resistance of S49.1 and WEHI7.2 lymphoid cell to cell death and DNA fragmentation induced by glucocorticoids and multiple chemotherapeutic drugs. Cancer Res 52: 5407–5411

Nakayama K-I, Nakayama K, Negishi I, Kuida K, Shinkai Y, Louie MC, Fields LE, Lucas PJ, Stewart V, Alt FW, Loh DY (1993) Disappearance of the lymphoid system in Bcl-2 homozygous mutant chimeric mice. Science 261: 1584–1588

Neilan JG, Lu Z, Afonso CL, Kutish GF, Sussman MD, Rock DL (1993) An African swine fever virus gene with similarity to the proto-oncogene bcl-2 and the Epstein-Barr virus gene BHRF1. J Virol 67: 4391–4394

Nuñez G, London L, Hockenbery D, Alexander M, McKearn JP, Korsmeyer SJ (1990) Deregulated Bcl-2 gene expression selectively prolongs survival of growth factor-deprived hemopoietic cell lines. J Immunol 144: 3602–3610

Nuñez G, Hockenbery D, McDonnell TJ, Sorensen CM, Korsmeyer SJ (1991) Bcl-2 maintains B cell memory. Nature 353: 71–73

Ohmori T, Podack ER, Nishio K, Takahashi M, Miyahara Y, Takeda Y, Kubota N, Funayama Y, Ogasawara H, Ohira T, Ohta S, Saijo N (1993) Apoptosis of lung cancer cells caused by some anti-cancer agents (MMC, CPT-11, ADM) is inhibited by BCL-2. Biochem Biophys Res Commun 192: 30–36

Oltvai ZN, Milliman CL, Korsmeyer SJ (1993) Bcl-2 heterodimerizes in vivo with a conserved homolog, Bax, that accelerates programed cell death. Cell 74: 609–619

Ray CA, Black RA, Kronheim SR, Greenstreet TA, Sleath PR, Salvesen GS, Pickup DJ (1992) Viral inhibition of inflammation: cowpox virus encodes an inhibitor of the interleukin-1β converting enzyme. Cell 69: 597–604

Reed JC, Tsujimoto Y, Alpers JD, Croce CM, Nowell PC (1987) Regulation of bcl-2 proto-oncogene during normal human lymphocyte proliferation. Science 236: 1295–1299

Reed JC, Miyashita T, Cuddy M, Cho D (1992) Regulation of p26-Bcl-2 protein levels in human peripheral blood lymphocytes. Lab Invest 67: 443–449

Russell JH, Masakowski V, Rucinsky T, Phillips G (1982) Mechanisms of immune lysis III. Characterization of the nature and kinetics of the cytotoxic T lymphocyte-induced nuclear lesion in the target. J Immunol 128: 2087–2094

Sentman CL, Shutter JR, Hockenbery D, Kanagawa O, Korsmeyer SJ (1991) bcl-2 inhibits multiple forms of apoptosis but not negative selection in thymocytes. Cell 67: 879–888

Shi Y, Sahai BM, Green DR (1989) Cyclosporin A inhibits activation-induced cell death in T-cell hybridomas and thymocytes. Nature 339: 625–626

Siegal RM, Katsumata M, Miyashita T, Louie DC, Greene MI, Reed JC (1992) Inhibition of thymocyte apoptosis and negative antigenic selection in bcl-2 transgenic mice. Proc Natl Acad Sci USA 89: 7003–7007

Sigal NH, Dumont FJ (1992) Cyclosporin A, FK-506, and rapamycin: pharmacologic probes of lymphocyte signal transduction. Annu Rev Immunol 10: 519–560

Staruch MJ, Sigal NH, Dumont FJ (1991) Differential effects of the immunosuppressive macrolides FK-506 and rapamycin on activation-induced T-cell apoptosis. Int J Immunopharmacol 13: 677–685

Strasser A, Harris AW, Cory S (1991a) bcl-2 transgene inhibits T cell death and perturbs thymic self-censorship. Cell 67: 889–899

Strasser A, Whittingham S, Vaux DL, Bath ML, Adams JM, Cory S, Harris AW (1991b) Enforced BCL2 expression in B-lymphoid cells prolongs antibody responses and elicits autoimmune disease. Proc Natl Acad Sci USA 88: 8661–8665

Strasser A, Harris AW, Corcoran LM, Cory S (1994) Bcl-2 expression promotes B- but not T-lymphoid development in scid mice. Nature 368: 457–460

Tao W, Teh S-J, Melhado I, Jirik F, Korsmeyer SJ, Teh H-S (1994) The T cell receptor repetoire of CD4⁻8⁺ thymocytes is altered by overexpression of the BCL-2 protooncogene in the thymus. J Exp Med 179: 145–153

Tsujimoto Y, Croce CM (1986) Analysis of the structure, transcripts, and protein products of bcl-2, the gene involved in human follicular lymphoma. Proc Natl Acad Sci USA 83: 5214–5218

Tsujimoto Y, Finger LR, Yunis J, Nowell PC, Croce CM (1984) Cloning of the chromosomes breakpoint of neoplastic B cells with the t(14;18) chromosome translocation. Science 226: 1097–1099

Vanhaesebroeck B, Reed JC, De Valck D, Grooten J, Miyashita T, Tanaka S, Beyaert R, Van Roy F, Fiers W (1993) Effect of bcl-2 proto-oncogene expression on cellular sensitivity to tumor necrosis factor-mediated cytotoxicity. Oncogene 8: 1075–1081

Vaux DL, Cory S, Adams JM (1988) Bcl-2 gene promotes haemopoietic cell survival and cooperates with c-myc to immortalize pre-B cells. Nature 335: 440–442

Vaux DL, Aguila HL, Weissman IL (1992a) Bcl-2 prevents death of factor-deprived cells but fails to prevent apoptosis in targets of cell mediated killing. Int Immunol 7: 821–824

Vaux DL, Weissman IL, Kim SK (1992b) Prevention of programmed cell death in Caenorhabditis elegans by human bcl-2. Science 258: 1955–1957

Veis DJ, Sentman CL, Bach EA, Korsmeyer SJ (1993a) Expression of the Bcl-2 protein in murine and human thymocytes and in peripheral T lymphocytes. J Immunol 151: 2546–2554

Veis DJ, Sorenson CM, Shutter JR, Korsmeyer SJ (1993b) Bcl-2-deficient mice demonstrate fulminant lymphoid apoptosis, polycystic kidneys, and hypopigmented hair. Cell 75: 229–240

Wagner AJ, Small MB, Hay N (1993) Myc-mediated apoptosis is blocked by ectopic expression of Bcl-2. Mol Cell Biol 13: 2432–2440

Walton MI, Whysong D, O'Conner PM, Hockenbery D, Korsmeyer SJ, Kohn KW (1993) Constitutive expression of human Bcl-2 modulates nitrogen mustard and camptothecin induced apoptosis. Cancer Res 53: 1853–1861

Watanabe-Fukunaga R, Brannan CI, Copeland NG, Jenkins NA, Nagata S (1992) Lymphoproliferation disorder in mice explained by defects in Fas antigen that mediates apoptosis. Nature 356: 314–317

Williams GT, Smith CA (1993) Molecular regulation of apoptosis: genetic controls on cell death. Cell 74: 777–779

Wyllie AH, Kerr JFR, Currie AR (1980) Cell death: significance of apoptosis. Int Rev Cytol 68: 251–306

Yonish-Rouach E, Resnitzky D, Lotem J, Sachs L, Kimchi A, Oren M (1991) Wild-type p53 induces apoptosis of myeloid leukaemic cells that is inhibited by interleukin-6. Nature 352: 345–347

Yuan J, Horvitz HR (1990) The Caenorhabditis elegans genes ced-3 and ced-4 act cell autonomously to cause programmed cell death. Dev Biol 138: 33–41

Yuan J, Sharam S, Ledoux S, Ellis HM, Horvitz HR (1993) The C. elegans cell death gene ced-3 encodes a protein similar to mammalian interleukin-1β-converting enzyme. Cell 75: 641–652

Lymphocyte Death, *p53,* and the Problem of the "Undead" Cell

D.J. Harrison, S.E.M. Howie, and A.H. Wyllie

1 Apoptosis and the Immune System

Apoptosis is involved in many facets of the development and homeostatic regulation of the immune system (Cohen 1993), including the elimination of self-reactive thymocytes, selection of B lymphocytes within the germinal centres, development of memory T cells and killing of target cells by activated cytotoxic T lymphocytes (Jenkinson et al. 1989; Finkel et al. 1991; Berke 1991; Liu et al. 1989; Akbar et al. 1993).

Apoptosis in lymphoid tissue was originally described in the thymus and this organ has remained a standard model for the study of apoptosis (Wyllie 1980). Approximately 97% of the precursor cells which enter the thymus die in situ by apoptosis during the processes of positive and negative selection. The small proportion of survivors which escape death pass into the periphery to form the circulating pool of naive T lymphocytes. Any alteration in the apoptotic ability of cells normally destined to die in this process would clearly have profound effects on the immune response (Sellins and Cohen 1987) because of the possible release of both nonuseful (which would have been deleted during negative

Cancer Research Campaign Laboratories, Department of Pathology, University Medical School, Teviot Place, Edinburgh, EH8 9AG, Scotland, UK selection) and autoreactive T cells (which would have been deleted during positive selection). The same principles would apply to the postantigen exposure driven somatic mutation and selection of B lymphocytes in germinal centres.

As well as immunologically specific deletion of lymphocytes during development and immune responses, lymphoid tissue, in common with other tissues, can suffer DNA damage which if unchecked could result in mutation and tumorigenesis. Due to the highly proliferative nature of lymphoid tissue during antigen driven immune responses it is likely that aberrant recombination events may occur not necessarily involving immune system-specific genes. Therefore, in addition to specifically immunological termination signals and pathways there are mechanisms of response which ensure that a cell which sustains DNA damage cannot pass a mutation to its daughter cells. A problem peculiar to B lymphocytes is that somatic mutation of immunoglobulin genes, which involves multiple DNA strand breaks, is a normal part of the physiological maturation of these cells and the characteristic development of increased antigen specificity. One pathway now recognised as critical to the protection of the organism against DNA damage in cells is that involving the tumour suppressor gene *p53* (LANE 1992).

2 *p53*: A Paradigm of Tumour Suppressor Genes

Wild-type *p53* has been characterised as a tumour suppressor gene (LEVINE et al. 1991) and alterations in the gene constitute the commonest molecular defect associated with human cancers. Mutations and deletions of *p53* have been described in human leukemias and lymphomas (HOLLSTEIN et al. 1991; MALKIN et al. 1990; SRIVASTAVA et al. 1990; CARDER et al. 1993; SAID et al. 1992; BAKER et al. 1990). These observations underscore the importance of *p53* in preventing the acquisition of mutations and passage through successive cell divisions *p53* mRNA is expressed at low levels in nearly all cells suggesting that it has a general and vital role to play in many different cell lineages (LEVINE et al. 1991). The protein p53 becomes phosphorylated in a cell cycle-dependent fashion by cdc2 kinase (BISCHOFF et al. 1990; STURZBECHER et al. 1990) and levels of phosphorylation are maximal during mitosis. The protein is able to enter and leave the nucleus (MARTINEZ et al. 1991; SHAULSKY et al. 1990) and it is known to alter the transcription of a wide variety of genes (FIELDS and JANG 1990; FARMER et al. 1992; KASTAN et al. 1992; RAGIMOV et al. 1993; MERCER et al. 1990; MOMAND et al. 1992; ZAMBETTI et al. 1992; MACK et al. 1993).

Cotransfection of wild-type *p53* with a variety of oncogenes can prevent or ameliorate cell transformation (FINLAY et al. 1989; ELIYAHU et al. 1989). Transfection by *p53* alone can suppress the growth of a number of different cell lines, either by inducing cell cycle growth arrest in late G1 or by triggering apoptosis (BAKER et al. 1990; LIN et al. 1992; YONISH-ROUACH et al. 1991; SHAW et al. 1992). This apparent paradox, that the same gene can cause growth arrest or death, can be rationalised by noting that in either event the risk of daughter cells receiving incorrectly coded DNA is minimised. Thus p53 has been coined as the "guardian of the genome" (LANE 1992, 1993). The effects of *p53* are achieved by the transcriptional regulation of a number of genes, principal amongst which may be *WAF*-1 (also known as *CIP*-1 or *CDI*-1) (XIONG et al. 1993; EL-DEIRY et al. 1993, 1994; DULIC et al. 1994). The 21 kDa product of *WAF*-1 binds to cdk2 and cyclin thereby causing cell cycle arrest in late G1 by activating the product of the *Rb* gene which shuts off E2F-1 (GU et al. 1993; SERRANO et al. 1993). It is not yet known whether the p21 *WAF*-1 pathway is also involved in the initiation of apoptosis (EL-DEIRY et al. 1994).

3 Structure and Function of p53

Separate functional domains of p53 have been identified by mutational analysis. The protein is usually present as an oligomer (KRAISS et al. 1988), and tetramerisation under certain conditions has been shown (FRIEDMAN et al. 1993). The COOH-terminal region is the oligomerisation domain and it contains a DNA binding site (FOORD et al. 1991; FUNK et al. 1992) and the NH_2-terminal region is responsible for transcription modulation (FIELDS and JANG 1990). A number of stimuli cause cell cycle arrest associated with stabilisation of the p53 protein rather than altered transcription of the gene (KUERBITZ et al. 1992). The common feature is DNA damage induced by, for example, UV light, γ-irradiation, the topoisomerase II inhibitor etoposide (VP16) and chemotherapeutic agents such as bleomycin (MALTZMAN and CZYZYK 1984; KASTAN et al. 1991; FRITSCHE et al. 1993; HALL et al. 1993; ZHAN et al. 1993). Precisely how p53 recognises DNA breaks is unknown at present but the protein catalyses repair of both single- and double-stranded DNA breaks (OBEROSLER et al. 1993; BAKALKIN et al. 1994; NELSON and KASTAN 1994). That DNA breaks are critical to the induction of p53 is shown by the effect of 3-aminobenzamide (LU and LANE 1993). This is an inhibitor of ADP polyribosyl transferase and thus of DNA repair. In the presence of this chemical the elevation of p53 following genotoxic injury is prolonged. In a DNA repair deficient mouse lacking the ERCC-1 gene, there is also evidence of increased p53 expression under conditions of unrepaired endogenous DNA injury possibly caused as an incidental consequence of normal oxidative metabolism (MCWHIR et al. 1993). When bound to DNA p53 can repress the activity of promoters whose initiation is dependent on the presence of a TATA box, by interaction of p53 with TATA binding proteins and a variety of other transcription factors (DUTIA et al. 1993; RAGIMOV et al. 1993; MACK et al. 1993). In addition p53 can down-regulate the expression of other genes whose expression is controlled by growth factors such as the early cell cycle genes, c-*fos* and c-*jun,* and also β-actin.

4 Generation of *p53*-Deficient Mice

One approach to study the physiological role of p53 has been to generate mice with a germline mutation in the *p53* gene (DONEHOWER et al. 1992; CLARKE et al. 1993; LOWE et al. 1993; TSUKUDA et al. 1993). This can be achieved by interrupting the wild-type *p53* gene with a plasmid containing the neomycin resistance gene in embryonal stem (ES) cells and then injecting the targeted cells into blastocysts. The blastocysts are then implanted in a pseudopregnant mouse. The host cells and the targeted ES cells both contribute to the progeny and hence they are chimaeric. If ES-derived cells are present in the gonads then germline transmission of the targeted event can take place resulting in progeny in which every cell is heterozygous. By selective cross-breeding it is then possible to generate mice homozygous for the disrupted *p53* gene. Several groups have produced *p53* defective animals in this way and the surprising initial finding was that the majority of mice homozygous for *p53* deficiency developed normally and were viable after birth. A small proportion of homozygous female embryos are anencephalic and die soon after delivery (unpublished observations). This suggested that either the main role of p53 was to respond to pathological stimuli and that its effect was not necessary for normal cell function, or that there was redundancy of the gene with other genes perhaps taking over the role in *p53* deficient animals. There was, however, higher than expected incidence of malignant neoplasms, both spontaneously occurring and after treatment with carcinogenic chemicals (DONEHOWER et al. 1992, 1993; PURDIE et al. 1994). The commonest tumour was a high grade T cell lymphoma arising in the thymus. One group has also reported the presence of B lymphomas in about 15% of cases (DONEHOWER et al. 1993) but we have seen no B lymphomas at all. It is possible that this reflects minor strain differences between the mice used or, alternatively, the different strategy used to generate p53 deficiency (compare MOSER et al. 1992). In Bradley's group the introduction of a *neo* gene in exon 5 rather than in 2 as in our own study might have allowed the transcription of part of the gene distal to the insertion of the interrupting sequence with unknown biological consequences (DONEHOWER et al. 1992).

5 p53 and the Regulation of Thymocyte Death

Thymocytes are probably the most frequently used cells for the study of apoptosis. Much of our understanding of the morphology and biochemical events associated with apoptosis have come from studies of these cells. The thymocyte, both in vivo and in vitro, readily undergoes apoptosis in response to a diverse range of stimuli, including heat shock, corticosteroids, γ-irradiation, calcium ionophore and phorbol ester, and aging.

Using thymocyte suspensions from mice, several groups have investigated the role of p53 in the death of these cells under a variety of conditions (CLARKE et al. 1993; LOWE et al. 1993). With dexamethasone there was the expected dose-dependent increase in apoptosis in thymocytes regardless of the *p53* genotype. The same response was seen with aging, in which there was a steady increase in the frequency of apoptosis with time in vitro. However when cells were exposed to a γ-irradiation source there was a striking effect of *p53* status on the occurrence of cell death. Thymocytes with wild-type *p53* showed the expected dose-dependent induction of apoptosis up to 14 Gy. Thymocytes from mice which were homozygous for the deficient *p53* allele were totally resistant to irradiation; there was no increase in the rate of cell death as the dose was increased. This indicates that p53 is essential for triggering apoptosis in thymocytes following DNA strand breakage by irradiation. A similar dependency was found for apoptosis triggered by strand breaks caused by the topoisomerase II inhibitor etoposide, although a lesser degree of apoptosis which was *p53*-independent did occur (CLARKE et al. 1993). Not only were the homozygous *p53*-deficient thymocytes totally resistant to death caused by DNA strand breakage but thymocytes from mice heterozygous for *p53* deficiency showed an intermediate sensitivity to irradiation, indicating a linear gene dosage effect between the wild-type and homozygote *p53*-deficient results. This effect was also seen in intact thymuses in vivo (LOWE et al. 1993). When calcium ionophore and phorbol ester were applied to the cells as a model of cell signaling there was no protective effect associated with carrying deficient *p53*.

Thus it is clear that in thymocytes p53 is essential for the initiation of death by apoptosis after DNA strand breakage caused by pathological stimuli but not by some other stimuli including signal transduction pathways and steroid treatment.

6 p53 and the Regulation of Peripheral T Lymphocyte Death

One reason why the thymocyte has proved so useful for the study of apoptosis is the predictable and relatively high rate of cell death which can be counted easily over the course of an experiment. The peripheral T lymphocyte population is more resilient and therefore the accurate assessment of T lymphocyte death in vitro is more difficult. To investigate the role of p53 in the death of these cells after a number of stimuli we have developed a technique for directly estimating the rate of apoptosis in lymph node tissue sections. This has been achieved employing a semiautomated microscope/image analyser (the HOME system) (BRUGAL et al. 1992) to count morphologically identified apoptotic cells. This gives an underestimate of the number of cells but it is rapid and reproducible and, in our hands, more accurate than in situ end labeling techniques or flow cytometry when the rate of apoptosis is low. Using this approach we have treated mice with a number of stimuli known to cause apoptosis. After varying periods of time animals were killed and lymph nodes were removed for quantitation of cell number, phenotypic analysis and apoptosis count.

Deletion of T lymphocytes can be achieved by ligation of CD4 on the cell surface by antibody or HIV peptides (NEWELL et al. 1990; BANDA et al. 1992; HOWIE et al. 1994). As a signal transduction pathway leading to apoptosis animals were treated with the anti-CD4 monoclonal antibody L3T4 (COBBOLD et al. 1984; HORNEFF et al. 1993). Several groups have previously shown that this causes a fall of around 50% in the CD4 positive T lymphocyte counts within 2 days. The effect is not seen when a control rat immunoglobulin of the same class is substituted for the anti-CD4 antibody. By counting the frequency of apoptosis in tissue sections we demonstrated that the induction of apoptosis was maximal 4 h after injection at 1.33% and that it remained higher than normal (which was 0.06%) for several days. Furthermore, by integration of the curve of frequency of apoptosis against time and since we knew what the fall in total cell count had been it was possible to arrive at an estimate for the length of time during which apoptosis could be recognised morphologically. By adding to this value the lag time between injection of anti-CD4 antibody (4 h) a final estimate of the duration of apoptosis is on the order of 5½ h, which compares favourably with other calculations of the duration of apoptosis before phagocytosis and degradation of the apoptotic fragments. No evidence of complement activation or inflamation was seen histologically, and there was no evidence of lymphocyte accumulation in other lymphoid or nonlymphoid tissues. It seems probable then that the fall in CD4 positive T cell number in lymph nodes can be accounted for simply by apoptosis (HOWIE et al. 1994). The precise mechanism for this effect is unknown but it involves signaling through the p56^{lck} tyrosine kinase associated with CD4 on the inner side of the cytoplasmic membrane and may also involve an additional trigger signal step mediated through Fas (CD95) antigen. This illustrates the ability of even small changes in the measureable rate of apoptosis to have a major effect on population size.

Using this model no effect was found on the fall of CD4 positive T lymphocytes according to *p53* status. The baseline rate of apoptosis in T cell areas of the lymph nodes was identical and there was a similar increase after anti-CD4 antibody treatment.

When animals were irradiated and lymph nodes were harvested at 4 h for direct counting of apoptosis in tissue sections there was a clear *p53* effect demonstrable (manuscript in preparation). Wild-type animals showed more than a 100-fold increase in the frequency of apoptosis whereas lymph nodes from homozygous *p53* deficient mice showed only a sixfold increase. There was a gene dosage effect with the heterozygous falling between the wild-type and homozygous deficient animals. However, the values for the heterozygotes were very much closer to the homozygous deficient than to the wild type, in sharp contrast to the results obtained with thymocytes. This indicates that the gene dosage effect of *p53* differs between thymic and peripheral T lymphocytes; peripheral mature T lymphocytes require a relatively higher level of p53 to initiate death following DNA strand breakage. An alternative explanation is that the effects of irradiation are altered when the whole animal is exposed rather than just isolated cells. Against this possibility is the clear gene dose effect which was noted when the survival of thymocytes irradiated in vivo was compared to thymocytes in vitro (Lowe et al. 1993). The increase in apoptosis seen in the homozygous mice is of interest. This is probably due to a surge of endogenous corticosteroids as a result of stress engendered while the mice were transported to and from the irradiation chamber (Munck et al. 1984). Mock exposed animals also showed this small but significant rise regardless of genotype.

We have found no difference in the cell cycle activity of freshly isolated T lymphocytes and no consistent differences have been noted following mitogen stimulation in vitro. Since most of these experiments have been performed using partly outbred mice it is quite possible that minor differences may have been overlooked. Several groups have reported the ease with which it is possible to culture a variety of cell types from *p53*-deficient mice (Harvey et al. 1993; Tsukada et al. 1993). This is particularly true of fibroblasts which rapidly immortalise and become aneuploid indicating the importance of p53 in this cell type for the maintenance of genomic stability. This does not appear to apply to freshly isolated peripheral blood mononuclear cells, and we have been unable to grow immortalised T lymphocyte lines so far, even in the presence of interleukin-2 (IL-2) conditioned medium. The role of p53 in different cell types may be to some extent lineage-dependent. Similar findings to those described above have been reported in a number of other in vitro systems for myeloid leukemia cell lines and haematopoietic cells in which p53 causes death rather than growth arrest (Yonish-Rouach et al. 1991; Ryan et al. 1993; Lotem and Sachs 1993).

7 p53 and Death Regulation in B Lymphocytes

Although *p53*-null mice have a virtually 100% incidence of malignancy a proportion of heterozygous mice were found to die without obvious evidence of tumour. Rather these mice showed nothing more than reactive lymphoid follicular hyperplasia although no specific source of infection was found and there was no histological evidence of autoimmune disease (Purdie et al. 1994).

Recent work from the laboratories of Greaves and Griffiths (manuscript in preparation) has investigated the role of p53 in murine IL-7-dependent B lymphocyte precursors. These cells, present in bone marrow, are exquisitely sensitive to the cytotoxic effect of γ-irradiation unlike more mature B cells (Griffiths et al., in press). Pre-B cells isolated from the bone marrow of mice homozygous for deficient *p53* demonstrated the same inability to undergo apoptosis following irradiation as was found with both thymocytes and peripheral T lymphocytes. Since these cells, unlike the thymocyte, can be maintained in culture in vitro by the addition of IL-7 it was possible to investigate the longer term effects of failing to undergo apoptosis after genotoxic injury. Clonogenic cells, the "undead" cells, were selected in medium supplemented with IL-7 but also containing 6-thioguanine. Clones which grew under these conditions had clearly become resistant to the effects of 6-thioguanine, thus reflecting mutation of the *HPRT* gene. Even more remarkable was the incidence of such mutations. In pre-B cells exposed to 10 Gy γ-radiation there was an approximately thousandfold increase in the mutation at the *HPRT* locus. Since only one marker gene was studied the overall incidence of mutation is probably much higher. The cloning efficiency in the experiments was significantly less than 100% indicating that the mutational load in some cells was lethal. These findings are clearly of great relevance for understanding carcinogenesis in a number of cell types. Failure of p53 to initiate cell death, perhaps because of a debilitating mutation of *p53* or its downstream effectors or overriding of the death signal by another survival signal such as *bcl-2*, would lead to cells surviving with mutations, possibly in critical growth regulatory genes, which should under normal circumstances be deleted. A proportion of these "undead" cells may then progress to form a tumour.

8 Has p53 a Physiological Role in Lymphocyte Apoptosis?

p53 mRNA and protein are undetectable in freshly isolated peripheral lymphocytes by northern analysis or immunoprecipitation (MILNER and MILNER 1981; REED et al. 1986; LUBBERT et al. 1989). They have been found at very low levels by flow cytometric antibody studies in mature, but not precursor, T lymphocytes (KASTAN et al. 1991). The protein is induced in thymocytes after γ-irradiation but even then it is detectable by immunohistochemistry in only around 6% of cells (CLARKE et al. 1993). After mitogenic lectin stimulation peripheral blood mononuclear cells do express *p53* mRNA (REED et al. 1986) but whether this is translated to protein is a moot point. In B lymphocytes an increase in *p53* mRNA preceded changes in κ light chain expression associated with differentiation and maturation of these cells (ALONI-GRINSTEIN et al. 1993). This cannot be regarded as convincing evidence for the involvement of *p53* in lymphocyte physiology since alterations of mRNA rarely relate to alteration in protein level or function. Somatic mutation occurs as part of the normal maturation and selection of B lymphocytes and involves DNA strand breakage. At this stage many cells are deleted due to the lack of a survival signal. Is it possible that p53 plays a role in positively selecting cells for deletion under these circumstances, not as a unique factor or even necessarily a particularly important one? Such a role would be consistent the incidence of B cell lymphomas noted in some murine models of *p53* deficiency.

The gene dosage effect noted in the heterozygote *p53*-deficient mice resistance to irradiation is interesting. Whereas thymocytes showed an effect midway between wild-type and homozygote deficient mature T cells showed a relatively higher resistance than expected. This might suggest that p53 only causes apoptosis after a certain threshold of DNA strand breakage has occurred. If so, and it is very difficult to suggest how such a mechanism could be tuned finely enough, is somatic mutation at or below the threshold level above which apoptosis occurs? It has been estimated that as few as ten double-strand DNA breaks can cause apoptosis in pre-B cells which is a level conceivably found during immunoglobulin gene rearrangement. Thus p53 may be effectively giving a "weak" death signal under these circumstances but its effect is ameliorated by other positive survival factors. Modulation of the death initiation pathway regulated by p53 has been shown in several myeloid leukemia cell lines by growth factors, principally IL-6. The bcl-2 family of genes may also act downstream of *p53* and act in a regulatory role. Absence of functional p53 in these circumstances would not be predicted as having a major effect on the selection of an immune repertoire. Expression genes involved in signal transduction from the cytoplasmic membrane, especially Fas and its ligand (ITOH et al. 1991, 1993; ALDERSON et al. 1993; SUDA et al. 1993), are probably more important physiological initiators of lymphocyte death. Defects in Fas found in the *lpr* mouse strain are associated with the development of severe autoimmune disease (WATANABE-FUKUNAGE et al. 1992). Since Fas is expressed in thymocytes (OGASAWARA et al. 1993) and mature T lymphocytes (TRAUTH et al. 1989) this suggests that defects in Fas result in escape of self-reactive cells into the periphery (ZHOU et al. 1993; RUSSELL et al. 1993). There is no evidence currently that Fas and p53 interact with one another.

9 Conclusions

p53 is of major importance in triggering death in a variety of lymphoid cell types after DNA strand breakage. Cells with nonfunctional or abnormal p53 do not respond appropriately to this kind of clastogenic insult and as a result survive with an increased tendency for inappropriate amplification, recombination events, tetraploidisation or mutations when, under normal circumstances, they should be deleted (Fig. 1) (RUIZ and WAHL 1990; LIVINGSTONE et al. 1992; YIN et al. 1992; HARTWELL 1992). This is very likely a critical step in the generation of mutations in

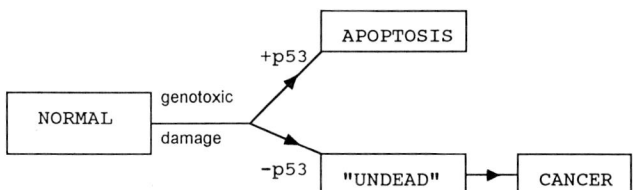

Fig. 1. A summary of the role of p53 in determining the outcome of genotoxic damage in lymphocytes, and perhaps other cell types as well. In fibroblasts the dominant pathway in the presence of p53 may be GI arrest rather than apoptosis; however, teleologically this model still applies since the end result is the prevention of genetic mutation being passed to daughter cells

lymphoid tumours (SELLINS and COHEN 1987). In contrast the majority of evidence at present does not support a role for p53 in normal population homeostasis, nor in lymphocyte death triggered through cell surface receptor ligation. However, further work is necessary to elucidate other signals which can countermand the effect of p53, since this may include growth factors such as IL-6 (YONISH-ROUACH et al. 1991; ZHU et al. 1994), perhaps leading to therapeutic strategies for the management of lymphoproliferative diseases.

Acknowledgements. This work was supported by the Cancer Research Campaign, the Medical Research Council and Scottish Hospitals Endowment Trust.

References

Akbar AN, Salmon M, Savill J, Janossy G (1993) A possible role for bcl-2 in regulating T-cell memory— a 'balancing act' between cell death and survival. Immunology Today 14: 526-532

Alderson MR, Armitage RJ, Maraskovsky E, Tough TW, Houx E, Schooley K, Ramsdell F, Lynch DH (1993) Fas Transduces activation signals in normal human T lymphocytes. J Exp Med 178: 2231–2235

Aloni-Grinstein R, Zar-Bar I, Alboum I, Goldfinger N, Rotter V (1993) Wild type p53 functions as a control protein in the differentiation pathway of the B-cell lineage. Oncogene 8: 3297–3305

Baker SJ, Markowitz S, Fearon ER. Willson JKV, Vogelstein B (1990) Suppression of human colorectal carcinoma cell growth by wild-type p53. Science 249: 912–915

Bakalkin G, Yakovleva T, Selivanova G, Magnusson KP, Szekely L, Kiseleva E, Klein G, Terenius L, Wiman KG (1994) p53 binds single stranded DNA ends and catalyzes DNA renaturation and strand transfer. Proc Natl Acad Sci USA 91: 413–417

Banda NK, Bernier J, Kurahara DK, Kurrle R, Haigwood N, Sekaly RP, Finkel TH (1992) Crosslinking CD4 by human immunodeficiency virus gp120 primes T cells for activation-induced apoptosis. J Exp Med 176: 1099–1106

Berke G (1991) Lymphocyte-triggered internal target disintegration. Immunol Today 12: 396–399

Bischoff JR, Friedman PN, Marshak DR, Prives C, Beach D (1990) Human p53 is phosphorylated by p60-cdc2 and cyclin B-cdc2. Proc Natl Acad Sci USA 87: 4766–4770

Brugal G, Dye R, Krief B, Chassery JM, Tanke H, Tucker JH (1992) HOME: highly optimized microscope environment. Cytometry 13: 109–116

Carder P, Wyllie AH, Purdie CA, Morris RG, White S, Piris J, Bird CC (1993) Stabilised p53 facilitates aneuploid clonal divergence in colorectal cancer. Oncogene 8: 1397–1401

Clarke AR, Purdie CA, Harrison DJ, Morris RG, Bird CC, Hooper ML, Wyllie AH (1993) Thymocyte apoptosis induced by p53-dependent and independent pathways. Nature 362: 849–852

Cobbold SP, Jayasuriya A, Nash A, Prospero TD, Waldmann H (1984) Therapy with monoclonal antibodies by elimination of t-cell subsets in vivo. Nature 312: 548–551

Cohen JJ (1993) Apoptosis. Immunol Today 14: 126-130

Donehower LA, Harvey M, Slagle BL, McArthur MJ, Montgomery CA Jr, Butel JS, Bradley A (1992) Mice deficient for p53 are developmentally normal but susceptible to spontaneous tumours. Nature 356: 215–221

Dutia A, Ruppert JM, Aster JC, Winchester E (1993) Inhibition of DNA replication factor RPA by p53. Nature 365: 79–82

Dulic V, Kaufmann WK, Wilson SJ, Tlsty TD, Lees E, Harper JW, Elledge SJ, Reed SI (1994) p53-dependent inhibition of cyclin-dependent kinase activities in human fibroblasts during radiation induced G1 arrest. Cell 76: 1013–1023

El Deiry WS, Tokiino T, Velculescu VE, Levy DB, Parsons R, Trent JM, Lin D, Mercer WE, Kinzier KM, Vogelstein B (1993) WAF1 a potential mediator of p53 tumor suppression. Cell 75: 817–825

El-Deiry WS, Harper JW, O'Conner PM, Velculescu VE, Canman CE, Jackman J, Pietenpol JA, Burrell M, Hill DE, Wang Y, Wiman KG, Mercer WE, Kastan MB, Kohn KW, Elledge SJ, Kinzler KW, Vogelstein V (1994) Waf1/Cip1 is induced in p53-mediated G1 arrest and apoptosis. Cancer Res 54: 1169–1174

Eliyahu D, Michaloviz D, Eliyahu S, Pinhasi-Kimhi O, Oren M (1989) Wild-type p53 can inhibit oncogene-mediate focus formation. Proc Natl Acad Sci USA 86: 8763–8767

Farmer G, Bargonetti J, Zhu H, Friedman P, Prywes R, Prives C (1992) Wild type p53 activates transcription in vitro. Nature 358: 83–86

Fields S, Jang SK (1990) Presence of a potent transcription activating sequence in the p53 protein. Science 249: 1046–1049

Finkel TH, Kubo RT, Cambier JC (1991) T-cell development and transmembrane signalling changing biological responses through an unchanging receptor. Immunol Today 12: 79–85

Finlay CA, Hinds PW, Levine AJ (1989) The p53 proto-oncogene can act as a suppressor of transformation. Cell 57: 1083–1093

Foord O, Navot N, Rotter V (1991) Isolation and characterization of DNA sequences that are specifically bound by wild-type p53 protein. Mol Cell Biol 13: 1378–1384

Friedman PN, Chen X, Bargonetti J, Prives C (1993) The p53 is an unusually shaped tetramer that binds directly to DNA. Proc Natl Acad Sci USA 90: 3319–3323

Fritsch M, Haessler C, Brandner G (1993) Induction of nuclear accumulation of the tumour-suppressor protein p53 by DNA damaging agents. Oncogene 8: 307–318

Funk WD, Pak DT, Karas RH, Wright WE, Sha JW (1992) A transcriptionally active DNA-binding site for human p53 protein complexes. Mol Cell Biol 12: 2866–2871

Griffiths SD, Goodhead DT, Marsden SJ, Wright EG, Krajewski S, Reed JC, Korsmeyer SJ, Greaves M (1994) IL7-dependent B lymphocyte precursor cells are ultra-sensitive to apoptosis. J Exp Med (in press)

Gu Y, Turck CW, Morgan DO (1993) Inhibition of CDK2 activity in vivo by an associated 20 K regulatory subunit. Nature 366: 707–710

Hall PA, McKee PH, Menage HP, Dover R, Lane DP (1993) high levels of p53 protein in UV-irradiated normal human skin. Oncogene 8: 203–207

Hartwell L (1992) Defects in a cell cycle may be responsible for the genomic instability of cancer cells. Cell 71: 543–546

Harvey M, Sands AT, Weiss RS, Hegi ME, Wiseman RW, Pantazis P, Giovanella BC, Tainsky MA, Bradley A, Donehower LA (1993) In vitro growth characteristics of embryo fibroblasts isolated from p53-deficient mice. Oncogene 8: 2457–2467

Hollstein M, Sidransky D, Vogelstein B, Harris CC (1991) p53 mutations in human cancers. Science 253: 49–53

Horneff G, Guse AH, Sculze-Koops H, Kalden JR, Burmester GR, Emmrich F (1993) Human CD4 modulation in vivo induced by antibody treatment. Clin Immunol Immunopathol 66: 80–90

Howie SEM, Sommerfield AJ, Gray E, Harrison DJ (1994) Peripheral T lymphocyte depletion by apoptosis after CD4 ligation in vivo: selective loss of CD44-ve and "activating" memory T cells. Clin Exp Immunol 95: 195–200

Itoh N, Yonehara S, Ishii A, Yonehara M, Mizushima S, Sameshima M, Hase A, Seto Y, Nagata S (1991) The polypeptide encoded by the cDNA for human cell surface antigen Fas can mediate apoptosis. Cell 66: 233–243

Itoh N, Rsujimoto Y, Nagata S (1993) Effect of bcl-2 on Fas antigen-mediate cell death. J Immunol 151: 621–627

Jenkinson EJ, Kingston R, Smith CA, Williams GT, Owen JJ (1989) Antigen-induced apoptosis in developing T cells: a mechanism for negative selection of the T cell receptor repertoire. Eur J Immunol 19: 2175–2177

Kastan MB, Onyekwere O, Sidransky D, Vogelstein B, Craig RW (1991) Participation of p53 protein in the cellular response to DNA damage. Cancer Res 51: 6304–6311

Kastan MB, Zhan Q, El-Deiry WS, Carrier F, Jacks T, Walsh WV, Plunckett BS, Vogelstein B, Fornace AJ Jr (1992) A mammalian cell cycle checkpoint pathway utilizing p53 and GADD45 in defective in Ataxia-Telangiectasia. Cell 71: 587–597

Kraiss S, Quaiser A, Oren M, Montenarh M (1988) Oligomerization of oncoprotein p53. J Virol 62: 4737–4744

Kuerbitz SJ, Plunkett BS, Walsh WV, Kastan MB (1992) Wild type p53 is a cell cycle checkpoint determinant following irradiation. Proc Natl Acad Sci USA 89: 7491–7495

Lane DP (1992) p53, guardian of the genome. Nature 358: 15–16

Lane DP (1993) A death in the life of p53. Nature 362: 786–787

Levine AJ, Momand J, Finlay CA (1991) The p53 tumour suppressor gene. Nature 351: 453–456

Lin D, Shields MT, Ullrich SJ, Appella E, Mercer WE (1992) Growth arrest induced by wild type p53 protein blocks cells prior to or near the restriction point in late G1 phase. Proc Natl Acad Sci USA 89: 9210–9214

Liu YJ, Johnson GD, Gordon J, Maclennan ICM (1989) Germinal centres in T cell dependent antibody responses. Immunol Today 13: 17–21

Livingstone LR, White A, Sprouse J, Livanos E, Jacks T, Tlsty TD (1992) Altered cell cycle arrest and gene amplification potential accompany loss of wild-type p53. Cell 70: 923–935

Lotem J, Sachs L (1993) Hematopoietic cells from mice deficient in wild-type p53 are more resistant to induction of apoptosis by some agents. Blood 82: 1092–1096

Lowe SW, Schmitt EM, Smith SW, Osborne BA, Jacks T (1993) p53 is required for radiation-induced apoptosis in mouse thymocytes. Nature 362: 847–849

Lu S, Lane DP (1993) Differential induction of transcriptionally active p53 following UV or ionizing radiation: defects in chromosome instability syndromes? Cell 75: 765–778

Lubbert M, Miller CW, Kahan J, Koeffler P (1989) Expression, methylation and chromatin structure of the p53 gone in untransformed and human T-cell leukemia virus type I-transformed human T-lymphocytes. Oncogene 4: 643–651

Mack DH, Vartikar J, Pipas JM, Laimins LA (1993) Specific repression of TATA-mediated but not initiator-mediated transcription by wild-type p53. Nature 363: 281–285

Malkin D, Li FP, Strong LC, Fraumeni JF Jr, Nelson CE, Kim DH, Kasssel J, Gryka MA, Bischoff FZ, Tainsky MA et al (1990) Germ line p53 mutations in a familial syndrome of breast cancer, sarcomas, and other neoplasms. Science 250: 1233–1238

Maltzman W, Czyzyk L (1984) UV irradiation stimulates levels of p53 cellular tumor antigen in nontransformed mouse cells. Mol Cell Biol 4: 1689–1694

Martinez J, Georgoff I, Martinez J, Levine AJ (1991) Cellular localization and cell cycle regulation by a temperature-sensitive p53 protein. Genes Dev 5: 151–159

McWhir J, Selfridge J, Harrison DJ, Squires S, Melton DW (1993) Mice with DNA repair gene (ERCC-1) deficiency have elevated levels of liver p53, nuclear abnormalities and die before weaning. Nat Genet 5: 217–224

Mercer WE, Shields MT, Amin M, Sauve GJ, Appella E, Romano JW, Ullrich SJ (1990) Negative growth in a glioblastoma tumor cell line that conditionally express human wild-type p53. Proc Natl Acad Sci USA 87: 6166–6170

Milner J, Milner S (1981) SV40-53K antigen: a possible role for 53K in normal cells. Virology 112: 785–788

Momand J, Zambetti GP, Olson DC, George D, Levine A (1992) The mdm-2 oncogene product forms a complex with the p53 protein and inhibits p53-mediated transactivation. Cell 69: 1237–1245

Moser Ar, Dove WF, Roth KA, Gordon JI (1991) J Cell Biol 116: 1517–1526

Munck A, Guyre PM, Holbrook NJ (1984) Physiological functions of glucocorticoids in stress and their relation to pharmacological actions. Endocr Rev 5: 25–44

Nelson WG, Kastan MB (1994) DNA strands breaks: the DNA template alterations that trigger p53-dependent DNA damage response pathways. Mol Cell Biol 14: 1815–1823

Newell MK, Haughn LJ, Maroun CR, Julius MH (1990) Death of mature T cells by separate ligation of CD4 and the T-cell receptor for antigen. Nature 347: 286–289

Oberoster P, Hloch P, Ramsperget U, Stahl H (1993) p53 catalyzed annealing of complementary single-stranded nucleic acids. EMBO J 12: 2389–2396

Ogasawara J, Watanabe-Fukunaga T, Adachi M, Matsuzawa A, Kasugai T, Kitamura Y, Itoh N, Suda T, Nagata S (1993) Lethal effect of the anti-fas antibody in mice. Nature 364: 806–809

Purdie CA, Harrison DJ, Peter A, Dobbie L, White S, Howie SEM, Saltter DM, Bird CC, Wyljie AH, Hooper ML, Clarke AR (1994) Tumour incidence, spectrum and ploidy in mice with a large deletion in the p53 gene. Oncogene 9: 603–609

Ragimov N, Kravskopf A, Naot N, Rotter V, Oren M, Aloni Y (1993) Wild-type but not mutant p53 can repress transcription initiation in vitro by interfering with the binding of basal transcription factors to the TATA motif. Oncogene 8: 1183–1193

Reed JC, Alpers JD, Nowell PC, Hoover RG (1986) Sequential expression of protooncogenes during lectin-stimulated mitogenesis of normal human lymphoctyes. Proc Natl Acad Sci USA 83: 3982–3986

Ruiz JC, Wahl GM (1990) Chromosomal destabilization during gene amplification. Mol Cell Biol 10: 3056–3066

Russell JH, Rush a, Weaver C, Wang R (1993) Mature T cells of autoimmune lpr/lpr mice have a defect in antigen-stimulated suicide. Proc Natl Acad Sci USA 90: 4409–4413

Ryan JJ, Danish R, Gottleib CA, Clarke MF (1993) Cell cycle analysis of p53-induced cell death in murine erythroleukemia cells. Mol Cell Biol 13: 711–719

Said JW, Barrera R, Shintaku IP, Nakamura H, Koeffler HP (1992) Immunohistochemical analysis of p53 expression in malignant lymphomas. Am J Pathol 141: 1343–1348

Sellins KS, Cohen JJ (1987) Gone induction by Gamma-radiation leads to DNA fragmentation in lymphocytes. J Immunol 139: 3199–3206

Serrano M, Hannon GJ, Beach D (1993) A new regulatory motif in cell-cycle control causing specific inhibition of cyclin D/CDK4. Nature 366: 704–707

Shaulsky G, Ben-Ze'ev A, Rotter V (1990) Subcellular distribution of the p53 protein during the cell cycle of Balb/c 3T3 cells. Oncogene 5: 1707–1711

Shaw P, Bovey R, Tardy S, Sahli R, Sordat B, Costa J (1992) Induction of apoptosis by wild-type p53 in a human colon tumor derived cell line. Proc Natl Acad Sci USA 89: 4495–4499

Srivastava S, Zou Z, Pirollo K, Blattner W, Change EH (1990) Nature 348: L 747–749

Sturzbecher HW, Maimets T, Chumakov P, Brain R, Addison C, Simanis V, Rudge K, Philip R, Grimaldi M, Court W, Jenkins JR (1990) p53 interacts with p34^{cdc2} in mammalian cells: implications for cell cycle control and oncogenesis. Oncogene 5: 795–800

Suda T, Takahashi T, Glostein P, Nagata S (1993) Molecular cloning and expression of the fas ligand, a novel member of the tumor necrosis family. Cell 75: 1169–1178

Trauth BC, Klas C, Peters AMJ, Matzku S, Moller P, Falk W, Debatin KM, Kramer PH (1989) Monoclonal antibody-mediated tumor regression by induction of apoptosis. Science 245: 301–305

Tsukada T, Tomooka Y, Takai S, Veda Y, Nishikawa S, Yagi T, Tokunaga T, Takeda N, Suda Y, Abe S, Matsuo I, Ikawa Y, Aizawa S (1993) Enhanced proliferative potential in culture of cells from p53-deficient mice. Oncogene 8: 3313–3322

Watanabe-Fukunaga R, Brannan CI, Copeland NG, Jenkins NA, Nagata S (1992) Lymphoproliferation disorder in mice explained by defects in Fas antigen and mediates apoptosis. Nature 356: 314–317

Wyllie AH (1980) Glucocorticoid induced thymocyte apoptosis is associated with endogenous endonuclease activation. Nature 284: 555–556

Xiong Y, Hannon GJ, Zhang H, Casso D, Kobayashi R, Beach D (1993) p21 is a universal inhibitor of cyclin kinases. Nature 366: 701–704

Yin Y, Tainsky MA, Bischoff FZ, Strong LC, Wahl GM (1992) Wild-type p53 restores cell cycle control and inhibits gene amplification in cells with mutant p53 alleles. Cell 70: 937–948

Yonish-Rouach, Resnitzky D, Lotem J, Sachs L, Kimchi A, Oren M (1991) Wild-type p53 induces apoptosis of myeloid leukaemic cells that is inhibited by interleukin-6. Nature 352: 345–347

Zambetti GP, Bargonetti J, Walker K, Prives C, Levine AJ (1992) Wild-type p53 mediates positive regulation of gene expression through a specific DNA sequence element. Genes Dev 6: 1143–1152

Zhan Q, Carrier F, Fornace AJ (1993) Induction of Cellular p53 activity by DNA-damaging agents and growth arrest. Mol Cell Biol 13: 4242–4250

Zhou T, Bluethmann H, Eldridge J, Berry K, Mountz JD (1993) Origin of CD4$^-$CD8$^-$ B220$^+$ T cells in MRL-lpr/lpr mice. J Immunol 150: 3651–3667

Zhu YM, Bradbury DA, Russell NH (1994) Wild-type p53 is required for apoptosis induced by growth factor deprivation in factor-dependent leukemic cells. Br J Cancer 69: 468–472

Death Genes in T Cells

J. Woronicz, B. Calnan, and A. Winoto

1 Introduction

Apoptosis is a general phenomenon which occurs in many organisms and in many cell types. It is characterized by chromosomal condensation, cleavage of DNA into nucleosome size fragments and subsequent cell death (Cohen et al. 1992; Ellis et al. 1991; Oppenheim 1991; Wyllie et al. 1984). Apoptosis occurs in more than 95% of thymocytes during development as a consequence of negative selection or lack of positive selection. Apoptosis is also likely to occur during peripheral tolerance in mature T cells (for review, see Cohen et al. 1992; Korsmeyer 1992). In vitro, T cell apoptosis can be initiated using antibodies for the CD3 T cell receptor complex, steroids, irradiation, and antibody specific for the Fas antigen (see below). The many cell death processes differ in their requirements and pathways. For example, the p53 gene is required for irradiation-induced T cell death but not for anti-CD3- or glucocorticoid-induced death (Lowe et al. 1993). Apoptosis by anti-CD3 and glucocorticoid requires gene transcription whereas Fas-mediated cell death does not (Cohen and Eisenberg 1992).

Apoptosis mediated through anti-CD3 might be related to negative selection. Cross-linking the T cell receptor with anti-CD3 antibody mimics ligands for the T cell receptor and can initiate apoptosis in immature T cells (thymocytes) and T cell hybridomas (Odaka et al. 1990; Shi et al. 1990; Ucker et al. 1989; Woronicz et al. 1994). Mature T cells usually do not die upon anti-CD3 stimulation unless they are exposed to a high dose of interleukin-2 (IL-2) beforehand (Critchfield et al. 1994; Lenardo 1991).

Department of Molecular and Cell Biology, Division of Immunology, University of California, Berkeley, CA 94720-3200, USA

The process of apoptosis in immature T cells and T cell hybridomas requires new gene synthesis and is sensitive to cyclosporin A (CsA), an immuo-suppressive drug widely used to alleviate transplant rejection (GAO et al. 1988; JENKINS et al. 1988; MERCEP et al. 1989; SHI et al. 1989). Induction of cell death by glucocorticoid also requires gene transcription, presumably as a result of activation by the glucocorticoid receptor (McCONKEY et al. 1989; WYLLIE 1980). Interestingly, signals from glucocorticoid and from CD3 can cancel each other out (ZACHARCHUK et al. 1990). T cell hybridomas and thymocytes stimulated with both agents do not die. These data suggest an interplay of apoptotic pathways initiated by glucocorticoid and by anti-CD3. In this review, we will provide a molecular characterization of Nur77, an orphan steroid receptor involved in T cell receptor-mediated apoptosis, and give a brief synopsis of the other apoptotic genes and how their pathways may relate to each other.

2 The Role of *nur77* Gene Family in Anti-CD3-Induced Apoptosis

To isolate genes crucial for apoptosis, several groups have used the subtractive hybridization technique to identify genes that are involved in the cell death process (LIU et al. 1994; OWENS et al. 1991; SCHWARTZ and OSBORNE 1993; WORONICZ et al. 1994). One of the genes induced during apoptosis is *nur77*, which encodes an orphan steroid receptor protein (LIU et al. 1994; WORONICZ et al. 1994). *nur77* or N10 (mouse cDNA) or NGFI-B (rat cDNA) was also previously isolated as a serum-induced immediate early gene in fibroblasts and as a nerve growth factor (NGF)-induced gene in neuronal cells (HAZEL et al. 1988; RYSECK et al. 1989; WATSON and MILBRANDT 1989). Its mRNA is expressed at low levels in thymus, brain, heart and lung. Similar to most members of the steroid receptor gene family, *nur77* encodes a protein with a strong transactivation domain, a DNA binding domain and a putative ligand binding domain. Unlike most steroid receptors, however, Nur77 does not have a known ligand. Indeed, no ligand may be required for transactivation by this receptor as shown by transient transfection experiments (PAULSEN et al. 1992; WILSON et al. 1991, 1992). In addition, Nur77 also contains a unique A box DNA binding domain, which is immediately downstream of the two zinc fingers (PAULSEN et al. 1992; WILSON et al. 1992). This A box DNA binding domain is also found in two other known orphan steroid receptors (WILSON et al. 1992, 1993). Another feature that sets Nur77 apart from the other steroid family members is its mode of DNA binding. In contrast to most steroid receptor proteins, Nur77 binds as a monomer to a nonpalindromic DNA element (WILSON et al. 1992, 1993). Its DNA binding site was determened as AAAAGGTCA by genetic selection in yeast (WILSON et al. 1991) and is similar to half of the estrogen receptor binding site (AGGTCA). Not much information, however, is known regarding the downstream genes regulated by Nur77. Only one gene, steroid 21-

hydroxylase, which is expressed in the nervous system, has been identified as a target gene for Nur77 (WILSON et al. 1993).

nur77 is part of a gene family. Another orphan steroid receptor, Nurr1, is highly related to nur77. Its predicted protein shares 92% identity with the Nur77 DNA binding domain (LAW et al. 1992; SCEARCE et al. 1993). Its protein can bind to the *nur77* DNA element, and introduction of the cDNA into NIH 3T3 cells can transactivate a reporter gene containing the *nur77* DNA element. *Nurr1* mRNA is expressed abundantly in brain and is hardly detectable in other tissues. Its expression in apoptotic T cells is yet to be determined.

In T cell hybridomas, a difference in kinetics of *nur77* mRNA induction can be seen between apoptotic and nonapoptotic cells (WORONICZ et al. 1994). The *nur77* message is induced very rapidly, as early as 30 min after stimulation with anti-CD3 or ionomycin and phorbol ester. In nonapoptotic cells—anti-CD3-stimulated EL4 cells or phorbol ester-treated hybridoma—*nur77* mRNA is rapidly down-regulated and is essentially undetectable after 3 h postinduction (WORONICZ et al. 1994). This is similar to its induction kinetics in serum-induced fibroblasts and NGF-induced neuronal cells (DAVIS et al. 1991; MILBRANDT 1988; RYSECK et al. 1989). In contrast, the mRNA level for *nur77* stays at a high level in dying anti-CD3 treated T cell hybridoma for at least 12 h postinduction (WORONICZ et al. 1994).

A more striking difference between dying and growing T cells was seen when the Nur77 protein activity was determined using a gel mobility shift assay. As another Nur77-like protein also exists that can bind to the same DNA element (LAW et al. 1992; SCEARCE et al. 1993), this assay detects all the Nur77 protein family members. In nonapoptotic EL4 cells, phorbol ester-stimulated T cell hybridomas or nonstimulated immature thymocytes, very little Nur77 family DNA binding activity could be detected. In contrast, a strong protein/DNA band could be seen in anti-CD3-treated apoptotic T cell hybridomas or apoptotic thymocytes (WORONICZ et al. 1994, see Table 1). The induction of *nur77* mRNA and protein is due to calcium signals as calcium ionophore (ionomycin) alone can induce the Nur77 family protein activity.

Table 1. Correlation of the Nur77 protein activity with anti-CD3 apoptosis

Cells	Signals	Apoptosis	Nur77 protein
T hybridomas	None	–	–
	Anti-CD3	++++	++++
	PMA	–	–
	Ionomycin	++++	++++
	Anti-CD3/dexamethasone	–	+/–
	Anti-CD3/cyclosporin A	–	–
EL4 lymphoma	None	–	–
	Anti-CD3	–	–
Thymocytes	None	–	–
	Anti-CD3	++++	++++

– indicates that the cells did not undergo apoptosis or the Nur77 activity was not detectable by the gel shift analysis; ++++ indicates occurrence of apoptosis or detection of the Nur77 protein activity by the gel shift analysis; +/– indicates little Nur77 protein activity; PMA, phorbol myristate acetate.

To show that the *nur77* gene family is required for apoptosis, we have used a dominant negative mutant version of Nur77 (Woronicz et al. 1994). The mutant was constructed and tested first in transient transfection experiments. At a one to one ratio to the wild-type protein, the Nur77 dominant negative mutant can block most of the Nur77 protein from functioning. Stable introduction of the *nur77* dominant negative mutant into T cell hybridomas protected the cells from anti-CD3-induced apoptosis. As the dominant negative mutant can block other members of the Nur77 family as well, these data only show that the *nur77* gene family is required for apoptosis (Woronicz et al. 1994). The requirement of Nur77 in activation-induced apoptosis was also shown independently using an anti-sense approach (Liu et al. 1994).

The immunosuppressive drug CsA and glucocorticoid are known to inhibit anti-CD3 apoptosis (Mercep et al. 1989; Shi et al. 1989; Zacharchuk et al. 1990). CsA is also reported in some instances to inhibit negative selection in vivo (Gao et al. 1988; Jenkins et al. 1988; Urdahl et al. 1994). Administration of CsA resulted in the escape of the autoreactive T cells to the peripheral lymphoid organs. Experiments using these two drugs suggest that the Nur77 family is the central control protein in activation-induced apoptosis. In the presence of CsA, gel mobility shift assay indicated that the Nur77 family protein was totally inactive at all time points (Woronicz and Winoto, manuscript in preparation, Table 1). Similarly, addition of dexamethasone to anti-CD3-treated T cell hybridomas also resulted in down-regulation of the Nur77 family protein activity (Table 1). This latter effect is manifested at the level of transcription as the *nur77* promoter is glucocorticoid-sensitive (B. Calnan, unpublished data). Based on these data, we put forward a simple model to explain the interrelationship between the glucocorticoid and anti-CD3 apoptosis pathways. We hypothesize that a protein X is induced by anti CD3 signals. This protein usually will mediate the prolonged induction of the transcription of the *nur77* gene family. In the presence of glucocorticoid, however, this protein X associates with the activated glucocorticoid receptor to form a repressor complex. The GR/X repressor complex can then inhibit the transcrip-

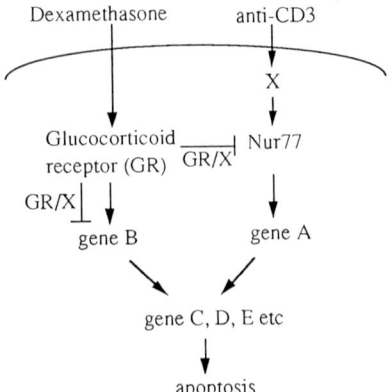

Fig. 1. A model of apoptosis by anti-CD3 and glucocorticoid. Dexamethasone, in the absence of other signals, activates glucocorticoid receptor (*GR*), which then translocates to the nucleus to activate gene B and starts a cascade of apoptosis genes (C, D, E, etc.) Anti-CD3, in the absence of dexamethasone, activates protein X, which stimulates high level expression of the *nur77* gene family. The Nur77 protein family then stimulates gene A and initiates a cascade of apoptosis genes (C, D, E, etc.) When the two signals are combined, GR associates with protein X to repress *nur77* gene family expression and expression of gene B. This leads to shutting down of the two apoptosis pathways

tional activation of both the *nur77* gene family and the downstream genes, effectively shutting off both apoptosis pathways (Fig. 1).

3 The Role of Other Genes (c-*myc*, *bcl*-2 family, *fas*, *PD*-1, p53, *ced*-3) in T Cell Apoptosis

Several other genes have been implicated in the various forms of apoptotic death. They include genes for the cytoplasmic/nuclear proteins *p53* (LOWE et al. 1993; SHAW et al. 1992; YONISH et al. 1991), c-*myc* (EVAN et al. 1992; SHI et al. 1992), the *bcl*-2 gene family (BOISE et al. 1993; KORSMEYER 1992; OLTVAI et al. 1993), *ced*-3 (*ice*, MIURA et al. 1993; YUAN et al. 1993) and the cell surface proteins *PD*-1 (ISHIDA et al. 1992) and *fas* (ITOH et al. 1991).

   p53. Overexpression of the tumor suppressor gene *p53* can induce apoptosis of a human colon tumor-derived cell line and myeloid leukemia cells (SHAW et al. 1992; YONISH et al. 1991). In the *p53* knock-out mutant mice, the lymphocytes are resistant to radiation but not to glucocorticoid- or anti-CD3-induced cell death (LOWE et al. 1993). p53 is a transcription factor, which regulates Waf1 (Cip1), a p21 kDa protein that can associate with cdk/cyclin complexes to inhibit the cell cycle (ELDEIRY et al. 1993; HARPER et al. 1993). Thus p53 may mediate apoptosis by affecting the progression of the cell cycle.

   c-*myc.* The c-*myc* gene was initially defined as a proto-oncogene frequently translocated to the immunoglobulin and T cell receptor loci in a variety of B and T cell tumors. Extensive studies in many laboratories identified c-*myc* as a transcription factor. It heterodimerizes with another factor, Max, to bind DNA (LUSCHER and EISENMAN 1990a, b). Both the c-Myc and Max proteins contain a dimerization domain composed of helix-loop-helix and leucine zipper motif, whereas only c-Myc contains a transactivation domain for transcription (for review, see BLACKWOOD et al. 1991, 1992; KRETZNER et al. 1992a, b). Hence a Max-Max homodimer will only bind DNA but cannot activate transcription. The basic regions of c-Myc and Max contribute to the DNA binding activity to a consensus DNA sequence CACGTG. Two other c-Myc family proteins, Mad and Mxi, were identified later (EVAN et al. 1992; KATO et al. 1992; KRETZNER et al. 1992a, b; LUSCHER and EISENMAN 1990a, b; PRENDERGAST et al. 1991; PRENDERGAST and ZIFF 1992; ZERVOS et al. 1993). They also contain the helix-loop-helix and leucine zipper protein motifs but can only bind to Max. It is generally believed that the ratio of c-Myc/Max and Mad/Max or Mxi/Mad proteins in the cells determines the outcome and function of the c-*myc* gene family. The c-Myc/Max protein can activate transcription, whereas the Max/Mad or Max/Mxi proteins repress transcription.

   In T cells, c-*myc* is an immediate early gene family that is rapidly up-regulated when T cells are stimulated through their T cell receptor complex (for review, see CRABTREE 1989). Its message is present at a low level in immature and mature T cell populations and can be stimulated with chemical agents that activate kinase

C and calcium influx. Recent evidence indicates that c-Myc plays a dual role in driving both proliferation and programmed cell death. In T cell hybridomas, in which stimulation by anti-CD3 antibody leads to apoptosis, addition of an anti-sense c-*myc* oligonucleotide can inhibit cell death (SHI et al. 1992). In serum-starved fibroblasts over-expression of c-*myc* also leads to cell death (EVAN) et al. 1992). In Chinese hamster ovary (CHO) cells that contain a heat inducible c-*myc*, heat shock treatment results in apoptosis mediated by c-Myc which can be rescued by Bcl-2 (BISSONNETTE et al. 1992; WURM et al. 1986). Introduction of a dominant negative c-Myc into T cell hybridomas can also protect the cells from anti-CD apoptosis (B. Calnan, unpublished data).

bcl-2. *bcl*-2 was originally described as a translocated oncogene to the immunoglobulin locus in B cell lymphomas (KORSMEYER 1992). It was found subsequently that overexpression of *bcl*-2 can protect an immature B cell line from undergoing apoptosis upon growth factor withdrawal (Blackwood et al. 1991, 1992; HOCKENBERY et al. 1990). Several studies showed that overexpression of *bcl*-2 can rescue cell death by irradiation, glucocorticoid, anti-CD3 and from IL-1β convertase (ICE)-mediated apoptosis (CHIOU et al. 1994; SENTMAN et al. 1991; STRASSER et al. 1991; YUAN et al. 1993). Its effect on negative selection, however, is not entirely clear (SIEGEL et al. 1992; STRASSER et al. 1991). *bcl*-2 deficient mice showed massive apoptosis of peripheral B and T cells after 3 weeks of age (NAKAYAMA et al. 1993; VEIS et al. 1993), indicating that Bcl-2 is crucial in maintaining the peripheral lymphocyte population.

Several other *bcl*-2-like genes (*bcl*-x and *bax*) have been identified and isolated (BOISE et al. 1993; OLTVAI et al. 1993). Two forms of Bcl-x proteins were identified due to alternative splicing of its mRNA: Bcl-x_L and Bcl-x_S. The Bcl-x_L protein is similar to Bcl-2 in that its overexpression protects cells from apoptosis (BOISE et al. 1993). Overexpression of either Bcl-x_S and Bax, by contrast, sensitizes the cell to programmed cell death. Bax protein was found to heterodimerize with Bcl-2, indicating that the ratio of Bcl-2/Bax heterodimer and Bax/Bax homodimer might determine the life or death of a cell (OLTVAI et al. 1993).

ced-3. The *ced*-3 gene was initially identified in *C. elegans* as a genetic locus required for apoptosis of certain cell lineages (ELLIS et al. 1991; YUAN et al. 1993). Cloning of *ced*-3 showed that its predicted protein is homologous to the mammalian ICE, a cysteine protease (CERRETTI et al. 1992; MIURA et al. 1993; NETT et al. 1992). Overexpression of the mouse *ice* gene can result in apoptosis of a rat fibroblast cell line (MIURA et al. 1993). Thus, a cysteine protease might be involved in T cell apoptosis as well.

PD-1. Using subtractive hybridization, Honjo's group identified a novel gene *PD*-1, which is induced in anti-CD3 apoptosis of T cell hybridomas and in IL-3 withdrawal of an IL-3-dependent pro-B cell line (ISHIDA et al. 1992). *PD*-1 has homology to the immunoglobulin superfamily but its function in apoptosis is still unknown.

fas. Antibody specific for the Fas antigen (or Apo1) can initiate cell death in a variety of cell types (ITOH et al. 1991; TRAUTH et al. 1989), Fas is a cell surface protein with homology to the TNF (tumor necrosis factor) receptor. Mice that are *lpr/lpr* contain a mutation in the Fas antigen, which presumably contribute to the autoreactive phenotype of these mice (Iтон et al. 1991; WANTANABE et al. 1992a, b). The Fas ligand was isolated recently and, as expected, is a member of the TNF family (Suda et al. 1993). The *gld* mutant mice show the same phenotype as *lpr/lpr* mice and contain a mutation in the Fas ligand (TAKAHASHI et al. 1994).

4 Conclusions

An understanding of the molecular mechanisms of apoptosis is unravelling at a rapid pace. Several genes involved in apoptosis have been identified and cloned. The *bcl-2* gene family seems to play a central role in most forms of apoptosis. Bcl-2 gain of function can block apoptosis mediated by most means (i.e., glucocorticoid, anti-CD3, irradiation). Bcl-2 can also block the pathways mediated by c-*myc*, *fas*, *ced-3* and *p53* (BISSONNETTE et al. 1992; CHIOU et al. 1994; ITOH et al. 1993; YUAN et al. 1993). Increasing the expression of *bcl-2*, however, does not block apoptosis mediated by cytotoxic T cells (VAUX et al. 1992). Its effect on negative selection is also not entirely clear at this point (SENTMAN et al. 1991; SIEGEL et al. 1992; STRASSER et al. 1991,1994).

It is not known what the interrelationship between the *nur77* gene family and the *bcl-2* gene family is. The *nur77* gene family is not likely to activate transcription of the *bcl-2* family as the mRNA levels of *bcl-2*, *bax* or *bcl-x* stay constant during anti-CD3 apoptosis in T cell hybridomas (S. Allbright, A. Winoto, unpublished data). Also, Nur77 is not involved in glucocorticoid or Fas-mediated cell death, or death of cytotoxic T cells due to withdrawal of IL-2 (J. Woronicz, B. Calnan, A. Winoto, unpublished data). Thus, Nur77 may be uniquely involved in the T cell receptor-mediated programmed cell death of thymocytes.

Many central questions regarding apoptosis are still left unresolved. For example: What is the endonuclease activated during apoptosis? Is *ced-3* involved in T cell apoptosis? Does *ced-3* inactivate or activate a crucial protein during apoptosis? Is the *nur77* gene family part of the intracellular signaling process in negative selection during T cell development? Is *nur77* a master gene for programmed cell death in thymocytes? Does *nur77* activate Ced-3 or other proteases? Does Nur77 affect the balance of Bcl-2/Bax and Bax/Bax protein dimer stochiometry? What are the roles of Bax and Bcl-x in T cell development? What is the physiological role of Fas-mediated apoptosis? With the advent of genetic manipulation and powerful molecular biological techniques, the coming years should promise to be exciting ones, when these questions will be answered, and the molecular pathway of T cell programmed cell death will be determined.

References

Bissonnette RP, Echeverri F, Mahboubi A, Green DR (1992) Apoptotic cell death induced by c-myc is inhibited by bcl-2. Nature 359: 552–554

Blackwood EM, Luscher B, Kretzner L, Eisenman RN (1991) The Myc: Max protein complex and cell growth regulation. Cold Spring Harb Symp Quant Biol 56: 109–117

Blackwood EM, Kretzner L, Eisenman RN (1992) Myc and Max function as a nucleoprotein complex. Curr Opin Genet Dev 2: 227–235

Boise LH, Gonzalez GM, Postema CE, Ding L, Lindsten T, Turka LA, Mao X, Nunez G, Thompson CB (1993) bcl-x, a bcl-2-related gene that functions as a dominant regulator of apoptotic cell death. Cell 74: 597–608

Cerretti DP, Kozlosky CJ, Mosley B, Nelson N, Van NK, Greenstreet TA, March CJ, Kronheim SR, Druck T, Cannizzaro LA (1992) Molecular cloning of the interleukin-1 beta converting enzyme. Science 256: 97–100

Chiou S-K, Rao L, White E (1994) Bcl-2 blocks p53-dependent apoptosis. Mol Cell Biol 14: 2556–2563

Cohen JJ, Duke RC, Fadok VA, Sellins KS (1992) Apoptosis and programmed cell death in immunity. Annu Rev Immunol 10: 267–293

Cohen PL, Eisenberg RA (1992) The lpr and gld genes in systemic autoimmunity: life and death in the Fas lane. Immunol Today 13: 427–428

Crabtree GR (1989) Contingent genetic regulatory events in T lymphocyte activation. Science 243: 355–361

Critchfield JM, Racke MK, Zuniga PJ, Cannella B, Raine CS, Goverman J, Lenardo MJ (1994) T cell deletion in high antigen dose therapy of autoimmune encephalomyelitis. Science 263: 1139–1143

Davis IJ, Hazel TG, Lau LF (1991) Transcriptional activation by Nur77, a growth factor-inducible member of the steroid hormone receptor superfamily. Mol Endocrinol 5: 854–859

Eldeiry WS, Tokino T, Velculescu VE, Levy DB, Parsons R, Trent JM, Lin D, Mercer WE, Kinzler KW, Vogelstein B (1993) Waf-1, a potential mediator of p53 tumor suppression. Cell 75: 817–825

Ellis RE, Yuan JY, Horvitz HR (1991) Mechanisms and functions of cell death. Annu Rev Cell Biol 7: 663–698

Evan GI, Wyllie AH, Gilbert CS, Littlewood TD, Land H, Brooks M, Waters CM, Penn LZ, Hancock DC (1992) Induction of apoptosis in fibroblasts by c-myc protein. Cell 69: 119–128

Gao EK, Lo D, Cheney R, Kanagawa O, Sprent J (1988) Abnormal differentiation of thymocytes in mice treated with cyclosporin A. Nature 336: 176–179

Harper JW, Adami GR, Wei N, Keyomarsi K, Elledge SJ (1993) The p21 Cdk-interacting protein Cip 1 is a potent inhibitor of G1 cyclin-dependent kinases. Cell 75: 805–816

Hazel TG, Nathans D, Lau LF (1988) A gene inducible by serum growth factors encodes a member of the steroid and thyroid hormone receptor superfamily. Proc Natl Acad Sci USA 85: 8444–8448

Hockenbery D, Nunez G, Milliman C, Schreiber RD, Korsmeyer SJ (1990) Bcl-2 is an inner mitochondrial membrane protein that blocks programmed cell death. Nature 348: 334–336

Ishida Y, Agata Y, Shibahara K, Honjo T (1992) Induced expression of PD-1 a novel member of the immunoglobulin gene superfamily upon programmed cell death. EMBO J 11: 3887–3895

Itoh N, Yonehara S, Ishii A, Yonehara M, Mizushima S, Sameshima M, Hase A, Seto Y, Nagata S (1991) The polypeptide encoded by the cDNA for human cell surface antigen Fas can mediate apoptosis. Cell 66: 233–243

Itoh N, Tsujimoto Y, Nagata S (1993) Effect of bcl-2 on Fas antigen-mediated cell death. J Immunol 151: 621–627

Jenkins MK, Schwartz RH, Pardoll DM (1988) Effects of cyclosporine A on T cell development and clonal deletion. Science 241: 1655–1658

Kato GJ, Lee WM, Chen LL, Dang CV (1992) Max: functional domains and interaction with c-Myc. Genes Dev 6: 81–92

Korsmeyer SJ (1992) Bcl-2: a repressor of lymphocyte death. Immunol Today 13: 285–288

Kretzner L, Blackwood EM, Eisenman RN (1992a) Myc and Max proteins possess distinct transcriptional activities. Nature 359: 426–429

Kretzner L, Blackwood EM, Eisenman RN (1992b) Transcriptional activities of the Myc and Max proteins in mammalian cells. In: Potter M, Melchers F (eds) Mechanisms in B cell neoplasia 1992. Springer, Berlin Heidelberg New York, pp 435–443 (Current topics in microbiology and immunology, vol 182)

Law SW, Conneely OM, DeMayo FJ, O'Malley BW (1992) Identification of a new brain-specific transcription factor NURR1. Mol Endocrinol 6: 2129–2135

Lenardo MJ (1991) Interleukin-2 programs mouse alpha beta T lymphocytes for apoptosis. Nature 353: 858–861

Liu Z-G, Smith SW, McLaughlin KA, Schwartz LM, Osborne B (1994) Apoptotic signals delivered through the T-cell receptor of a T-cell hybrid require the immediate-early gene nur77. Nature 367: 281–284

Lowe SW, Schmitt EM, Smith SW, Osborne BA, Jacks T (1993) p53 is required for radiation-induced apoptosis in mouse thymocytes (see comments). Nature 362: 847–849

Luscher B, Eisenman RN (1990a) New light on Myc and Myb, part I. Myc. Genes Dev 4: 2025–2035

Luscher B, Eisenman RN (1990b) New light on Myc and Myb, part II. Myb. Genes Dev 4: 2235–2241

McConkey DJ, Hartzell P, Nicotera P, Orrenius S (1989) Calcium-activated DNA fragmentation kills immature thymocytes. FASEB J 3: 1843–1849

Mercep M, Noguchi PD, Ashwell JD (1989) The cell cycle block and lysis of an activated T cell hybridoma are distinct processes with different Ca^{2+} requirements and sensitivity to cyclosporine A. J Immunol 142: 4085–4092

Milbrandt J (1988) Nerve growth factor induces a gene homologous to the glucocorticoid receptor gene. Neuron 1: 183–188

Miura M, Zhu H, Rotello T, Hartweig EA, Yuan J (1993) Induction of apoptosis in fibroblast by IL-1beta converting enzyme, a mammalian homolog of the C. Elegans cell death gene ced-3. Cell 75: 653–660

Nakayama K, Nakayama K, Negishi I, Kuida K, Loh D (1993) Disappearance of the lymphoid system in bcl-2 homozygous mutant chimeric mice. Science 261: 1584–1588

Nett MA, Cerretti DP, Berson DR, Seavitt J, Gilbert DJ, Jenkins NA, Copeland NG, Black RA, Chaplin DD (1992) Molecular cloning of the murine IL-1beta converting enzyme cDNA. J Immunol 149: 3254–3259

Odaka C, Kizaki H, Tadakuma T (1990) T cell receptor-mediated DNA fragmentation and cell death in T cell hybridomas. J Immunol 144: 2096–2101

Oltvai ZN, Milliman CL, Korsmeyer SJ (1993) Bcl-2 heterodimerizes in vivo with a conserved homolog, Bax, that accelerates programmed cell death. Cell 74: 609–619

Oppenheim RW (1991) Cell death during development of the nervous system. Annu Rev Neurosci 14: 453–501

Owens GP, Hahn WE, Cohen JJ (1991) Identification of mRNAs associated with programmed cell death in immature thymocytes. Mol Cell Biol 11: 4177–4188

Paulsen RE, Weaver CA, Fahrner TJ, Milbrandt J (1992) Domains regulating transcriptional activity of the inducible orphan receptor NGFI-B. J Biol Chem 267: 16491–16496

Prendergast GC, Lawe D, Ziff EB (1991) Association of Myn, the murine homolog of max, with c-Myc stimulates methylation-sensitive DNA binding and ras cotransformation. Cell 65: 395–407

Prendergast GC, Ziff EB (1992) A new bind for Myc. Trends Genet 8: 91–96

Ryseck RP, MacDonald BH, Mattei MG, Ruppert S, Bravo R (1989) Structure, mapping and expression of a growth factor inducible gene encoding a putative nuclear hormonal binding receptor. EMBO J 8: 3327–3335

Scearce LM, Laz TM, Hazel TG, Lau LF, Taub R (1993) RNR-1 a nuclear receptor in the NGFI-B/Nur77 family that is rapidly induced in regenerating liver. J Biol Chem 268: 8855–8861

Schwartz LM, Osborne BA (1993) Programmed cell death apoptosis and killer genes. Immunol Today 14: 582–590

Sentman CL, Shutter JR, Hockenbery D, Kanagawa O, Korsmeyer SJ (1991) bcl-2 inhibits multiple forms of apoptosis but not negative selection in thymocytes. Cell 67: 879–888

Shaw P, Bovey R, Tardy S, Sahli R, Sordat B, Costa J (1992) Induction of apoptosis by wild-type p53 in a human colon tumor-derived cell line. Proc Natl Acad Sci USA 89: 4495–4499

Shi YF, Sahai BM, Green DR (1989) Cyclosporin A inhibits activation-induced cell death in T-cell and thymocytes. Nature 339: 625–626

Shi YF, Szalay MG, Paskar L, Sahai BM, Boyer M, Singh B, Green DR (1990) Activation-induced cell death in T-cell hybridomas is due to apoptosis. Morphologic aspects and DNA fragmentation [published erratum]. Immunol (1990) 145(11): 3945. J Immunol 144: 3326–3333

Shi Y, Glynn JM, Guilbert LJ, Cotter TG, Bissonnette RP, Green DR (1992) Role for c-myc in activation-induced apoptotic cell death in T cell hybridomas. Science 257: 212–214

Siegel RM, Katsumata M, Miyashita T, Louie DC, Greene MI, Reed JC (1992) Inhibition of thymocyte apoptosis and negative antigenic selection in bcl-2 transgenic mice. Proc Natl Acad Sci USA 89: 7003–7007

Strasser A, Harris AW, Cory S (1991) bcl-2 transgene inhibits T cell death and perturbs thymic self-censorship. Cell 67: 889–899

Strasser A, Harris AW, Cory S (1994) Positive and negative selection of T cells in T-cell receptor transgenic mice expressing a bcl-2 transgene. Proc Natl Acad Sci USA 91: 1376–1380

Suda T, Takahashi T, Golstein P, Nagata S (1993) Molecular cloning and expression of the Fas ligand a novel member of the tumor necrosis factor family. Cell 75: 1169–1178

Takahashi T, Tanaka M, Brannan C, Jenkins NA, Copeland NG, Suda T, Nagata S (1994) Generalized lymphoproliferative disease in mice, caused by a point mutation in the fas ligand. Cell 76: 969–976

Trauth BC, Klas C, Peters AM, Matzku S, Moller P, Falk W, Debatin KM, Krammer PH (1989) Monoclonal antibody-mediated tumor regression by induction of apoptosis. Science 245: 301–305

Ucker DS, Ashwell JD, Nickas G (1989) Activation-driven T cell death. I Requirements for de novo and translation and association with genome fragmentation. J Immunol 143: 3461–3469

Urdahl KB, Pardoll DM, Jenkins MK (1994) Cyclosporin A inhibits positive selection and delays negative selection in alpha-beta transgenic mice. J Immunol 152: 2853–2859

Vaux DL, Aguila HL, Weissman IL (1992) Bcl-2 prevents death of factor deprived cells but fails to prevent apoptosis in targets of cell mediated killing. Int Immunol 4: 821–824

Veis DJ, Sorenson CM, Shutter JR, Korsmeyer SJ (1993) Bcl-2-deficient mice demonstrate fulminant lymphoid apoptosis polycystic kidneys and hypopigmented hair. Cell 75: 229–240

Watanabe FR, Brannan CI, Copeland NG, Jenkins NA, Nagata S (1992a) Lymphoproliferation disorder in mice explained by defects in Fas antigen that mediates apoptosis. Nature 356: 314–317

Watanabe FR, Brannan CI, Itoh N, Yonehara S, Copeland NG, Jenkins NA, Nagata S (1992b) The cDNA structure, expression, and chromosomal assignment of the mouse Fas antigen. J Immunol 148: 1274–1279

Watson MA, Milbrandt J (1989) The NGFI-B gene, a transcriptionally inducible member of the steroid receptor gene superfamily: genomic structure and expression in rat brain after seizure induction. Mol Cell Biol 9: 4213–4219

Wilson TE, Fahrner TJ, Johnston M, Milbrandt J (1991) Identification of the DNA binding site for NGFI-B by genetic selection yeast. Science 252: 1296–1300

Wilson TE, Paulsen RE, Padgett KA, Milbrandt J (1992) Participation of non-zinc finger residues in DNA binding by two nuclear orphan receptors. Science 256: 107–110

Wilson TE, Fahrner TJ, Milbrandt J (1993a) The orphan receptors NGFI-B and steroidogenic factor 1 establish monomer binding as a third paradigm of nuclear receptor-DNA interaction. Mol Cell Biol 13: 5794–5804

Wilson TE, Mouw AR, Weaver CA, Milbrandt J, Parker KL (1993b) The orphan nuclear receptor NGFI-B regulates expression of the gene encoding steroid 21-hydroxylase. Mol Cell Biol 13: 861–868

Woronicz JD, Calnan B, Ngo V, Winoto A (1994) Requirement for the orphan steroid receptor Nur77 in apoptosis of T-cell hybridomas. Nature 367 (in press)

Wurm FM, Gwinn KA, Kingston RE (1986) Inducible overproduction of the mouse c-myc protein in mammalian cells. Proc Natl Acad Sci USA 83: 5414–5418

Wyllie AH (1980) Glucocorticoid-induced thymocyte apoptosis is associated with endogenous endonuclease activation. Nature 284: 555–556

Wyllie AH, Morris RG, Smith AL, Dunlop D (1994) Chromatin cleavage in apoptosis: association with condensed chromatin morphology and dependence on macromolecular synthesis. J Pathol 142: 67–77

Yonish RE, Resnitzky D, Lotem J, Sachs L, Kimchi A, Oren M (1991) Wild-type p53 induces apoptosis of myeloid leukaemic cells that is inhibited by interleukin-6. Nature 353: 345–347

Yuan J, Shaham S, Ledoux S, Ellis HM, Horvitz HR (1993) The C. Elegans cell death gene ced-3 encodes a protein similar to mammalian interleukin-1beta converting enzyme. Cell 75: 641–652

Zacharchuk CM, Mercep M, Chakraborti PK, Simons SJ, Ashwell JD (1990) Programmed T lymphocyte death Cell activation- and steroid-induced pathways are mutually antagonistic. J Immunol 145: 4037–4045

Zervos AS, Gyuris J, Brent R (1993) Mxil, a protein that specifically interacts with Max to bind Myc-Max recognition sites. Cell 72: 223–232

Molecular Events in Thymocyte Apoptosis

S.W. Smith[1], K.A. McLaughlin[2], and B.A. Osborne[1,2]

1 Introduction

Progenitor T cells arise in the bone marrow and are transported to the thymus gland where they continue to develop. During this period of development, immature cells of the T cell lineage begin to express cell surface markers such as CD4 and CD8. Prior to these events, the T cell receptor (TCR) undergoes rearrangements resulting in the potential expression of upwards of 10^9 different receptors. Therefore, it is important that the immune system has the ability to eliminate T cells that recognize self-antigens with high affinity. At the CD4$^+$CD8$^+$ stage of development, thymocytes encounter antigen-presenting cells which determine their ultimate fate. These double positive cells may go on to differentiate into a cytotoxic T cell, a helper T cell, or be deleted. One of the remaining questions in thymic development is the paradox of how immature T cells are both positively and negatively selected in an MHC restricted manner.

Two models have been suggested to explain this phenomenon: (1) the instructive model, which argues that recognition of self-peptide presented on different antigen presenting cells results in either positive or negative selection;

[1] Department of Veterinary and Animal Sciences, 309 Paige Lab, University of Massachusetts, Amherst, MA 01003, USA
[2] Department of Molecular and Cellular Biology, Morril Science Center, University of Massachusetts, Amherst, MA 01003, USA

(2) the affinity-avidity model, which implies that the type of selection corresponds to the strength with which the T cell recognizes self-antigen. In this model, thymocytes that recognize peptide-MHC with a low affinity or at a low concentration differentiate into single positive mature T cells. Thymocytes that recognize the peptide-MHC complex with high affinity or at high concentrations, die. Recent experiments have supported the affinity-avidity model of positive selection (ASHTON-RICKARDT et al. 1994; HOGQUIST et al. 1994; SEBZDA et al. 1994). Since a large number of immature thymocytes are eliminated by negative selection or non-functional rearrangement of the TCR, only about 5%–10% of cells differentiate into mature T cells (SCOLLAY et al. 1980). This deletion is known to occur by a form of programmed cell death called apoptosis (SWAT et al. 1991; SMITH et al. 1989; MURPHY et al. 1990). Apoptosis is characterized by chromatin condensation along the nuclear envelope, cleavage of the chromatin into 180–200 base pair fragments, and blebbing of the plasma membrane.

In addition to engagement of the TCR, apoptosis can be induced in thymocytes by a variety of different stimuli. Curiously, these cells are more susceptible to the induction of death by treatments that would not affect other cells. For example, corticosteroids levels that do not kill other cells and nonlethal levels of irradiation will induce apoptosis in thymocytes (WYLLIE 1980; NELIPOVICH et al. 1988).

These observations raise the question of whether the cell death program is partially engaged at this point in development, suggesting that the cells require a signal to survive. Specifically, a signal to differentiate, given during positive selection, may change the state of the cell temporarily, making it more resistant to cell death. Alternatively, it might be that the cells need to receive a death signal. Perhaps the high avidity signal the cells receive during negative selection actively engages the death program. Lastly, it is also conceivable that both these hypotheses are at work simultaneously. For example, it is known that the death program is still active in peripheral T cells maintaining homeostasis (KABELITZ et al. 1993). These data suggest the death program is available throughout the life of a T cell. However, it is also known that cells that do not receive a positive selection signal do not appear to make it into the periphery (BLACKMAN et al. 1990; MOMBAERTS et al. 1992; SHINKAI et al. 1992). The signal provided during positive selection might not be strong enough to engage the suicide program. Once terminal differentiation begins, there might be a change in the intracellular communication or an increase in some protective factor (possibilities to be discussed later in this chapter), which decrease the susceptibility of the cell. These changes might act to protect the cell from engagement of the death program during one developmental stage while keeping the program available for later use.

In the following paragraphs, this review will concentrate on what is known about the molecular mechanism of apoptosis in thymocytes. We will discuss the types of proteins that may be involved and review what is salient regarding genes that have been shown to be associated or directly involved in cell death induced via different mechanisms. Also, the issue of whether these different stimuli employ separate pathways to bring about the demise of the cells or if the pathways converge at some common point will be addressed.

2 Are There Multiple Mechanisms for Thymocyte Apoptosis?

Table 1 illustrates the ever-increasing number of agents that can induce apoptosis in thymocytes. At first glance, these agents are diverse and seemingly unrelated. At least three different hypotheses may be proposed to account for death in all these instances. The first hypothesis involves all the unrelated death inducers producing some common death signal. A second hypothesis proposes that the various inducers engage separate induction pathways but these pathways converge to form a single pathway leading to cell death. Lastly, different pathways of apoptotic death may occur by mechanisms unique to each inducing agent, implying that genes in each pathway are nonoverlapping. The third hypothesis seems the most unlikely because it would be inefficient for the cell to maintain so many different mechanisms to produce the same end result.

In looking for genes that might fit the criteria defined by the first hypothesis, *bcl*-2 is a gene that one might consider. As detailed below, however, *bcl*-2, while a critical regulator of many types of cell death, does not appear to be involved in all forms of death in T lymphocytes. *bcl*-2 was identified as a gene that is up-regulated in B cell follicular lymphomas due to a translocation that placed the gene in front of a strong immunoglobulin enhancer (TSUJIMOTO et al. 1985). More recently, it was shown that *bcl*-2 acts to prevent death (VAUX et al. 1988; HOCKENBERY et al. 1990). It is expressed as a membrane protein and is thought to play a role in preventing oxidative stress by somehow inhibiting free radical damage to membranes (HOCKENBERY et al. 1993). Studies examining the normal expression of *bcl*-2 in the thymus show a high level of expression in CD4$^-$CD8$^-$CD3$^-$ cycling thymocytes and single positive mature thymocytes, but a low level of expression in the double positive population (GRATIOT-DEANS et al. 1993; MOORE et al. 1994). Overexpression of *bcl*-2 inhibits cell death caused by

Table 1.   Inducers of apoptosis in thymocytes

Inducers	References
Antigen presentation	ASHWELL et al. 1987
Antibodies to TCR/CD3	SMITH et al. 1989
Calcium ionophores	WYLLIE et al. 1984
Phorbol esters	KIZAKI et al. 1989
Agents to elevate cAMP	McCONKEY et al. 1990
Glucocorticoids	WYLLIE 1980
Apo-I/Fas	TRAUTH et al. 1989
TNF-α	HERNANDEZ-CASELLES and STUTMAN 1993
γ-irradiation	SELLINS and COHEN 1987
Adenosine	KIZAKI et al. 1988
Epipodophyllotoxin	YE et al. 1993
Etoposide	CLARKE et al. 1993
5-fluorouracil	LOWE et al. 1993b
Adriamycin	LOWE et al. 1993b
2,3,7,8 tetrachlorodibenzo-*p*-dioxin	McCONKEY et al. 1988

TCR, T cell receptor.

irradiation, corticosteroids, and anti-CD3. However, the ability of this gene to inhibit negative selection is controversial. In most instances, in transgenic mice expressing *bcl*-2 in thymocytes, superantigens induce clonal deletion (STRASSER et al. 1991; SENTMAN et al. 1991; SIEGAL et al. 1992). These data suggest *bcl*-2 is not involved in negative selection.

If the Bcl-2 protein did inhibit all types of cell death, then a possible common thread resulting from the different inducers of apoptosis may be the generation of free oxygen radicals or other downstream events. If this idea is applied to the hypotheses listed above, then it may suggest that thymocytes are, in general, ill prepared to deal with free radicals. This vulnerability may reflect the existence of low enzyme levels in thymocytes that help scavenge free radicals or convert them to H_2O.

One idea that argues against the common signal hypothesis and more favorably for the convergence hypothesis is that there appear to be genes that are unique to specific pathways. The majority of these genes appear to be putative transcription factors, which suggests a molecular program for the induction of apoptosis. Although the generation of free radicals may be important in many cases, the early events leading to death are quite distinct. This argues strongly for the second hypothesis, that various inducers initially send the cells into the death program via different routes, but then the pathways converge and share distal events. We will present data that support this hypothesis and discuss the genes that have been identified as being involved or associated with apoptosis.

3 Primary Response Genes

Primary response genes are genes that are expressed very early after mitogenic stimulation and do not require de novo protein synthesis. In fact, they are superinduced with agents that inhibit protein synthesis, such as cycloheximide, suggesting posttranslational control of the expression levels of these genes. This group of genes includes a number of transcription factors and well known proto-oncogenes. Those who have studied cellular differentiation have recognized that this important class of genes appear to be the switches that determine differentiation, proliferation and growth arrest (NGUYEN et al. 1993; SELVAKUMARAN et al. 1992). Six primary response genes, c-*fos*, c-*jun*, c-*myc*, *nur*77, *RP*8 and *egr*-1, have been found to be associated with lymphoid apoptosis. c-fos and c-jun, regulate transcription by participation in AP-1 complexes, have been found to be important in cytokine deprivation in mature T cells (COLOTTA et al. 1992), and two others, c-*myc* and *nur*77 (EVAN et al. 1992; SHI et al. 1992; LIU et al. 1994; WORONICZ et al. 1994), have been found to be important in TCR mediated apoptosis in thymocytes. *RP*8 and *egr*-1, have not been identified as being essential for death to occur, but are up-regulated in response to certain inducers of apoptosis and shall be discussed briefly in this review. The number of primary response genes identified as being important suggests that these genes are likely to be at the forefront of the decision for the thymocyte to live or die.

c-*myc* was the first primary response gene to be identified as being involved in thymocyte apoptosis. c-Myc is a nuclear phosphoprotein that can bind DNA and activate transcription (SCHWEINFEST et al. 1988; KADDURAH-DAOUK et al. 1987). It is known to heterodimerize with a proteins like Max and Mad (BLACKWOOD and EISENMAN 1991; PRENDERGRAST et al. 1991; AYER et al. 1993), but also has been shown to bind cell cycle proteins like Rb in vitro (RUSTGI et al. 1991). Little is known about the role of c-Myc in cell death.

Two recent papers suggest that deregulation of c-*myc* influences the cell death decision. The first study examined the effect of over-expression of c-Myc in a rat-I fibroblast line. The authors were able to show that overexpression of c-Myc in serum deprived fibroblasts caused apoptosis (EVAN et al. 1992). The second study looked specifically at a T cell hybridoma line and demonstrated that if c-Myc expression is blocked through the use of antisense oligonucleotides, apoptosis could not be induced by stimulation through the TCR. However, this apoptotic block did not interfere with the signals through the TCR that influenced interleukin-2 (IL-2) production (SHI et al. 1992).

We, as well as others, have demonstrated that another primary response gene, *nur77*, also known as *NGFIB*, *NIO*, *TisI* and *Nak*-1, is required for apoptosis in T cells (LIU et al. 1994; WORONICZ et al. 1994). Nur77, a member of the steroid/ thyroid hormone receptor superfamily, is a zinc finger phosphoprotein that was identified in 3T3 cells as a gene expressed when the cells transcend the G_0/G_1 border (LAU and NATHANS 1987) and in PCI2 cells as a nerve growth factor inducible gene (WATSON and MILBRANDT 1989). Nur77 recently has been shown to bind to the estrogen receptor-like half site and activate transcription (WILSON et al. 1992). This receptor thus far has been described as an orphan receptor, meaning that no ligand has been identified as yet. We have shown that with *nur77* antisense expression, protein expression is reduced and death inhibited in a T cell line (LIU et al. 1994). Others have shown that overexpression of a dominant negative mutant of Nur77 inhibited apoptosis (WORONICZ et al. 1994). This protein was shown to bind its recognition element during the induction of death in a T cell hybridoma, thereby suggesting that the protein is actively participating in the regulation of transcription of certain genes during the apoptotic program (WORONICZ et al. 1994). Additionally we have shown that *nur77* expression is induced in thymocytes as they undergo negative selection in thymic organ cultures (S. Smith, L. Spain, L. Berg, B.A. Osborne, unpublished results). Taken together, the data suggest that *nur77* acts as a transcription factor and future studies to identify downstream genes should reveal other components of this cell death pathway. To date, the only recognized gene activated by Nur77 is steroid 21-hydroxylase in adrenocortical cells (WILSON et al. 1993).

Two primary response genes that are expressed during thymocyte apoptosis but have not been shown to be required are *egr*-1 and *RP*-8. The expression of *egr*-1 has been associated with TCR stimulation and cell death in thymocytes (S.W. Smith, unpublished). Egr-1, also known as NGFI-A, Krox24, zif/268 and

Tis8, is a nuclear phosphoprotein transcription factor with a TIIIA type zinc finger domain. It binds to the recognition sequence GCG(G/T)GGGCG, which is found in the promoter regions of numerous primary response genes such as c-*jun* and *nur77*, and within the *egr*-1 promoter (CHRISTY and NATHANS 1989; LEMAIRE et al. 1990; WILLIAMS and LAU 1993). The protein encoded by this gene has been implicated in activation (LEMAIRE et al. 1990). Egr-1 also has been demonstrated to be up-regulated in immature B cells upon stimulation through the immunoglobulin receptor (SEYFERT et al. 1990; MITTELSTADT and DeFRANCO 1993). RP8, a putative transcription factor that is superinduced by cycloheximide, is up-regulated within an hour after stimulation causing this gene to fall in our definition of primary response genes. RP8 was identified by J.J. Cohen and colleagues as a 1.3 kb transcript induced by γ-irradiation and dexamethasone stimulation. The sequence of the gene is not similar to any known family of genes, but it does appear to contain a zinc finger domain (OWENS et al. 1991). It is interesting that this gene appears to be shared by two very different pathways and its induction is an early event in the initiation of cell death. These considerations qualify this gene for inclusion on a list of those that may be important in death induced by radiation and dexamethasone. Clarification of its exact role awaits further investigation.

4 Steroid/Thyroid Hormone Receptor Superfamily

The steroid/thyroid hormone receptor family is a rapidly growing superfamily of genes characterized by their zinc finger and ligand binding domains. This family has been implicated in such diverse processes as development and differentiation, hormonal control, reproduction, and cholesterol biosynthesis (EVANS 1988). These nuclear receptors bind only one of two DNA recognition elements. They are able to establish fine specificity and a multitude of functions by competition, heterodimerization, and homodimerization with various members of this large family of proteins. These proteins can expand their function further by recognition of inverted or direct repeats and interactions with other transcription factors (FULLER 1991; BEATO 1989). These genes and their ligands have become known for their association or linkage to death in a wide variety of circumstances. Examples of nuclear receptor involvement can be seen in a wide variety of organisms and tissue types. The intersegmental muscle death in the moth *Manduca sexta* is triggered by a fall in the hormone ecdysone (TRUMAN and SCHWARTZ 1984), and prostate gland regression following castration is associated with a fall in androgen (MONPETITE et al. 1986; ISAACS 1984).

Most notably, two members of this family have been implicated directly in apoptotic mechanisms in thymocytes; *nur77*, as discussed earlier, and the glucocorticoid receptor (GR). The GR protein is localized in a stable complex with hsp90 in the nuclear membrane in an uninduced cell. It is activated by its ligand, corticosteroid, and in T cell lines or thymocytes will elicit apoptosis. Death by corticosteroids can be inhibited by calcium signals in thymocytes, such as those given during TCR stimulation, and by oxysterols, such as 25-hydroxycholesterol, in T cell leukemic lines (BAKOS et al. 1993). Dexamethasone induced death is enhanced by retinoic acid and by many protease inhibitors, such as E-64, leupeptin, acetyl-leucyl-normethionyl, diisopropyl fluorophosphate, and phenyl-methylsulfonyl fluoride (SARIN et al. 1993).

5 Cell Cycle and DNA Repair

The control of the cell cycle is a very interesting part of the apoptotic process. Following induction of apoptosis, thymocytes typically arrest at the G_1/S border (ASHWELL et al. 1987). However, research has shown that arrest at G_1 is not a prerequisite and that the cells can die at many different points in the cycle (COTTER et al. 1992). The role of cell cycle arrest may differ from cell type to cell type and depend on status of differentiation. What is known is that the cell cycle arrest occurs as an early event in cell death and that certain inhibitors will block death without relieving the cell cycle block (SARIN et al. 1993; IWATA et al. 1992). The involvement of proteins that regulate the cell cycle make some sense in that the chromatin condensation and breakdown of the nuclear envelope during apoptosis closely resemble situations that occur during mitotsis. Recently, an early activation of p34^{cdc2}, a serine-threonine kinase critical in controlling entry into mitosis, has been identified as being required for the apoptotic death that occurs in cytotoxic T lymphocyte target cells (SHI et al. 1994).

Proto-oncogenes and tumor suppressor genes that control the cell cycle in one fashion or another also have been found to be important in apoptosis. c-myc and p53 both have been implicated in apoptosis in thymocytes; however they are not activated by the same inducers. c-myc, as previously discussed, is essential for TCR induced apoptosis, while p53 has been shown to be required with other triggers for apoptosis. p53, a tumor suppressor gene, is believed to be responsible for arresting the cell in G_1 allowing DNA repair to occur (LANE 1992). Cells that do not contain functional p53 do not suffer growth arrest when irradiated (KASTAN et al. 1992). Many transformed cell lines have shed light upon the role of p53 in cell death. When a temperature sensitive p53 mutant was introduced into a murine myeloid leukemic cell line deficient for p53 protein, the expression of the wild-type protein in these cells induced an apoptotic death (YONISH-ROUACH et al. 1991). Furthermore, when p53 is expressed in this line, it does not arrest the cells at G_1 as might be expected. Instead, the cells arrest at multiple points in the cycle. This has suggested to some that death, in this case, may be occurring due to conflicting signals (YONISH-ROUASCH et al. 1993; CHIOU et al. 1994).

We, in collaboration with Scott Lowe and Tyler Jacks, have shown that in mice lacking the p53 gene due to targeted disruption, thymocytes are not able to enter the cell death program upon irradiation, but die normally by corticosteroids or agents that mimic TCR engagement (Lowe et al. 1993a; Clarke et al. 1993). Chemotherapeutic agents that also cause DNA damage, such as 5-fluorouracil, etoposide, or adriamycin, also are incapable of inducing apoptosis in cell lines from these mice suggesting a similar early mechanism of death centered around the p53 protein (Lowe et al. 1993b; Clarke et al. 1993). Bcl-2 recently has been shown to inhibit death in E1A transformed cells transfected with the *p53* temperature sensitive mutant (Chiou et al. 1994). While the exact role played by *p53* in cell death is still unknown, one current model is that expression of wild-type *p53* holds the cell in an arrested state until DNA perturbations that effect growth are repaired or adjusted; if DNA repair does not occur, the cell is induced to die.

Poly(ADP-ribose) polymerase is an enzyme that has been implicated in several of the same pathways as p53. It is a DNA repair enzyme important in lymphoid cell differentiation (Johnstone and Williams 1982) and activated when the 116 kDa pro-enzyme is proteolytically cleaved to release an 85 kDa fragment. Compounds that inhibit this cleavage inhibit the activation of Poly(ADP-ribose) polymerase and inhibit the endonucleolytic cleavage of the DNA (Kaufmann et al. 1993). This enzyme has been shown to be activated by irradiation, various chemotherapeutic agents, and dexamethasone. It functions by covalently attaching poly(ADP-ribose) polymers to broken strands of DNA and various nuclear proteins (Benjamin and Gill 1980). NAD^+ is the substrate for this enzyme and it has been postulated that it is the depletion of this substrate that leads to death (Hoshino et al. 1992). An inhibitor of the synthesis of the enzyme, 3-amino-benzamide, inhibits the reduction in NAD^+ pools seen during death and blocks death as measured by vital dyes without inhibiting the DNA fragmentation (Hoshino et al. 1993). Further investigation is required to determine if this model is correct.

6 Proteases and Proteolysis

Recently, a neglected but very important group of proteins with a wide range of activities have received a great deal of attention by cell death researchers. Proteases, the activators, regulators, and housekeepers of the cell, have been implicated in the programmed death of many cell types induced by a variety of stimuli. Many years have passed since granzymes and serine proteases were first shown to be released from cytotoxic T cells (CTLs) and enter target cells through pores punched in the membrane by perforin (Henkart 1985). This CTL targeted killing induces many of the same characteristics as classic apoptosis only it occurs at a much more rapid rate and does not seem to require macromolecular synthesis. Recently a mouse has been generated which is deficient in the granzyme B protease by targeted disruption. The CTLs from this animal are hindered in their ability to cause rapid cleavage of DNA in target cells, although death measured by lysis does not seem impaired by much. This suggests strongly that this serine protease is an important part of the signaling pathway to DNA fragmentation (HEUSEL et al. 1994). Proteases are important in thymocyte induced cell death as well. As mentioned previously, the agents that block the proteolytic cleavage of poly(ADP-ribose), such as tosyl-L-lysine chloromethyl ketone, tosyl-L-phenylalanine chloromethyl ketone, N-ethylmaleimide, and iodoacetamide, inhibit death (KAUFMANN et al. 1993). Inhibitors of calpain, cysteine, and serine proteases block the death a of a T cell hybridoma, 2B4, by anti-CD3 and anti-Thy-I (SARIN et al. 1993). Only leupeptin reversed the cell cycle block and inhibited death, suggesting two divergent pathways for cell cycle arrest and death. All the inhibitors used on this cell line enhanced dexamethasone mediated cell death suggesting an important role for proteases in the cross-talk between the signaling pathways of corticosteroid mediated and TCR induced cell death (SARIN et al. 1993).

Furthermore, a very exciting discovery is the identification of a protease required for cell death in the small free-living soil nematode *C. elegans*. This protease, encoded by a gene known as *ced-3*, has homology with a mammalian cysteine protease, IL-1β converting enzyme (ICE) (YUAN et al. 1993). ICE is required for the cleavage of the inactive pro-IL-Iβ 33 kDa protein into an active 17.5 kDa form. The overexpression of *ced-3* or ICE in rat-I fibroblasts induces cell death (MIURA et al. 1993).

An interesting proteolytic pathway that also appears to play a role in lymphocyte apoptosis involves the polyubiquination of proteins. This pathway is a nonlysosomal ATP-dependent multistep process in which a 76 amino acid ubiquitin molecule is, covalently attached to a protein. In most cases, this tag targets the protein for degradation by other enzymes in the pathway. The targets may be abnormal proteins or normal proteins that undergo rapid turnover like cyclin B, p53, c-Myc, and c-Fos (MURRAY et al. 1989, CIECHANOVER et al. 1991; CHOWDERY et al. 1994). Studies have shown that in lymphocytes induced to die by low levels of radiation, there are increased levels of ubiquitin RNA and increased ubiquitinated nuclear proteins (DELIC et al. 1993). The timing of this event appears to coincide with the cleavage of DNA into nucleosomal size pieces suggesting a possible role for ubiquitin in the changes of the state of the DNA. Ubiquitin has been shown to be essential in this process by the addition of antisense oligonucleotides to cells prior to irradiation, demonstrating that inhibition of ubiquitin protein levels inhibits irradiation induced apoptosis (DELIC et al 1993). Interestingly, ubiquitin levels have also been shown to be up-regulated in the intersegmental muscles of the tobacco hawkmoth *Manduca sexta* at the posteclosion stage, when the muscle cells are no longer needed and are undergoing programmed cell death (SCHWARTZ 1991).

7 Cell Surface Antigens

Surface molecules are important in a wide variety of activities such as adhesion, homing, identification, and signal transduction. We already have discussed the role of the TCR complex in transducing a signal in thymocytes which is responsible for a life or death decision, but there are many other molecules that may participate in this process. For example, costimulation may be required for death of thymocytes. In vitro studies of thymocytes have shown that the death program is engaged much more efficiently if costimulation is provided. In these studies, antibodies to the CD28 molecule and the Thy-I antigen were needed in conjunction with antibodies to the TCR complex for death to occur (PUNT et al. 1994; NAKASHIMA et al. 1993).

Still other cell death pathways can be engaged through receptors that are distinct from the TCR complex. A very interesting example of this is the Apo-I/Fas molecule, This receptor is a 48 kDa protein expressed at high levels on activated T cells and thymocytes with an intermediate level of TCR expression (DEBATIN et al. 1994). The ligand for Apo-I/Fas has recently been identified. It is called FasL and it is a member of the tumor necrosis factor (TNF) family (SUDA et al. 1993). It maps to chromosome 1 and is responsible for the observed phenotype in the *gld* mice (TAKAHASHI et al. 1994). Antibodies to Apo-I/Fas will induce cell death, and mice deficient in Apo-I have autoimmune problems, although thymocyte development appears to occur normally (DEBATIN et al. 1990; WATANABE-FUKUNAGA et al. 1992). These data suggest that Apo-I/Fas is important in T cell homeostasis and may play a role in thymocyte deletion.

The TNF receptor (TNFR) is a member of the same family as Apo-I/Fas. The cytokines TNF-α and lymphotoxin-α were first shown to induce apoptotic-like death in many transformed cell lines (CARSWELL et al. 1975). Now it is clear from a targeted deletion of lymphotoxin that these cytokines have a role in lymph node development and in inflammation (DE TOGNI et al. 1994). Both Apo-I/Fas and TNFR contain cytoplasmic domains that are required for induction of apoptotic activities triggered through these ligands (TARTAGLIA et al. 1993).

Receptors also have been implicated in the removal of dying cells and debris. It is important that dying cells be removed from areas where death is occurring and this is accomplished very efficiently by phagocytes. However, dying cells need a marker to be distinguished from the viable cells. The vitronectin receptor appears to serve this function for dying neutrophils. Antibodies to the vitronectin receptor will block the removal of the neutrophils from a dying population by macrophages (SAVILL et al. 1990). It is likely that thymocytes also have a surface molecule that mediates removal of dead cells.

Two other putative receptors have been identified recently as being associated with cell death in thymocytes. One of them, *PD*-1, is induced a few hours after stimulation through the TCR and has homology to the immunoglobulin family of genes (ISHIDA et al. 1992). The other one, *RP*-2, is thought to be a receptor because of its homology to integral membrane proteins. Its 1.8 kb mRNA is induced by dexamethasone and γ-irradiation within 2 h (Owens et al. 1991). No definitive role has been assigned to either of these genes as yet. One might speculate that their expression enhances or transduces new signals or serves as an identification tag for removal.

8 Conclusions

Very little is known about the signaling pathways leading to thymocyte apoptosis. Most of the work done on thymocyte death has been performed on T cell hybridomas, due to the difficulty in working with thymocytes in vitro. A good example is the expression experiments needed to establish a gene in the apoptotic pathway. These transfections have been done using hybridomas because of the difficulty in transfecting thymocytes. The disparity between a transformed cell and a normal thymocyte might produce misleading results. Of course, the ideal experimental system is to have a transgenic or knock-out mouse for the gene of interest, but these are not always available.

From the known available data concerning the molecular events that occur during apoptosis, a preliminary model of thymocyte death can be generated. Figure 1 presents a simplistic model for the molecular mechanism of apoptosis in thymocytes. The guiding hypothesis in the design of this model is that separate, early signaling events converge on a common pathway. We propose that the irradiation and the corticosteroid pathways converge early because the two pathways seem to share many genes in common. Bcl-2 can inhibit death induced by both of these agents and the two pathways share the induction of *RP*-2 and *RP*-8, as well as the activation of poly(ADP-ribose) polymerase.

The TCR pathway has an interesting relationship to the corticosteroid induced pathway: they are mutually antagonistic of each other (Zacharchuk et al. 1990; Iseki et al. 1991). The mechanism(s) for this inhibition are not understood at present; however one could postulate that members of the steroid/thyroid hormone family may be engaged in intracellular cross-talk. Evidence for cross-talk extends further when one uses inhibitors of the TCR pathway. Retinoic acid, a vitamin A metabolite important in development, inhibits TCR mediated death but enhances the corticosteroid pathway (Iwata et al. 1992). The protease inhibitors used by Henkart and colleagues also inhibit TCR induced cell death and enhance dexamethasone mediated death (Sarin et al. 1993).

Cell cycle arrest is portrayed in this model as a side pathway that is engaged as an early event. It appears to be separate from the actual cell death program because many inhibitors of death will reverse death without reversing the cell cycle arrest (Iwata et al. 1992; Sarin et al. 1993). The genes that have been found to be required or associated with death are also shown in Fig. 1. After the cell has made a commitment to die through the activation of genes unique to each pathway, the pathways converge. The cell then undergoes the morphological

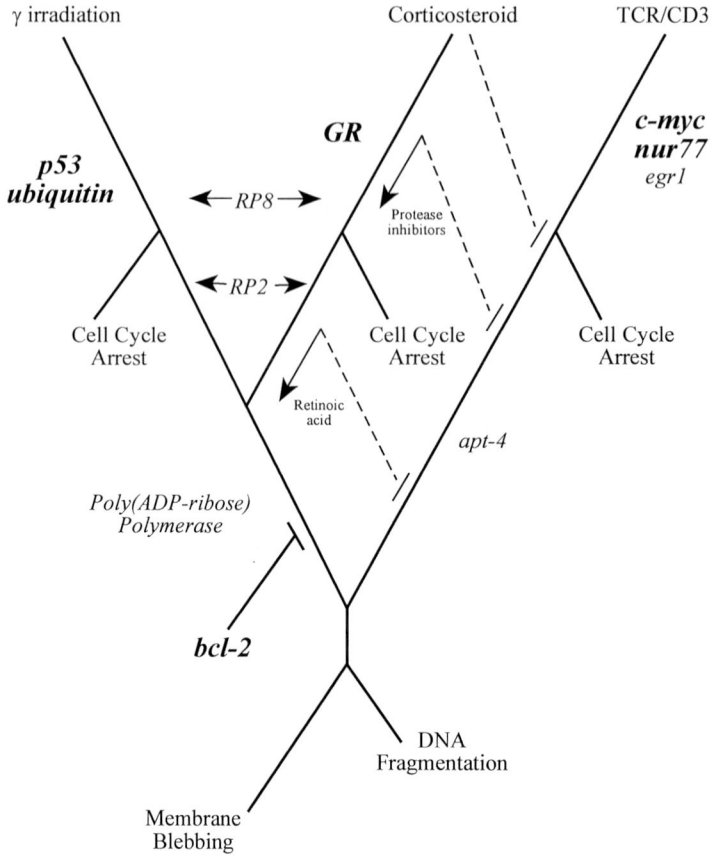

DEATH

Fig. 1. Converging pathways of thymocyte apoptosis. Apoptosis is induced by unique signaling events which converge and share a common pathway. Genes associated with apoptosis are placed next to the pathways. The genes that are known to be essential for death are printed in *bold type*. *Dotted lines* represent agents that inhibit the T cell receptor (TCR)/CD3 signaling pathway, such as corticosteroids, protease inhibitors and retinoic acid. *Solid lines* that run parallel to the corticosteroid pathway depict agents that enhance corticosteroid-induced death

changes characteristic of apoptosis, such as DNA fragmentation. The signals to undergo membrane blebbing and DNA fragmentation are separate and are illustrated as diverging because many cases exist in which one can see death without DNA fragmentation and, in one case, DNA fragmentation has been detected without death (HEUSEL et al. 1994; HOSHINO et al. 1992). The molecular events discussed in this review are just the initial pieces of the puzzle available to reveal the steps leading to apoptosis. It will be important, in future experiments, to identify the downstream genes of the known transcription factors to understand exactly how they function in the cell death pathway.

Acknowledgements. The authors wish to thank R. Goldsby, L. Grimm and B. Ganguly for their useful comments and L. Korpiewski for help in preparing the manuscript. This work was supported by NIH RO1 GM 47922.

References

Ashton-Rickardt PG, Bandeira A, Delaney JR, van Kaer L, Pircher H-P, Kinkernagel RM, Tonegawa S (1994) Positive and negative thymocyte selection induced by different concentrations of a single peptide. Science 263: 1615–1618

Ashwell JD, Cunningham RE, Noguchi PD, Hernandez D (1987) Cell growth cycle block of T cell hybridomas upon activation with antigen. J Exp Med 165:173–194

Ayer DE, Cratzner L, Eisenman RN (1993) Mad: a heterodimeric partner for Max that antagonizes Myc transcriptional activity. Cell 72: 211–222

Bakos JT, Johnson BH, Thompson EB (1993) Oxysterol-induced cell death in human leukemic T-cells correlates with oxysterol binding protein occupancy and is independent of glucocorticoid-induced apoptosis. J Steroid Biochem Mol Biol 46: 415–426

Beato M (1989) Gene regulation by steroid hormones. Cell 56: 335–344

Benjamin RC, Gill DM (1980) ADP-ribosylation in mammalian cell ghosts. Dependence of poly(ADP-ribose)synthesis on strand breakage in DNA. J Biol Chem 255: 10493–10501

Blackman M, Kappler J, Marrack P (1990) The role of the T cell receptor in positive and negative selection of developing T cells. Science 248:1335–1351

Blackwood EM, Eisenman RN (1991) Max: a helix-loop-helix protein that forms a sequence specific DNA binding complex with myc. Science 251: 1211–1217

Carswell EA, Old LJ, Kassel RL, Green S, Fiore N, Williamson B (1975) An endotoxin-induced serum factor that causes necrosis of tumors. Proc Natl Acad Sci USA 72: 366–370

Chiou S-K, Rao L, White E (1994) Bcl-2 blocks p53-dependent apoptosis. Mol Cell Biol 14: 2556–2563

Chowdary DR, Dermody JJ, Jha KK, Ozer HL (1994) Accumulation of p53 in a mutant cell line defective in the ubiquitin pathway. Mol Cell Biol 14: 1997–2003

Christy B, Nathans D (1989) DNA binding site of growth-factor-inducible protein zif268. Proc Natl Acad Sci USA 86: 8737–8741

Ciechanover A, DiGiuseppe JA, Bercovich B, Orian A, Richter JD, Schwartz AL, Brodeur GM (1991) Degradation of nuclear oncoproteins by the ubiquitin system in vitro. Proc Natl Acad Sci USA 88: 139–143

Clarke AR, Purdie CA, Harrison DJ, Morris RG, Bird CC, Hooper ML, Wyllie AH (1993) Thymocyte apoptosis induced by p53-dependent and independent pathways. Nature 362: 849–852

Colotta F, Polentarutti N, Sironi M, Mantovani A (1992) Expression and involvement of c-fos and c-jun proto-oncogenes in programmed cell death induced by growth factor deprivation in lymphoid cell lines. J Biol Chem 267: 18278–18283

Cotter T, Glyn J, Echeverri F, Green DR (1992) The induction of apoptosis by chemotherapeutic agents occurs in all phases of the cell cycle. Anticancer Res 12: 773–780

Debatin KM, Goldmann CK, Bamford R, Waldmann TA, Krammer PH (1990) Monoclonal-antibody-mediated apoptosis in adult T-cell leukaemia. Lancet 335: 497–500

Debatin K-M, Suss D, Krammer PH (1994) Differential expression of APO I on human thymocytes implications for negative selection? Eur J Immunol 24: 753–758

Delic J, Morange M, Magdelenat H (1993) Ubiquitin pathway involvement in human lymphocyte γ-irradiation-induced apoptosis. Mol Cell Biol 13: 4875–4883

De Togni P, Goellner J, Ruddle NH, Streeter PR, Fick A, Mariathasan S, Smith SC, Carlson R, Shornick LP, Strauss-Schoenberger J, Russell JH, Karr R, Chaplin DD (1994) Abnormal development of peripheral lymphoid organs in mice deficient in lymphotoxin. Science 264: 703–707

Evan GI, Wyllie AH, Gilbert CS, Littlewood TD, Land H, Brooks M, Waters CM, Penn LZ, Hancock DC (1992) Induction of apoptosis in fibroblasts by c-myc protein. Cell 69: 119–128

Evans RM (1988) The steroid and thyroid hormone receptor superfamily. Science 240: 889–895

Fuller PJ (1991) Steroid receptor superfamily: mechanisms of diversity. FASEB J 5: 3092–3099

Gratiot-Deans J, Ding L, Turka LA, Nunez G (1993) bcl-2 proto-oncogene expression during human T cell development. J Immunol 151: 83–91

Hernandez-Caselles T, Stutman O (1993) Immune functions of tumor necrosis factor. I. Tumor necrosis factor induces apoptosis of mouse thymocytes and can also stimulate or inhibit IL-6-induced proliferation depending on the concentration of mitogenic costimulation. J Immunol 151: 3999–4012

Henkart PA (1985) Mechanism of lymphocyte-modiated cytotoxicity. Annu Rev Immunol 3: 31–58

Heusel JW, Wesselschmidt RL, Shresta S, Russel JH, Ley TJ (1994) Cytotoxic lymphocytes require granzyme B for rapid induction of DNA fragmentation and apoptosis in allogeneic target cells. Cell 76: 977–987

Hockenbery DM, Nunez G, Milliman C, Schreiber RD, Korsmeyer SJ (1990) Bcl-2 is an inner mitochondrial membrane protein that blocks programmed cell death. Nature 348: 334–336

Hockenbery DM, Oltvai ZN, Yin XM, Milliman CL, Korsmeyer SJ (1993) Bcl-2 functions in an antioxidant pathway to prevent apoptosis. Cell 75: 241–251

Hogquist KA, Jameson SC, Heath WR, Howard JL, Bevan MJ, Carbone FR (1994) T cell receptor antagonist peptides induce positive selection. Cell 76: 17–27

Hoshino J, Beckmann G, Kroger H (1993) 3-aminobenzamide protects the mouse thymocytes in vitro from dexamethasone-mediated apoptotic cell death and cytolysis without changing DNA strand breakage. J Steroid Biochem Mol Biol 44: 113–119

Iseki R, Mukai M, Iwata M (1991) Regulation of lymphocyte apoptosis: signals for the antagonism between activation and glucocorticoid-induced death. J Immunol 147: 4286–4292

Ishida Y, Agata Y, Shibahara K, Honio T (1992) Induced expression of PD-1, a novel member of the immunoglobulin gene superfamily, upon programmed cell death. EMBO J 11: 3887–3895

Issacs JT (1984) Antagonistic effect of androgen on prostatic cell death. Prostate 5: 545–557

Iwata M, Mukai M, Nakai Y, Iseki R (1992) Retinoic acids inhibit activation-induced apoptosis in T cell hybridomas and thymocytes. J Immunol 149: 3302–3308

Johnstone AP, Williams GT (1982) Role of DNA Breaks and ADP-ribosyl transferase activity in eukaryotic differentiation demonstrated in human lymphocytes. Nature 300: 368–470

Kaddurah-Daouk R, Greene JM, Baldwin AS, Kingston RE (1987) Activation and repression of mammalian gene expression by the c-myc protein. Genes Dev 1: 347–357

Kabelitz D, Pohl T, Pechhold K (1993) Activation-induced cell death (apoptosis) of mature peripheral T lymphocytes. Immunol Today 14: 338–339

Kastan MB, Zhan Q, El-Deiry WS, Carrier F, Jacks T, Walsh W, Plunkett BS, Vogelstein B, Fornace AJ (1992) A mammalian cell cycle checkpoint pathway utilizing p53 and GADD45 is defective in Ataxia-Telangiectasia. Cell 71: 587–597

Kaufmann SH, Desnoyers S, Ottaviano Y, Davidson NE, Poirrer GG (1993) Specific proteolytic cleavage of poly(ADP-ribose) polymerase: an early marker of chemotherapy induced apoptosis. Cancer Res 53(17): 3976–3985

Kizaki H, Shimada H, Ohsaka F, Sakurada T (1988) Adenosine, deoxyadenosine, and deoxyguanosine induce DNA cleavage in mouse thymocytes. J Immunol 141: 1652–1657

Kizaki H, Tadakuma T, Odaka C, Muramatsu J, Ishima Y (1989) Activation of a suicide process in thymocytes through DNA fragmentation by Ca ionophore and phorbol esters. J Immunol 143: 1790–1794

Kyprianou N, English HF, Davidson NE, Issacs JT (1991) Programmed cell death during regression of the MCF-7 human breast cancer following estrogen ablation. Cancer Res 51: 162–166

Lane DP (1992) p53, guardian of the genome. Nature 358: 15–16

Lau L, Nathans D (1987) Expression of a set of growth-related immediate early genes in BALB/c 3T3 cells: coordinate regulation of c-fos or c-myc. Proc Natl Acad Sci USA 84: 1182–1186

Lemaire P, Vesque C, Schmitt J, Stunnenberg H, Frank R, Charnay P (1990) The serum-inducible mouse gene Krox-24 encodes a sequence-specific transcriptional activator. Mol Cell Biol 10: 3456–3467

Liu Z-G, Smith SW, McLaughlin KA, Schwartz LM, Osborne BA (1994) Apoptotic signals delivered through the T-cell receptor of a T-cell hybrid require the gene nur77. Nature 367: 281–284

Lowe SW, Schmitt EM, Smith SW, Osborne BA, Jacks T (1993a) p53 is required for radiation-induced apoptosis in mouse thymocytes. Nature 362: 847–849

Lowe SW, Ruley HE, Jacks T, Houseman DE (1993b) p53 dependent apoptosis modulates the cytotoxicity of anticancer agents. Cell 74: 957–967

McConkey DJ, Hartzel P, Duddy SK, Hakansson H, Orrenius S (1988) 2,3,7,8 tetrachlorodibenzo-p-dioxin kills immature thymocytes by Ca^{2+}-mediated endonuclease activation. Science 242: 256–258

McConkey DJ, Hartzel P, Offenius S (1990) Agents that elevate CAMP stimulate DNA fragmentation in thymocytes. J Immunol 145: 1227–1230

Mittelstadt PR, DeFranco AL (1993) Induction of early response genes by crosslinking membrane Ig on B lymphocytes. J Immunol 150: 4822–4832

Miura M, Zhu H, Rotello R, Hartweig EA, Yuan J (1993) Induction of apoptosis in fibroblasts by IL-Iβ-converting enzyme, a mammalian homolog of the C. elegans cell death gene ced-3. Cell 75: 653–660

Mombaerts P, Iacomini J, Johnson RS, Herrup K, Tonegawa S, Papaioannou VE (1992) Rag-1-deficient mice have no mature B and T lymphocytes. Cell 68: 869–877

Monpetit M, Lawless K, Tenniswood M (1986) Androgen repressed messages in the rat ventral prostate. Prostate 8: 25–36

Moore NC, Anderson G, Williams GT, Owen JJT, Jenkinson EJ (1994) Developmental regulation of bcl-2 expression in the thymus. Immunology 81: 115–119

Murphy KM, Heimberger AB, Loh DY (1990) Induction by antigen of intrathymic apoptosis of CD4+CD8+ TCR[lo] thymocytes in vivo. Science 250: 1720–1722

Murray AW, Soloman MJ, Kirshner MW (1989) The role of cyclin synthesis and degradation in the control of maturation promoting factor activity. Nature 339: 280–286

Nakashima I, Mei-Yi P, Hamaguchi M, Iwamoto T, Rahman SMJ, Zhang Y-H, Kato M, Ohkusu K, Katano Y, Yoshida T, Koga Y, Isobe K-I, Nagase F (1993) Pathway of signal delivery to murine thymocytes triggered by cocrosslinking CD3 and Thy-I for cellular DNA fragmentation and growth inhibition. J Immunol 151: 3511–3520

Nelipovich PA, Nikonova LV, Umansky SR (1988) Inhibition of poly(ADP-ribose) polymerase as a possible reason for activation of Ca^{2+}/Mg^{2+} endonuclease in thymocytes of irradiated rats. Int J Radiat Biol 53: 749–765

Nguyen HQ, Hoffman-Liebermann B, Liebermann DA (1993) The zinc finger transcription factor egr-1 is essential for and restricts differentiation along the macrophage lineage. Cell 72: 197–209

Owens GP, Hahn WE, Cohen JJ (1991) Identification of mRNAs associated with programmed cell death in immature thymocytes. Mol Cell Biol 11: 4177–4188

Prendergrast GC, Lawe D, Ziff EB (1991) Association of myn, the murine homolog of max, with c-myc stimulates methylation sensitive DNA binding and Ras cotransformation. Cell 65: 395–407

Punt JA, Osborne BA, Takahama Y, Sharrow SO, Singer A (1994) Negative selection of CD4+CD8+ thymocytes by T cell receptor-induced apoptosis requires a costimulatory signal that can be provided by CD28. J Exp Med 179: 709–713

Rustgi AK, Dyson N, Bernards R (1991) Amino-terminal domain of c-myc and N-myc proteins mediate binding to the retinoblastoma gene product. Nature 352: 541–544

Sarin A, Adams DH, Henkart PA (1993) Protease inhibitors selectively block T cell hybridoma and activated peripheral T cells. J Exp Med 178: 1693–1700

Savill J, Dransfield I, Hogg N, Haslett C (1990) Vitronectin receptor-mediated phagocytosis of cells undergoing apoptosis. Nature 343: 170–173

Schwartz LM (1991) The role of cell death genes during development. Bioessays 13: 389–395

Schweinfest CW, Fujiwara S, Lau LF, Papas TS (1988) c-myc can induce expression of G_o/G_I transition genes. Mol Cell Biol 8: 3080–3087

Scollay RG, Butcher EG, Weissman IL (1980) Thymus cell migration: quantitative aspects of cellular traffic from the thymus to the periphery in mice. Eur J Immunol 10: 210–218

Sebzda E, Wallace VA, Mayer J, Yeung RSM, Mak TW, Ohashi PS (1994) Positive and negative thymocyte selection induced by different concentrations of a single peptide. Science 263: 1615–1618

Sellins KS, Cohen JJ (1987) Gene induction by gamma-irradiation leads to DNA fragmentation in lymphocytes. J Immunol 139: 3199–3206

Selvakumaran M, Liebermann DA, Hoffman-Liebermann B (1992) Deregulated c-myb disrupts interleukin 6 or leukemia inhibitory factor-induced myeloid differentiation prior to c-myc: role in leukemogenesis. Mol Cell Biol 12: 2493–2500

Sentman CL, Shutter JR, Hockenbury D, Kanagawa O, Korsmeyer SJ (1991) Bcl-2 inhibits multiple forms of apoptosis but not negative selection in thymocytes. Cell 67: 879–888

Seyfert VL, McMahon S, Glenn W, Cao X, Sukhatme VP, Monroe JG (1990) egr-1 expression in surface Ig-mediated B cell activation. J Immunol 145: 3647–3653

Shi L, Nishioka WK, Th'ng J, Bradbury EM, Litchfield DW, Greenberg AH (1994) Premature p34[cdc2] activation required for apoptosis. Science 263: 1143–1145

Shi Y, Glynn JM, Guilbert LJ, Cotter TG, Bissonette RP, Green DR (1992) Role of cmyc in activation induced apoptotic cell death in T cell hybridomas. Science 257: 212–214

Shinkai Y, Rathbun B, Alt F, Ham K-P, Oltz EM, Stewart V, Mendelsohn M, Chaffon J, Datta M, Young F, Stall AM (1992) Rag-2-deficient mice lack mature lymphocytes owing to inability to initiate V(D)J rearrangement. Cell 68: 855–867

Siegal RM, Katsumata M, Miyashita T, Louie DC, Greene MI, Reed JC (1992) Inhibition of thymocyte apoptosis and negative selection in bcl-2 transgenic mice.Proc Natl Acad Sci USA 89: 7003–7007

Smith CA, Williams GT, Kingston R, Jenkinson EJ, Owen JJT (1989) Antibodies to CD3/T-cell receptor complex induce death by apoptosis in immature T cells in thymic cultures. Nature 337: 181–184

Strasser A, Harris AW, Cory S (1991) bcl-2 transgene inhibits T cell death and perturbs thymic self censorship. Cell 67: 889–899

Suda T, Takahashi T, Golstein P, Nagata S (1993) Molecular cloning and expression of the Fas ligand a, novel member of the tumor necrosis family. Cell 75: 1169–1178

Swat W, Ignatowicz L, von Boehmer H, Kisielow P (1991) Clonal deletion of immature CD4$^+$CD8$^+$ thymocytes in suspension culture by extrathymic antigen presenting cells. Nature 351: 150–153

Takahashi T, Tanaka M, Brannan CI, Jenkins NA, Copeland NG, Suda T, Nagata S (1994) Generalized lymphoproliferative disease in mice, caused by a point mutation in the Fas ligand. Cell 76: 969–976

Tartaglia L, Ayres T, Wong G, Goeddel D (1993) A novel domain within the 55 kD TNF receptor signals cell death. Cell 74: 845–850

Trauth BC, Klas C, Peters AMJ, Matzku S, Moller P, Fa W, Debatin K-M, Krammer PH (1989) Monoclonal antibody-mediated tumor regression by induction of apoptosis. Science 245: 301–305

Truman JM, Schwartz LM (1984) Steroid regulation of neuronal death in the moth nervous system. J Neurosci 4: 274–280

Tsujimoto Y, Gorham J, Cossman J, Jaffe E, Croce C (1985) The T(14;18) chromosome translocations involved in B cell neoplasms result from mistakes in VDJ joining. Science 229: 1390–1393

Vaux DL, Cory S, Adams JM (1988) bcl-2 gene promotes haemopoietic cell survival and cooperates wkth c-myc to immortalize pre-B cells. Nature 335: 440–442

Watanabe-Fukunaga R, Brannan CI, Copeland NG, Jenkins NA, Nagata S (1992) Lymphoproliferative disorder in mice explained by defects in Fas antigen that mediates apoptosis. Nature 356: 314–317

Watson MA, Milbrandt J (1989) The NGFI-B gene, a transcriptionally inducible member of the steroid receptor gene superfamily: genomic structure and expression in rat brain after seizure induction. Mol Cell Biol 9: 4213–4219

Williams GT, Lau LF (1993) Activation of the inducible orphan receptor gene nur77 by serum growth factors: dissociation of immediate early and delayed early responses. Mol Cell Biol 13: 6124–6136

Wilson TE, Paulsen RE, Padgett KA, Milbrandt J (1992) Participation. of non-zinc finger residues in DNA binding by two nuclear orphan receptors. Science 256: 107–110

Wilson TE, Mouw AR, Weaver CA, Milbrandt J, Parker KL (1993) The orphan receptor NGFI-B regulates expression of the gene encoding steroid 21-hydroxylase. Mol Cell Biol 13: 861–868

Woronicz JD, Cainan B, Ngo V, Winoto A (1994) Requirement for the orphan steroid receptor Nur77 in apoptosis of T-cell hybridomas. Nature 367: 277–281

Wyllie AH (1980) Glucocorticoid induced thymocyte apoptosis is associated with endogenous endonuclease activation. Nature 284: 555–556

Wyllie AH, Morris RG, Smith AL, Dunlop D (1984) Chromatin cleavage in apoptosis: association with condensed chromatin morphology and dependence on macromolecular synthesis. J Pathol 142: 67–77

Ye X, Georgoff I, Fleisher S, Coffman FD, Cohen S, Fresak K (1993) The machanism of epipodophyllotoxin-induced thymocyte apoptosis: possible role for a novel Ca^{2+} independent protein kinase. Cell Immunol 151: 320–335

Yonish-Rouach E, Resnitzky D, Lotem J, Sachs L, Kimchi A, Oren M (1991) Wild-type p53 induces apoptosis of myeloid leukaemic cells that is inhibited by interleukin 6. Nature 352: 345–347

Yonish-Rouach E, Grunwald D, Wilder S, Kimchi A, May E, Lawrence JJ, May P, Oren M (1993) p53-mediated cell death: relationship to cell cycle control. Mol Cell Biol 13: 1415–1423

Yuan J, Shaham S, Ledoux S, Ellis HM, Horvitz HR (1993) The C. elegans cell death gene ced-3 encodes a protein similar to mammalian interleukin-Iβ converting enzyme. Cell 75: 641–652

Zacharchuk CM, Mercep M, Chakraborti PK, Simons SS, Ashwell JD (1990) Programmed T lymphocyte death-cell activation- and steroid-induced pathways are mutually antagonistic. J Immunol 145 4037–4045

Tissue Transglutaminase: A Candidate Effector Element of Physiological Cell Death

M. Piacentini

Dedicato a mia Madre in occasione del suo settantesimo compleanno

1 Introduction

The metazoans possess a genetic program of cell death (defined morphologically as apoptosis) which plays a vital role in their development and maintainance of tissue homeostasis (WYLLIE et al. 1980; FESUS et al. 1991b). Over the last few years, it has become clear that cells not only control their proliferative and differentiative pathways, but also need specific positive stimuli to survive (FESUS et al. 1991b; RAFF et al. 1992). In the absence of the appropriate survival signals, cells enter an active program of cell death which requires the participation of functionally distinct sets of genes (ARENDS and WYLLIE 1991; FESUS et al. 1991b). Despite its widespread importance, little is yet known about the biochemical events leading to the physiological deletion of cells in tissues (ARENDS and WYLLIE 1991; FESUS et al. 1991b). While the mechanisms of initiation and regulation of apoptosis have attracted the attention of many researchers, by far less interest has been focused on the effector genes that determine the irreversible phenotypical changes and clearance of the dying cells. These final events allows naturally

Department of Biology, University of Rome "Tor Vergata", Via della Ricerca Scientifica, 00133 Rome, Italy occurring cell death to behave as a "social" phenomenon which does not produce damage or inflammation in tissues (ARENDS and WYLLIE 1991; FESUS et al. 1991b). Several independent laboratories have shown tissue transglutaminase (tTG) to be a potentially important player in the last stage of the cell death program (for review see PIACENTINI et al. 1994). This gene is specifically expressed in cells dying during mammalian development and in those undergoing apoptosis in various physiological and experimental settings (FESUS et al. 1991b; PIACENTINI et al. 1994). This chapter reviews recent studies concerning the expression and the possible role of tTG in apoptotic cells; particular emphasis is given to tTG expression in the cell death pathways described under normal and pathological conditions in the immune system.

2 Transglutaminases

Transglutaminases belong to a gene family coding for intracellular and extra-cellular enzymes catalyzing Ca^{2+}-dependent cross-linking reactions between polypeptide chains (FOLK 1980). The reaction determines the posttranslational modification of proteins by establishing $\varepsilon(\gamma$-glutamyl)lysine cross-linkings and/or by the covalent incorporation of di- and polyamines and histamine (FOLK 1980; FESUS et al. 1985; PIACENTINI et al. 1988). Diamines and polyamines may also participate in cross-linking reactions through the formation of N, N-bis(γ-glutamyl) polyamine bonds (FOLK 1980; PIACENTINI et al. 1988). The number of natural glutaminyl substrates identified in cells is very low, while several suitable acyl acceptor protein substrates have been characterized (GREENBERG et al. 1992). The transglutaminase-catalyzed cross-linking reaction leads to protein polymerization; the polypeptides included in the polymer can be destroyed only by proteolytic degradation of protein chains (FOLK and FINLAYSON 1977; GREEN 1977; FOLK 1980). In fact, endoproteases capable of hydrolyzing the cross-links formed by transglutaminases have not been described in vertebrates; even the lysosomes do not express enzymes able to split the $\varepsilon(\gamma$-glutamyl)lysine bonds (FOLK 1980; FESUS et al. 1989, 1991b).

Several distinct transglutaminase gene products have been characterized so far both in vertebrates and invertebrates (LORAND et al. 1981; WILLIAMS-ASHMAN 1984; ICHINOSE et al. 1986; IKURA et al. 1988; GENTILE et al. 1991; KLEIN et al. 1992; KIM et al. 1993). Although the various TG forms seem to be involved in apparently different phenomena (blood coagulation, wound healing, terminal differentiation and cell death by apoptosis), all of them converge toward the protection of cell and tissue integrity (FESUS et al. 1991b; PIACENTINI et al. 1994).

3 Tissue Transglutaminase and Apoptosis

Tissue transglutaminase, or type II transglutaminase, is a cytosolic protein of about 80 kDa (its molecular weight is slightly different among different species) which has been shown to selectively accumulate in cells undergoing death by apoptosis (FESUS et al. 1991b; PIACENTINI et al. 1994). The onset of apoptosis is generally associated with a large increase in the tTG mRNA level followed by an enhancement of enzyme synthesis and of cross-linking activity (FESUS et al. 1987, 1989; PIACENTINI et al. 1991a, b; KNIGHT et al. 1991; STRANGE et al. 1992). Upon apoptosis induction, tTG mRNA in a given cell population may reach ten fold higher level than found in controls (PIACENTINI et al. 1991a); however, in the single apoptotic cell, the increase of tTG mRNA level is much higher (PIACENTINI et al. 1994). The experimental models in which the enzyme has been shown to increase in apoptotic cells include the best characterized in vivo and in vitro cell death systems (see following sections; FESUS et al. 1991a; PIACENTINI et al. 1994).

As mentioned above, tTG catalyzes the formation of protein cross-links which are biologically irreversible. This irreversibility has created difficulties in our understanding of the physiology of tTG action inside living cells. In fact, it is conceivable to suppose that the assembly of irreversible protein polymers could not take place in proliferating cells. It is now clear that this is indeed the case, and the cross-linked proteins accumulate in nondividing, terminally differentiated cells, such as keratinocytes and chondrocytes, and in cells undergoing death by apoptosis (GREEN 1977; FESUS et al. 1991b; AESCHLIMANN et al. 1993; PIACENTINI et al. 1994). tTG is a Ca^{2+}-dependent enzyme that is not active at the Ca^{2+} levels normally detected in viable cells; however, the rise of Ca^{2+} concentration reported in cells undergoing apoptosis is sufficient to activate the enzyme (FOLK 1980; FESUS et al. 1987, 1989; KNIGHT et al. 1991; PIACENTINI et al. 1991a, b). The activation of tTG protein in dying cells results in the assembly of highly cross-linked intracellular protein nets which are stabilized by both spermidine-derived and ε(γ-glutamyl)lysine containing cross-links (FESUS et al. 1989; PIACENTINI et al. 1991 b). The large amount of cross-links confer to these intracellular polymeric structures striking physicochemical properties such as insolubility in SDS and chaotropic agents (FESUS et al. 1989; PIACENTINI et al. 1991b). Biochemical characterization of these cross-linked protein scaffolds revealed that they contain several known intracellular proteins (actin, annexin II, vinculin, fibronectin, involucrin), some unknown proteins (KNIGHT et al. 1993a; TARCSA et al. 1993) and DNA cleaved into oligonucleosomes (FESUS et al. 1989).

On such a premise one wonders whether the expression of tTG in a cell might be considered as a biochemical marker of a preapoptotic stage. The tTG gene is constitutively expressed in a few cell types localized in specific mammalian tissues (endothelial cells, smooth muscle cells and mesangial cells; THOMAZY and FESUS 1989). A simplistic conclusion drawn from these findings would be that the presence of the tTG protein in a cell cannot per se be considered as indicative of apoptosis. However, under physiological conditions tTG gene is not expressed in

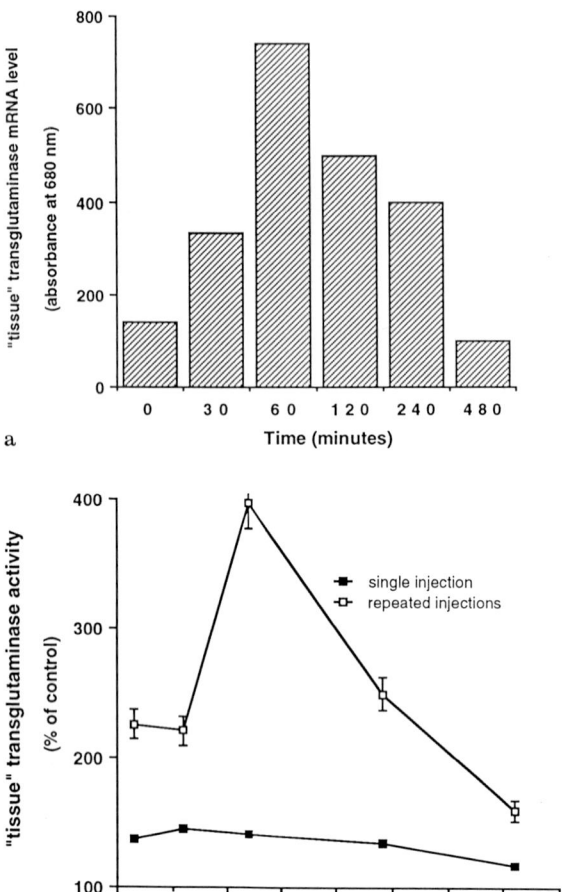

Fig. 1a, b. Effect of dexamethasone on tissue transglutaminase (tTG) mRNA (**a**) and enzyme activity (**b**) levels of normal thymocytes. **a** A suspension of freshly prepared rat thymocytes were incubated in complete medium in the presence of 10^{-6} *M* dexamethasone. At the reported time intervals, total RNA was extracted and the tTG mRNA level measured by northern bolt analysis, as previously described (PIACENTINI et al. 1991a). The amount of tTG mRNA, as determined by recalculating the absorbance values obtained from densitometric analysis of the northern blot bands (using total RNA), is shown. **b** Dexamethasone was administered intraperitonealy into adult mice (0.5 mg/kg body weight) in a single daily injection for 2 consecutive days. At the reported time intervals, the thymuses were collected, extensively washed in PBS and homogenized as previously described (MASTINO et al. 1992). tTG activity was measured as pmoles of [^{3}H] putrescine incorporated into protein/hour per mg protein and expressed as a percentage of values obtained in mice treated with control diluent. Data are the mean ± S.E.M. of triplicate determinations carried out in three different experiments.

the majority of cells and its mRNA is transcribed as a consequence of the induction of apoptosis (Fig. 1a; PIACENTINI et al. 1991a). It is not known what role, if any, could be played by tTG in those cells which constitutively express the enzyme. Nevertheless, these cells are localized in tissue areas exposed to environmental and functional stress (THOMAZY and FESUS 1989); hence, to avoid harmful conse-quences, they might have the apoptotic machinery ready to act whenever their integrity is affected. In keeping with this hypothesis are the findings indicating that, in some cases, apoptosis can take place in the absence of protein synthesis in anucleated cells (RAFF et al. 1992). Thus, it is conceivable to suppose that during evolution the ancestor tTG gene acquired cell type-specific regulation which allows preventive accumulation of the enzyme in cells particularly exposed to environmental stress. It must be recalled that the tTG protein can be posttranslationally regulated and is inactive in microenvironments which have low Ca^{2+} and high GTP levels (GREENBERG et al. 1992). These findings raise the important question of the regulation of the effector genes during the last stages of apoptosis. It is well known that apoptosis can be induced in different cells by a wide spectrum of nontoxic and toxic stimuli (ARENDS and WYLLIE 1991; FESUS et al. 1991b). This finding could imply that regulation of the putative killer genes has multiple accesses or that different lethal hits end in a common signaling pattern. It has been shown that de novo transcription of the tTG gene is induced by several factors, including transforming growth factor-β (TGF-β) retinoic acid, prostaglandin E2 (PGE2) and interleukin-6 (IL-6), which also modulate apoptosis (FESUS et al. 1991b; GREENBERG et al. 1992; SUTO et al. 1993; PIACENTINI et al. 1994). This multiple regulation might be typical of the effector elements of programmed cell death. The recent cloning and sequencing of the 5' flanking region of the tTG gene seems to confirm this hypothesis. In fact, the sequence analysis of a 2.3 kb fragment of the cloned genomic DNA revealed the presence of potential binding sites for several regulatory factors (SUTO et al. 1993), suggesting that transcription of a putative apoptotic effector gene can be controlled by a multifunctional promoter. Future studies should identify the consensus sequences for these factors by studying the genomic regulatory regions of genes involved in apoptosis.

Have the "cell death genes" acquired multiple functions during evolution? Recent findings indicate that internucleosomal cleavage of DNA, which so far has been considered the hallmark of apoptosis, might not be due to a specific endonuclease but could result by activation of DNase I (PEITSCH et al. 1993) expressed in a wide variety of cells. This is also the case with other putative apoptotic genes such as $p34^{cdc2}$, c-myc, SGP-2, and the vitronectin and the asialoglycoprotein receptors (SHI et al. 1994; EVAN et al. 1992; BUTTYAN et al. 1989; SAVILL et al. 1990; DINI et al. 1992) which are known to carry out additional functions.

4 Tissue Transglutaminase Expression and Apoptosis in the Immune System

4.1 Thymic Selection

Apoptosis is the cellular mechanism involved in elimination of immature thymo-cytes during both positive and negative selection of T lymphocytes (COHEN et al. 1992). In thymocytes the death program can be triggered by several exogenous stimuli such as glucocorticoids, removal of growth factors, exposure to γ-irradia-tion, antigen binding involving the CD3/T cell receptor (TCR), and cell surface treatment by anti-APO1/FAS antibody, thus suggesting that T cells are function-ally inclined to undergo cell death (COHEN et al. 1992). Figure 1 shows that tTG gene expression is rapidly induced in thymocytes undergoing apoptosis upon both in vivo and in vitro treatment with glucocorticoids.

So far, the intrinsic mechanisms and the biochemical mediators of naturally occurring cell death in the thymus have not been completely identified. In particular, very little is known about the intrinsic and/or environmental signals which can form the basis of the default death of immature (CD4+/CD8+) thymocytes in the thymus cortex. In fact, glucocorticoids seem not to be produced inside the thymus and is not clear how TCR-mediated signals could result in a survival signal during positive selection and a death impulse in the negative selection (PALMER et al. 1993). Increasingly important is the "dialogue" occurring in the thymus between the T and stromal cells (RITTER and BOYD 1993).

Table 1. Summary of tissue transglutaminase expression in cells undergoing apoptosis in the immune system

Cell	Apoptotic inducer	Tissue transglutaminase	Reference
In vivo			
Thymocytes	Glucocorticoid	Activity	Fig.1
Thymocytes	Prostaglandin E2	Protein and activity	MASTINO et al. 1992
Peripheral blood lymphocytes	HIV infection	Protein	Unpublished observation
In vitro			
Thymocytes	Glucocorticoids	mRNA	Fig. 1
CD4+ lymphocytes	HIV and prostaglandin E2	Protein and activity	MASTINO et al. 1993
CD4+ lymphocytes	Antigen	Protein and activity	AMENDOLA et al. 1994
CD4+ lymphocytes	HIV-gp120	Protein and activity	AMENDOLA et al. 1994
Monocytes/macrophage	HIV and M-CSF	Protein	BERGAMINI et al. 1994
U937	HIV	Protein	Unpublished observation
CBMC	Interleukin-2 withdrawal	Protein	Unpublished observation
Lymphoid lines	Anti-FAS antibody	Not expressed	L. Fesus, personal communication
Erythroleukemic cells	Natural Killer cells or cytotoxic lymphocytes	Activity	KNIGHT et al. 1993b

The nonlymphoid cells (epithelial, dendritic, macrophages and nurse cells) create specific microenvironments in which the developing thymocytes are selected on the basis of direct cell–cell interaction and/or by the action of soluble molecules (RITTER and BOYD 1993). Among these there are PGE2, which is produced by a variety of stromal cells (MASTINO et al. 1992). The administration of a synthetic analogue of PGE2 causes a selective and dramatic depletion of CD4$^+$/CD8$^+$ (double positive), CD3/ TCR$^{\alpha\beta}$ cells by apoptosis in the thymus of young mice (MASTINO et al. 1992). It is noteworthy that tTG enzyme activity is increased over the control level as early as 3 h upon PGE2 treatment; the PGE2 effect is thymus-specific, being the induction of tTG not detectable in any other organs (Table 1). The increased enzyme activity is due to a large induction of tTG in several cells localized in the thymus cortex showing the distinctive features of apoptosis (condensed chromatin and nuclear fragmentation) (MASTINO et al. 1992).

4.2 Activation-Induced Apoptosis

The first report demonstrating involvement of tTG in the immune system was that of NOVOGRODSKY et al. (1978), who showed increased enzymatic activity upon concanavalin A and phytohemagglutinin-induced proliferation of peripheral blood lymphocytes. This observation led to the hypothesis that tTG could have a role in cell proliferation, subsequent studies contradicted this early conclusion and showed that tTG expression is reduced in proliferating vs nonproliferating cells (PIACENTINI et al. 1991b). The increased expression of tTG during lymphocyte activation can now be reinterpreted on the basis of recent findings, showing that apoptosis is triggered in lymphocytes at various stages including antigen presentation. In fact, it has recently been shown that CD4$^+$ T cells, from established T cell clones, undergo apoptosis after exposure to antigen (KABELITS et al. 1993; AMENDOLA et al. 1994; CRITCHFIELD et al. 1994). The mechanism through which MHC-T cell receptor (TCR) interaction primes cell for apoptosis is not clear; however, steric perturbation of the CD4–TCR interaction may be responsible. (DIAMOND et al. 1990), since in immature thymocytes (SMITH et al. 1989) and mature murine T cells and hybridomas (NEWELL et al. 1990), the induction of apoptosis mediated by TCR stimulation is very likely a consequence of incomplete signal transduction (McCONKEY et al. 1990). Interesting features of lymphocyte apoptosis can be deduced by the involvement of tTG in the antigen-specific death (AMENDOLA et al. 1994) (Table 1): Whereas resting T cells do not show any tTG protein, after antigen exposure, the enzyme accumulates in the cytoplasm of a subset of stimulated cells. In T cell clones the synthesis of tTG protein precedes the appearance of the typical apoptotic phenotype, which occurs only when these "primed" cells receive additional CD3-transduced signals (AMENDOLA et al. 1994). Interestingly, cyclosporin A (CsA) markedly reduces the activation-induced apoptosis in T cell clones, leading to the accumulation of primed tTG positive cells. The CsA-dependent inhibition of antigen-induced apoptosis is likely due to blockade of the effector elements involved in activation.

In fact, CsA is an immunosuppressive agent that blocks T cell activation by preventing lymphokine production and by interfering with TCR-mediated Ca^{2+} signal transduction (BALDARI et al. 1991). Only a sustained increase in the intra-cellular Ca^{2+} level triggers the irreversible commitment to death by activating effector "killer" genes which in turn modify the structure of the cells toward that typical of apoptosis (PIACENTINI et al. 1994). These findings seem to suggest that accumulation of tTG in the cytoplasm, in the absence of the extreme apoptotic phenotype, could indeed highlight a preapoptotic stage.

AIDS can be considered as an activation-induced disease, and immunosup-pressors such as CsA have already been introduced in its therapy (AMEISEN and CAPRON 1991). The progressive loss of CD4+ helper T cells in the late stages of HIV infection is the major mechanism by which HIV induces immunodeficiency (FAUCI 1988; MEYAARD et al. 1992). Increasing importance is now being attributed to alternative pathogenic factors in HIV infection such as the binding of circulating gp120 to CD4+ receptor, cell-mediated immunoresponse and autoimmunity (NEWELL et al. 1990; LAURENT-CRAWFORD et al. 1991; TERAI et al. 1991; GOUGEON and MONTAGNIER 1992). T cells after preincubation with gp120 in vitro become refrac-tory to stimulation through the TCR (DIAMOND et al. 1990). Several mechanisms could account for gp120-induced inhibition. It has been recently reported that, when preceded by ligation of CD4, signaling through TCR results in T cell unresponsiveness which elicits cell death of mature T cells by apoptosis (GROUX et al. 1992). We have recently demonstrated that the binding of gp120 to CD4 molecules induces the expression of effector elements of programmed cell death such as tTG (Table 1). In this case also, the antigen-specific death subse-quent to the exposure to gp120 is prevented by CsA, indicating that the inhibition is due to the block of the effector elements expressed upon gp120 binding to CD4 receptor (AMENDOLA et al. 1994; BANDA et al. 1992; CEFAI et al. 1990). These observations are relevant for understanding an important aspect of AIDS therapy: blocking apoptosis in seropositive patients. However, before introduction of an anti-apoptotic therapeutic strategy in AIDS treatment, the important question of whether "primed" T cells (which underwent a number of passages in the cascade of events leading to apoptosis) still act as immunocompetent functional elements should be addressed.

4.3 Is Tissue Transglutaminase Expression Relevant in HIV-Induced Apoptosis?

HIV infects, in addition to T lymphocytes, other immunocompetent cells which express the CD4 receptor, such as monocytes macrophages (M/M) (FUACI 1988). Infection of cells of the M/M lineage by HIV plays an important role in the pathogenesis of AIDS (PANTALEO et al. 1993). These cells may be resistant to the cytopathic effect of HIV and could serve as a viral reservoir in the body. In vivo, M/M act under the control of different cytokines such as macrophage colony-stimulating factor (M-CSF) and granulocyte/macrophage colony-stimulating factor (GM-CSF). These cytokines affect both maturation and function of macrophages and modulate in vitro the replication of HIV in these cells. It has recently been shown that M-CSF causes HIV to be cytopathic for M/M in vitro by inducing syncytia formation and apoptosis (BERGAMINI et al. 1994). Interestingly, in these cells tTG is induced and activated, as indicated by the increased number of extensively cross-linked apoptotic bodies found in the cultures with respect to the controls. This enhanced tTG activity is paralleled by a large reduction in viral particle release into the medium, despite the fact that the amount of intracellular virions detected in the M-CFS-stimulated cells was much higher than that found in infected cultures with no apoptosis. Death of HIV-infected syncytia by apoptosis is not associated with virus release, thus suggesting that a specific cell death program is triggered in fused cells with the physiological goal of eliminating the potentially infective agents they have engulfed. It is well established that apoptotic cells do not lyse but fragment into membrane-sealed apoptotic bodies (WYLLIE et al. 1980). Apoptosis might therefore have a protective role in reducing the spread of infectious virions from dying infected cells. An interesting question is whether the virions could be trapped inside the dying cell by tTG-dependent cross-linking. It has recently been shown that the viral transmembrane glycoproteins gp41 and gp120, can very effectively act as in vitro substrates for tTG (MARINIELLO et al. 1993). Thus, it might be conceivable that the viral proteins can be incorporated together with the constitutively expressed cytoplasmic protein in the intracellular polymers assembled by the enzyme in cells undergoing apoptosis. If this "caging effect" proves to be effective, the best therapeutic strategy would be the targeted induction of apoptosis in the infected cells.

5 On the Role of Tissue Transglutaminase in Programmed Cell Death

As far as the role of tTG in apoptosis is concerned at least two potential interrelated functions can be envisaged: (1) tTG in cooperation with other effector elements might have a direct effect in killing and/or (2) tTG-dependent cross-linking could stabilize the apoptotic cells before their clearance. Interesting clues regarding these two hypotheses derive from transfection studies carried out in various mammalian cells. Cell lines (human neuroblastoma SK-N-BE(2), BALB-C 3T3 and L929 fibroblasts) transfected with a full length tTG cDNA all show a large reduction in their proliferative capacity paralleled by an increased rate of cell death (GENTILE et al. 1992; MELINO et al. 1994). The dying tTG-transfected cells exhibit both cytoplasmic and nuclear changes characteristic of cells undergoing apoptosis. Conversely, transfection of neuroblastoma cells with an expression vector containing segments of the human tTG cDNA in the antisense orientation results in a pronounced decrease of both spontaneous and induced apoptosis (MELINO et al. 1994). These findings indicate that the tTG-catalyzed irreversible cross-linking of intracellular protein might represent an important biochemical event in the induction of structural changes characteristic of cells dying by apoptosis. Recent studies indicate that overexpression of other putative effector elements (DNase I and proteases) can kill viable cells with the phenotypical features of apoptosis (PEITSCH et al. 1993; SHI et al. 1994). These findings are not surprising when related to the functions of these enzymes and to the somewhat artificial gene transfer approach. It must be considered that the intracellular level reached by the effector elements in the preapoptotic cells in naturally occurring cell death is far less than that obtained by transfection studies. In controlled physiological conditions it is very likely that different effector elements play complementary integrated functions, in different cell compartments, during the final stages of the cell death process. In fact, proteolytic activity could be required for the activation of specific regulatory proteins and p34^{cdc2}-dependent nuclear disruption, which in turn may allow cytosolic DNAses and/or endonuclease/s access to chromatin (MIURA et al. 1993; SHI et al. 1994; PEITSCH et al. 1993; EASTMAN 1994; HUGHES and CIDLOWSKI 1994). In this context Ca^{2+}-dependent tTG activation might play an important role in the condensation of cytoplasm and its subsequent controlled fragmentation. Interestingly, in cells overexpressing the tTG the higher apoptotic rate observed in the transfectants is associated with reduced leakage of intracellular macromolecules (Piacentini and Fesus, unpublished observations). Thus, tTG-dependent protein polymerization could temporarily stabilize the cytoplasm of dying apoptotic cells before phagocytosis. This phenomenon might aid in preventing the release of harmful intracellular components into the extracellular space (enzymes, DNA, RNA, viruses) and explain why apoptosis is an immuno-logically silent event not associated with inflammation and scar formation in the surrounding tissues (WYLLIE et al. 1980; ARENDS and WYLLIE 1991).

6 Conclusions

The original idea proposed in the 1970s by A. Wyllie and J.F. Kerr, that cell death by apoptosis is a genetically regulated event, has been strengthened by the identification of several genes participating in the process (FESUS et al. 1989; BUTTYAN et al. 1989; HOCKENBERY et al. 1990; SAVILL et al. 1990; ELLIS et al. 1991; YONISH-ROUACH et al. 1991; BURSCH et al. 1992; DINI et al. 1992; EVAN et al. 1992; SHI et al. 1994; MIURA et al. 1993; PEITSCH et al. 1992). In spite of the exponential increase in the number of studies on gene-dependent cell death, a single "killer" gene has not yet been identified in mammalian cells. A number of distinct enzymes might work in a coordinate fashion to achieve the irreversible, fast and clean removal of apoptotic bodies. It is also quite well established that, under both physiological and pathological conditions, only part of this integrated pathway of cell death can be triggered, resulting in distinct cell death pathways. We suggest that tTG, at least in some cell types, could be one of the potential candidate killer elements of the apoptotic program. By extensively cross-linking intracellular proteins, the enzymes could modify cell organization determining those irreversible ultrastructural changes typical of apoptotic cells. The fact that no diseases involving deregulated, overexpression of the tTG gene have been reported so far could imply that its induction is incompatible with cell survival.

Acknowledgements. I thank Drs Alessandra Amendola, Massimo Di Rao, Maria Grazia Farrace, Serafina Oliverio and Lucia Piredda for performing much of the work reported in this review. I also thank Prof. Francesco Autuori for support and encouragement, Prof. Gerry Melino and Laszlo Fesus for helpful discussion and Prof. Paolo Luly and Vittorio Colizzi for critical reading the manuscript. This work was supported by grants from Italian Ministery of Health "AIDS" Project, E.C. BIOMED 1, A I.R.C. and C.N.R. Bilateral project "Italy-Hungary".

References

Aeschlimann D, Wetterwald A, Fleisch H, Paulsson M (1993) Expression of tissue transglutaminase in skeletal tissue correlates with events of terminal differentiation of chondrocytes. J Cell Biol 120: 1461–1470

Ameisen JC, Capron A (1991) Cell dysfunction and depletion in AIDS: the programmed cell death hypothesis. Immunol Today 4: 102–105

Amendola A, Lombardi L, Oliverio S, Colizzi V, Piacentini M (1994) HIV-1 gp120-dependent induction of apoptosis in antigen specific human T cell clones is chracterized by tissue transglutaminase expression and is prevented by cyclosporin A. FEBS Lett 339: 258–264

Arends MJ, Wyllie AH (1991) Apoptosis. Mechanism and role in pathology. Int Rev Exp Pathol 32: 223–245

Baldari CT, Macchia G, Heguy A, Melli M, Telford J (1991) Cyclosporin A blocks calcium-dependent pathways of gene activation. J Biol Chem 266: 19103–19108

Banda NK, Bernier J, Kurahara DK, Kurrle R, Haigwood N, Sekaly RP, Finkel TH (1992) Crosslinking CD4 by human immunodeficiency virus gp120 primes T cells for activation-induced apoptosis. J Exp Med 176: 1099–1106

Bergamini A, Capozzi M, Piacentini M (1994) MCS-F stimulation induces cell death in HIV infected multinucleated monocyte. Immunol Lett 42: 35–40

Bursch W, Oberhammer F, Schulte-Hermann R (1992) Cell death by apoptosis and its protective role against disease. TIPS 13: 245–251

Buttyan R, Olsson CA, Pintar J, Chang C, Bandyk M, Poying NG, Sawczuk IS (1989) Induction of TRPM-2 gene in cells undergoing programmed cell death. Mol Cell Biol 9: 3473–3481

Cefai D, Debre P, Kaczorek M, Idziorek T, Autran B, Bismuth G (1990) Human immunodeficiency virus I glycoproteins gp120 and gp160 specifically inhibit the CD3/T cell-antigen receptor phosphoinositide transduction pathway. J Clin Invest 86: 2117–2124

Cohen JJ, Duke RC, Fadok VA, Sellins KS (1992) Apoptosis and programmed cell death in immunity. Annu Rev Immunol 10: 267–293

Critchfield JM, Racke MK, Zuniga-Pflucker JC, Cannella B, Raine CS, Goverman J, Lenardo MJ (1994) T cell deletion in high dose therapy of autoimmune encephalomyelitis. Science 263: 1139–1142

Diamond D, Sleckman B, Gregory T, Lasky L, Greestain J, Burakoff S (1990) Inhibition of CD4+ T-Cell function by the HIV-envelope protein gp120. J Immunol 141: 3715–3717

Dini L, Autuori F, Lentini A, Oliverio S, Piacentini M (1992) The clearance of apoptotic cells in the liver is mediated by the asialoglycoprotein receptor. FEBS Lett 296: 174–178

Eastman A (1994) Deoxyribonuclease II in apoptosis and the significance of intracellular acidification. Cell Death Differ 1: 7–9

Ellis RE, Yuan J, Horvitz HR (1991) Mechanism and functions of cell death. Annu Rev Cell Biol 7: 663–698

Evan, GI, Wyllie AH, Gilbert CS, Littlewood TD, Land H, Brooks M, Waters CM, Penn LZ, Hancock DC (1992) Induction of apoptosis in fibroblasts by c-myc protein. Cell 69: 119–128

Fauci AS (1988) The human immunodeficiency virus: infectivity and mechanisms of pathogenesis. Science 239: 617–622

Fesus L, Szucs EF, Barrett KE, Metcalfe DD, Folk JE (1985) Activation of transglutaminase and production of protein-bound γ-glutamylhistamine in stimulated mouse mast cells. J Biol Chem 260: 13771–13778

Fesus L, Thomazy V, Falus A (1987) Induction and activation of tissue transglutaminase during programmed cell death. FEBS Lett 224: 104–108

Fesus L, Thomazy V, Autuori F, Ceru' MP, Tarcsa E, Piacentini M (1989) Apoptotic hepatocytes become insoluble in detergents and chaotropic agents as a result of transglutaminase action. FEBS Lett 245: 150–154

Fesus L, Tarcsa E, Kedei E, Autuori F, Piacentini M (1991a) Degradation of cells dying by apoptosis leads to accumulation of ε(γ-glutamyl)lysine isodipeptide in culture fluid and blood. FEBS Lett 284: 109–112

Fesus L, Davies PJA, Piacentini M (1991b) Apoptosis: molecular mechanism in programmed cellular death. Eur J Cell Biol 56: 170–177

Folk JE (1980) Transglutaminases. Annu Rev Biochem 49: 517–431

Folk JE, Finlayson S (1977) The ε(γ-glutamyl)lysine crosslink and the catalytic role of transglutaminase. Adv Prot Chem 31: 1–133

Gentile V, Saydak M, Chiocca EA, Akande O, Birckbicheler PJ, Lee KN, Stein JP, Davies PJA (1991) Isolation and characterization of cDNA clones to mouse macrophage and human endothelial cell tissue transglutaminases. J Biol Chem 264: 478–483

Gentile V, Thomazy V, Piacentini M, Fesus L, Davies PJA (1992) Expression of tissue transglutaminase in balb-c 3t3 fibroblasts: effects on cellular morphology and adhesion. J Cell Biol 119: 463–474

Gougeon ML, Montagnier L (1992) New concepts in the mechanisms of CD4+ lymphocyte depletion in aids, and the influence of opportunistic infections. Res Microbiol 361–373

Green H (1977) Terminal differentiation of cultured human epidermal cells. Cell 11: 405–416

Geenberg CS, Birckbichler P, Rice RH (1992) Transglutaminases: multifunctional cross-linking enzymes that stabilize tissues. FASEB J 5: 3071–3077

Groux H, Torpier G, Monté D, Mouton Y, Capron A, Ameisen JC (1992) Activation-induced death by apoptosis in CD4+ T cells from human immunodeficiency virus-infected asymptomatic individuals. J Exp Med 175: 331–340

Hockenbery D, Zutter M, Hickey W, Nahm M, Korsmeyer SJ (1990) Bcl-2 protein is an inner mitochondrial membrane protein that blocks topographically programmed cell death. Nature 348: 334–336

Hughes FM, Cidlowski JA (1994) Apoptotic DNA degradation: evidence for novel enzymes. Cell Death Differ 1: 11–17

Ichinose A, Hendrickson LE, Fujikawa K, Davie DJ (1986) Amino acid sequence of the subunit of human factor XIII. Biochemistry 25: 6900–6906

Ikura K, Nasu T, Yokota T, Tsuchiya Y, Sasaki R, Chiba H (1988) Amino acid sequence of guinea pig liver transglutaminasi from its cDNA sequence. Biochemistry 27: 2898–2905

Kabelitz D, Pohl D, Pechhold K (1993) Activation-induced cell death (apoptosis) of mature peripheral T lymphocytes. Immunol Today 14: 338–345

Kim IG, Gorman JJ, Park SC, Chung SI, Steinert PM (1993) The deduced sequence of the novel protransglutaminase E (TGase3) of human and mouse. J Biol Chem 268: 12682–12690

Klein JD, Guzman E, Kuehn GD (1992) Purification and partial characterization of transglutaminase from physarum polycephalum. J Bacteriol 174: 2599–2604

Knight CRL, Rees RC, Griffin M (1991) Apoptosis: a potential role for cytosolic transglutaminase and its importance in tumor progression. Biochim Biophys Acta 1096: 312–320

Knight CRL, Hand D, Piacentini M, Griffin M (1993a) Characterization of the transglutaminase mediated large molecular weight polymer from rat liver; its relationship to apoptosis and its importance in carcinogenesis and tumour progression. Eur J Cell Biol 60: 210–217

Knight CRL, Rees RC, Platts A, Johnson T, Griffin M (1993b) Interleukin-2-activated effector lymphocytes mediate cytotoxicity by indicing apoptosis in human leukaemia and solid tumor target cells. Immunology 79: 535–541

Laurent-Crawford AG, Krust B, Muller S, Riviere Y, Rey-Cuillé MA, Bechet JM, Montagnier L, Hovanessien AG (1991) The cytopathic effect of HIV is associated with apoptosis. Virology 185: 829–839

Lorand L, Hsu LKH, Siefring GE, Rafferty NS (1981) Lens transglutaminase and cataract formation. Proc Natl Acad Sci USA 78: 1356–1362

Mariniello L, Esposito C, Di Pierro P, Cozzolino A, Pucci P, Porta R (1993) HIV transmembrane glycoprotein gp41 is an amino acceptor and donor substrate for transglutaminase in vitro. Eur J Biochem 215: 99–104

Mastino A, Piacentini M, Grelli S, Favalli C, Autuori F, Tentori L, Oliverio S, Garaci E (1992) Induction of apoptosis in thymocytes by prostaglandin PE2 in vivo. Dev Immunol 2: 263–271

Mastino A, Grelli S, Piacentini M, Oliverio S, Favalli C, Perno F, Garaci E (1993) Correlation between induction of lymphocyte apoptosis and prostaglandin PGE2 production by macrophages infected with HIV. Cell Immunol 152: 120–130

McConkey DJ, Orrenius S, Jondal M (1990) Cellular signaling in programmed cell death (apoptosis). Immunol Today 11: 120–121

Melino G, Annicchiarico-Petruzzelli M, Piredda P, Candi E, Gentile V, Davies PJA, Piacentini M (1994) "tissue" tansglutaminase and apoptosis: sense and antisense transfection studies in human neuroblastoma cells Mol Cell Biol 14: 6584–6596

Meyaard L, Otto SA, Jonker RR, Mijnster MJ, Keet RPM, Miedema F (1992) programmed death of T cells in HIIV infection. Science 257: 217–219

Miura M, Zhu H, Rotello R, Hartwieg EA, Yuani (1993) Induction of apoptosis in fibroblasts by IL-1B-converting enzyme, a mammalian homolog of the C. elegans cell death gene ced-3. Cell 75: 653–660

Newell MK, Haughn LJ, Maroun CR, Julius MH (1990) Death of mature T cells by separate ligation of CD4 and T cell receptor for antigen. Nature 347: 286–289

Novogrodsky A, Quittner S, Rubin L, Stenzel KH (1978) Transglutaminase activity in human lymophocytes: early activation by phytomitogens. Proc Natl Acad Sci USA 75: 1157–1161

Palmer DB, Hayday A, Owen MJ (1993) Is TCRbeta expression an essential event in early thymocyte development. Immunol Today 14: 460–462

Pantaleo G, Graziosi C, Denmarest JF, Butini L, Montroni M, Fox CH, Orenstein JM, Kotler DP, Fauci AS (1993) HIV infection is active and progressive in lymphoid tissue during the clinically latent stage of disease. Nature 362: 355–358

Peitsch MC, Polzar B, Stephan H, Crompton T, MacDonald HR, Mannherz HG, Tschopp J (1993) Characterization of the endogenous deoxyribonuclease involved in nuclear DNA degradation during apoptosis (programmed cell death). EMBO J 12: 371–377

Piacentini M, Martinet N, Beninati S, Folk JE (1988) Free and protein-conjugated polyamines in mouse epidermal cells. J Biol Chem 263: 3790–3794

Piacentini M, Autuori F, Dini L, Farrace MG, Ghibelli L, Piredda L, Fesus L (1991a) "tissue" trans-glutaminase is specifically expressed in neonatal rat liver cells undergoing apoptosis upon epidermal growth factor-stimulation. Cell Tissue Res 263: 227–235

Piacentini M, Fesus L, Farrace MG, Ghibelli L, Piredda P, Melino G (1991b) The expression of "tissue" transglutaminase in two human cancer cell lines is related with the programmed cell death (apoptosis). Eur J Cell Biol 54: 246–254

Piacentini M, Davies PJA, Fesus L (1994) Transglutaminase in cells undergoing apoptosis. In: Tomei LO, Cope FO (eds) Apoptosis II: the molecular basis of apoptosis in desease. Cold Spring Harbor Laboratory press, Cold Spring Harbor, pp 143–163

Raff MC, Barris BA, Burne JF, Coles HS, Ishizaki Y, Jacobson MD (1992) Programmed cell death and the control of cell survival: lessons from the nervous system. Science 262: 695–700

Ritter MA, Boyd RL (1993) Development in the thymus: it takes two to tango. Immunol Today 14: 462–469

Savill JS, Dransfield I, Hogg A, Haslett C (1990) Vitronectin receptor-mediated phagocytosis of cells undergoing apoptosis. Nature 343: 170–173

Shi L, Nishioka WK, Th'ng J, Bradbury EM, Litchfield DW, Grenberg AH (1994) Premature p34cdc2 activation required for apoptosis. Science 263: 1143–1145

Smith CA, Williams GT, Kingston R, Jenkinson EJ, Owen JJT (1989) Antibodies to CD3/T-cell receptor complex induce death by apoptosis in immature T cells in thymic cultures. Nature 337: 181–184

Strange R, Li F, Saurer S, Burkhard A. Friis RR (1992) Apoptotic cell death and tissue remodelling during mouse mammary gland involution. Development 115: 49–58

Suto N, Ikura K, Shinagawa R, Sasaki R (1993) Identification of promoter region of guinea pig liver transglutaminase gene. Biochim Biophys Acta 1172: 319–322

Tarcsa E, Kedei N, Thomazy V, Fesus L (1993) An involucrin-like protein in hepatocytes serves as a substrate of tissue transglutaminase during apoptosis. J Biol Chem 267: 25648–25652

Terai C, Korbluth RS, Pauza CD, Richmann DD. Carson DA (1991) Apoptosis as a mechanism of cell death in cultured T lymphoblasts acutely infected with HIV-1. J Clin Invest 87: 1710–1715

Thomazy V, Fesus L (1989) Differential expression of tissue transglutaminase in human cells. Cell Tissue Res 255: 215–224

Williams-Ashman HG (1984) Transglutaminases and the clotting of mammalian seminal fluid. Mol Cell Biochem 58: 51–61

Wyllie AH, Kerr JF, Currie AR (1980) Cell death: the significance of apoptosis. Int Rev Cytol 68: 251–306

Yonish-Rouach ED, Resnitzky S, Lotem J, Sachs S, Kimichi G, Oren M (1991) Wild-type p53 induces apoptosis of myeloid leukaemic cells that is inhibited by interleukin-6. Nature 352: 345–348

Chronic Activation of the Immune System in HIV Infection: Contribution to T Cell Apoptosis and Vβ Selective T Cell Anergy

M.L. Gougeon

1 Introduction

Since the initial isolation in 1983 of the human immunodeficiency virus (HIV) that is known to cause AIDS (acquired immune deficiency syndrome) (BARRE-SINOUSSI et al. 1983) and despite an extensive knowledge of the molecular characteristics of this virus, one of the main questions is still not answered: what causes the

Unité d'Oncologie Virale, Département SIDA et Rétrovirus, Institut Pasteur, 28 rue du Dr. Roux, 75724 Paris Cedex 15, France immune system collapse in HIV infection? Indeed the hallmark of AIDS is the progressive disappearance of CD4⁺ helper T lymphocytes which, in addition to being the targets of the virus (KLATZMANN et al. 1984), play a major role in the immune system. There is no question that the presence of the virus in lymphoid organs and its active replication associated with direct killing of the target is at least associated with CD4⁺ T cell depletion. However it has become clear for many investigators that direct mechanisms, even if they play a role, are inadequate to explain the extent of immunosuppression. HIV disease is now considered as a multifactorial process and the immunopathogenic mechanisms involved appear to be very complex (FAUCI 1988, 1993; GOUGEON and MONTAGNIER 1993; LEVY 1993).

In the present review some of the indirect mechanisms potentially involved in AIDS pathogenesis will be discussed, focusing on the causes and consequences of a persistent and inappropriate activation of the immune system leading to paralysis (anergy) or self-destruction (apoptosis) of patients' noninfected T cells.

2 Viral Infection and CD4 Cell Depletion

2.1 Viral Infection

Primary infection with HIV is generally followed a few weeks later by high levels of viremia, associated in the majority of individuals with clinical symptoms. Within weeks to months, a humoral and cellular immune response to HIV develops, which is correlated with a decreased viremia. The patients enter a phase of apparent clinical latency. The initial studies by HARPER et al. (1986) demonstrated, with the use of in situ hybridization for HIV-1-specific RNA, that only $1/10^4$–$1/10^5$ peripheral blood lymphocytes (PBLs) and lymph node cells could be identified as HIV-positive in vivo. However, owing to the development of more sensitive molecular techniques, it has become clear that the quantity of virus in the plasma and in the PBLs in HIV-infected individuals is much higher (HO et al. 1989; SCHNITTMAN et al. 1990; HSIA and SPECTOR 1991). In situ polymerase chain reaction (PCR) allowed detection of between 0.1% and 10% of blood mononuclear cells positive for HIV provirus in patients according to the stage of the disease (asymptomatic vs AIDS), and it is likely that most infected cells contain HIV provirus in a latent (or defective) form that was not detected earlier (BAGASRA et al. 1993; PATTERSON et al. 1993). These latently infected cells constitute a reservoir of persistent infection that cannot be targeted by the host's immune surveillance mechanism. Interestingly enough, simultaneous analysis of viral burden in the blood and lymphoid organs indicated that HIV is sequestered in lymph nodes both as an extracellular virus trapped in the follicular dendritic cell network of the germinal centers and as intracellular virus usually in a latent form (PANTALEO et al. 1993; EMBRETSON et al. 1993). Furthermore, during the apparently latent period between infection with HIV and the overt symptoms of AIDS, HIV is actively replicating in lymphoid organs (PANTALEO et al. 1993) despite a low viral burden and low replication in PBLs, indicating that a state of true microbiological latency does not exist. In fact HIV infection is active in lymphoid organs early in the clinical course of the disease and the degree of virus replication increases as HIV disease progresses.

2.2 HIV-Induced Immune Deficiency

A number of factors seem to be involved in HIV-induced immune deficiency. A direct cytopathic effect (CPE) of HIV on CD4$^+$ T cells might contribute to the immune CD4 T cell depletion. The cytopathic effect of HIV in CD4$^+$ T cells is manifested by ballooning of cells and formation of syncytia. The CPE of HIV-1 and HIV-2 on CD4$^+$ T cell cultures is associated with apoptosis (TERAI et al. 1991; LAURENT-CRAWFORD et al. 1991), and it was shown that apoptosis is triggered by the viral envelope glycoprotein gp120 (LAURENT-CRAWFORD et al. 1992). Another potential mechanism for CD4$^+$ cell loss is the covering of cells carrying the CD4 molecule with gp120. These uninfected cells are then recognized as virus-infected cells by NK (natural killer) effector cells or CTL (cytotoxic T lymphocytes) and subsequently destroyed, even though they are not infected by the virus (LANZAVECCHIA et al. 1988; WEINHOLD et al. 1989). Therefore, cytotoxic CD8$^+$ cells might kill normal CD4$^+$ cells and those infected with HIV (PANTALEO et al. 1990; RIVIERE et al. 1989). Anti-lymphocyte antibodies may also play a role in immune deficiency. Autoantibodies to the CD4 protein have been detected in HIV-infected individuals and might be involved in CD4$^+$ lymphocyte death (CHAMS et al. 1988).

2.3 HIV-Related Functional Helper Defects

Besides the direct effects of HIV on CD4$^+$ cells, viral proteins released by infected cells could interfere with the normal events in signal transduction. Indeed very early in the course of the disease, before CD4 cell depletion, functional defects of helper T cells are observed in patients' lymphocytes, characterized by the impairment of in vitro T cell receptor (TCR)-dependent activation in response to MHC-restricted recall antigens (SHEARER and CLERICI 1991) or to anti-CD3 monoclonal antibodies (MIEDEMA et al. 1988). In vitro experiments analyzing the effect of gp120-CD4 interaction on subsequent response of normal CD4$^+$ T cells to lymphocyte activation indicated the induction of an anergic state in these gp120-presignaled T cells (DI RIENZO et al. 1993). Thus extracellular signal mediated by gp120 via CD4 molecule may affect signal transduction. For example, the HIV-1 gp120 has been found to form an intracellular complex with CD4 and p56^{lck} in the endoplasmic reticulum (CRISE and ROSE 1990). The retention of this tyrosine kinase in the cytoplasm could affect the function of the cell. In addition, gp120 signaling is also supposed to be involved in the programming of noninfected CD4$^+$ T cells for cell death by apoptosis, as discussed below.

Paradoxically, concomitant to the inhibition of lymphocyte activation (anergy) of helper CD4⁺ T cells in patients, a chronic activation of the immune system is observed: expression of T cell activation antigens on CD4⁺ and CD8⁺ T cells, spontaneous B cell hyperactivation, lymph node hyperplasia early in the course of infection, increase cytokine expression, etc. The persistence of virus and viral replication throughout the course of HIV disease may play a primary role in the maintenance of this chronic activation and it has been proposed that superantigens (either of bacterial origin or encoded by HIV) may contribute to this activation. The potential influence of superantigens in AIDS pathogenesis is also discussed below.

3 Chronic Activation and Spontaneous Apoptosis

3.1 Apoptosis

Apoptosis (programmed cell death, PCD) is an active suicide mechanism that constitutes the principal form of cell death for lymphocytes. In general the cell undergoing apoptosis sustains profound structural changes and one of these is a nuclear collapse associated with condensation of chromatin which tends to marginate in crescents around the nuclear envelope. The nuclear collapse, visible by light or electron microscopy, indicates extensive damage to chromatin which is degraded into single and multiples oligonucleosomes. The DNA is cleaved in the internucleosomal linker region, and electrophoretic separation of DNA from apoptotic cells reveals a "ladder" pattern of bands averaging about 200 bp, 400 bp, 600 bp, etc., corresponding to oligonucleosomal fragments (WYLLIE et al. 1980, 1984). This fragmentation of DNA is enzymatic and generally occurs after activation of a calcium-dependent endogenous endonuclease (ARENDS et al. 1990).

There are several examples of PCD in T lymphocytes. This process is involved in the negative intrathymic selection of the T cell repertoire which leads to the clonal deletion of autoreactive T cells and to the establishment of self-tolerance (JENKINSON et al. 1989). Immature thymocytes undergo apoptosis in response to glucocorticoid hormones, calcium ionophores or antibodies to the TCR/CD3 complex (SMITH et al. 1989; KIZAKI et al. 1989). Mature T cells generally respond to TCR/CD3 stimulation by proliferation and differentiation. However under certain circumstances apoptosis is thought to mediate the death of antigen-activated mature T cells (WEBB et al. 1990; KAWABE and OCHI 1991; NEWELL et al. 1990).

3.2 Spontaneous Apoptosis During HIV-Infection and Immune Dysfunction

Our early studies (MONTAGNIER et al. 1989) indicated that, when PBLs from HIV-infected individuals were cultured in a survival medium lacking lymphocyte growth factors, there was a significant difference in the loss of viability of these lymphocytes as compared to those of control individuals. This premature cell death of patient's lymphocytes did not represent the death of infected cells. In addition it was observed early in the asymptomatic phase of the disease and was more pronounced in the AIDS stage.

Preliminary analysis by flow cytometry indicated the presence in cultures from patients' PBLs of a living population but weakly stained with orange acridine, suggesting that they might correspond to cells undergoing apoptosis. Apoptosis was monitored by the presence of oligonucleosomal DNA fragments and was confirmed by electron microscopy to demonstrate chromatin condensation (GOUGEON et al. 1991, 1992).

Both CD4$^+$ and CD8$^+$ T cells from HIV-infected individuals undergo spontaneous apoptosis after a few hours of culture (GOUGEON et al. 1993a, b): this was observed whether the purification of CD4$^+$ and CD8$^+$ T cells was carried out prior to culture or was done at the end of the culture before DNA extraction. This suggests that the triggering of apoptosis in each T cell subpopulation does not depend upon the presence of the other. Phenotyping of CD4$^+$ and CD8$^+$ T cells dying of apoptosis after 24 h of culture indicated that some of them expressed the activation markers CD45RO or DR. A recent study we performed with lymphocytes from West African individuals confirmed the ability of T cells from HIV-1-infected individuals to spontaneously die via apoptosis after a short-term culture and revealed that HIV-2 has the same indirect apoptotic effect on uninfected peripheral T cells. Furthermore we found that, both in HIV-1 and HIV-2-infected individuals, the extent of PCD correlates with progression of the disease (GOUGEON and MONTAGNIER et al. 1994; TOURE-BALDE et al. 1994).

Spontaneous apoptosis of peripheral CD4$^+$ and CD8$^+$ T lymphocytes has also been described in acute viral infections in both Epstein-Barr virus (EBV) (infectious mononucleosis) and varicella zoster virus (VZV) (chickenpox) infected individuals (UEHARA et al. 1992; AKBAR et al. 1993). These patients have expanded circulating T cell populations expressing CD45RO and HLA-DR and these activated cells have been shown to undergo apoptosis upon in vitro culture. Since apoptosis of these cells in viral infections can be prevented by cytokines, interleukin-2 (IL-2) in EBV-infected patients (AKBAR et al. 1993) or IL-1α + IL-2 in HIV-infected patients (GOUGEON et al. 1993a), it suggests that after T cell activation in vivo, the expanded population is destined to perish unless some factors, such as cytokines, promote their survival.

The T cell lymphocytosis associated with EBV and VZV infections is transient as the absolute number of circulating T lymphocytes and the relative proportion of CD4$^+$ and CD8$^+$ cells return to normal upon resolution of the disease. It probably occurs via a rapid clearance by apoptosis of the majority of activated T cells blasts in vivo, which allows a balance between cell death and survival. In the case of a chronic infection such as the retroviral HIV infection, persistence of immune activation generating suicide-sensitive CD45RO⁺ T cells, expressing low *bcl*-2 (AKBAR et al. 1993), would induce regular deletion of memory CD4⁺ and CD8⁺ T cells, contributing to the immune deficiency. The disappearance of such cells is particularly dramatic if the immune system from HIV-infected individuals is not capable of spontaneous regeneration. Indeed there is evidence that in HIV-infected individuals the thymus is severely damaged. Thus, even if the bone marrow precursor cells are still present in HIV-infected individuals, it is questionable whether the reconstitution of normal immune function would occur in the absence of an intact thymic microenvironment.

Persistence of immune activation may have other consequences. From a virological standpoint, virus spread is more efficient in activated cells and, in addition, cellular activation induces expression of virus in cells latently infected with HIV (ROSENBERG and FAUCI 1989; POLI and FAUCI 1990). From an immunological standpoint, persistent exposure of the immune system to activation may lead to immune dysfunction and either loss of the ability to respond to an antigen (anergy) or induction of an abnormal program of cell death, as discussed below.

4 Activation-Induced Apoptosis and CD4 Cell Deletion

4.1 Activation-Induced Apoptosis in Patients' T Lymphocytes

The asymptomatic phase of HIV infection is characterized by functional defects of CD4⁺Th cells, i.e., impairment in the in vitro proliferation to recall antigens, alloantigens, mitogens and anti-CD3 activation (SHEARER and CLERICI 1991; MIEDEMA et al. 1988). This functional impairment is followed by the slow decline of CD4⁺ T cells and the development of AIDS pathogenesis. Besides the HIV-related direct or indirect mechanisms described above and thought to be responsible for CD4 cell decline, recent developments in this area of research have suggested that the loss of CD4 cells in HIV-infected individuals is associated with activation-induced cell death by apoptosis.

Indeed, in addition to the spontaneous apoptosis described above, several groups have described that in vitro TCR-dependent or independent activation of patients' T cells commit in a fraction of them a cell suicide program (GOUGEON et al. 1991, 1992; GROUX et al. 1992; MEYAARD et al. 1992). Apoptosis depends on the intracellular availability of certain key proteins including a calcium-dependent endonuclease (WYLLIE et al. 1991; ARENDS et al. 1990). We have found that increasing the intracellular Ca²⁺ mobilization, by stimulating patients' T cells during several hours with ionomycin, induces apoptosis in a fraction of T cells (GOUGEON et al. 1993a). Ionomycin is supposed to activate the endogenous endonuclease and to induce apoptosis in primed cells such as immature thymocytes (McCONKEY et al. 1989). Polyclonal activators such as anti-CD3 antibodies are also able to induce apoptosis in patients' T cells (MEYAARD et al. 1992). It is noteworthy that both CD4⁺ and CD8⁺ T cells are susceptible to activation-induced cell death with these polyclonal activators. A more specific cell death of CD4⁺ T cells can be observed upon activation of patients' T cells with CD4 tropic stimuli such as MHC class II-dependent bacterial superantigens (GROUX et al. 1992; GOUGEON et al. 1993a; GOUGEON and MONTAGNIER 1994) or PWM mitogen (GROUX et al. 1992).

4.2 Putative Mechanisms of In Vivo Priming and Triggering of Apoptosis

Several nonexclusive mechanisms may contribute to in vivo triggering of PCD in the course of HIV infection. Since only a few cells are infected, a systemic effect of HIV proteins or disturbances of cytokine regulatory network are likely to be involved rather than direct infection. The observation that, in addition to CD4⁺ T cells, CD8⁺ T cells may undergo PCD raises the important question of the relevance of in vitro findings of PCD to in vivo CD4 cell depletion.

One hypothesis states that, during HIV infection, interaction of the gp120 envelope protein of HIV with CD4 receptor on T cells may result, in the absence of costimulatory signal, in the programmation of CD4⁺ lymphocytes for subsequent death by apoptosis when they meet their corresponding antigen (AMEISEN and CAPRON 1991). This hypothesis was supported by observations in the murine model showing that, when preceeded by ligation of CD4, signaling through TCRαβ results in CD4⁺ T cell unresponsiveness due to the induction of activation-dependent cell death by apoptosis (NEWELL et al. 1990). More recently, this experiment was reproduced in the human model with purified gp120 (BANDA et.al. 1992).

Additional mechanisms may contribute to PCD induction in HIV infection. T cell deletion may result from defective activation signals: optimal T cell activation requires the delivery of a cosignal, provided by the antigen presenting cell (APC), in addition to the antigenic signal delivered via the TCR. TCR stimulation occurring in the presence of an inappropriate cosignal (delivered by HIV-infected monocytes, for example) might lead to T cell deletion. This hypothesis is supported by results showing that cosignal through CD28 molecule (GROUX et al. 1992) or delivered by IL-1α (GOUGEON et al. 1993a) prevents cell death and restores partially the activation of patients' T lymphocytes.

If the first hypothesis allows one to explain priming for PCD of CD4⁺ T cells, the latter could also account for the death of CD8⁺ T cells. However, other questions may be raised: Do CD8⁺ T cells die in vitro as a consequence of their hyperactivation but would survive in patients? Do CD8⁺ T cells die in vivo but are replaced? Is the selective loss of CD4⁺ T cells during HIV infection also the consequence of impaired renewal of this population?

Finally, another hypothesis was formulated to explain CD4 cell deletion in AIDS pathogenesis after the discovery that superantigens are encoded by murine retroviruses (murine mammary tumor virus, MMTV) (CHOI et al. 1991). This theory proposes that HIV might cause, in conjunction with class II genes, cell anergy and deletion of non-infected CD4⁺ T cells by encoding a superantigen (JANEWAY 1991). Progression of CD4 T cell depletion would require cycles of mutation in the retroviral superantigen gene resulting in the elimination of CD4 T cells bearing different Vβs over time. Several reports suggested that superantigens may be involved in the pathogenesis of AIDS, as summarized below, but there is no direct evidence today that HIV encodes for a superantigenic activity.

5 Superantigens in HIV Infection: T Cell Anergy and Deletion

5.1 Bacterial Superantigens

Physiological activators of T lymphocytes are antigens which are recognized, in the context of MHC molecules, through their interaction with the variable V portions of the TCR α and β chains (DAVIS and BJORKMAN 1988). However, T cells recognize superantigens on the basis of the expressed Vβ alone, independently of the other variable TCR segments. The superantigens bind to MHC proteins and this complex, by engaging Vβ, can stimulate many T cells. Exogenous bacterial superantigens comprise a set of protein toxins produced by *Staphylococcus*, *Streptococcus* or *Mycoplasma* that are recognized, in the context of MHC class II molecules, by T cells expressing particular TCR Vβ gene families, causing strong T cell activation associated with toxic shock and autoimmune diseases (KAPPLER et al. 1989; MARRACK and KAPPLER 1990).

5.2 Retroviral Superantigens

It has been proposed that a superantigen of microbial origin encoded either by HIV or an unrelated microbe might be involved in the pathogenesis of HIV disease and particularly might contribute to CD4 cell depletion. Indeed, in vivo studies of acute confrontation with bacterial superantigens have shown that the Vβ-specific expansion of superantigen-reactive CD4⁺ T cells is followed by deletion and anergy of the corresponding Vβ subsets (MACDONALD et al. 1991). Moreover, a retroviral superantigen encoded by MMTVs was shown to be responsible for activation and subsequent deletion of CD4⁺ T cell subsets expressing corresponding Vβ elements (CHOI et al. 1991; KORMAN et al. 1992). Involvement of a superantigen (truncated gag protein) in the pathogenesis of another murine retroviral infection, MuL V, was also reported and was shown to be responsible for the murine acquired immunodeficiency syndrome (MAIDS) (HUGIN et al. 1991; KANAGAWA et al. 1992).

5.3 Vβ T Cell Receptor Repertoire Analysis of T Cells from HIV-Infected Individuals

Since HIV-1 is a retrovirus, and because the pathology of HIV infection and AIDS involves predominantly the same CD4⁺ T cells that are commonly involved in superantigen-associated phenomena, it was suggested that HIV might cause, in conjunction with class II genes, cell anergy and deletion of noninfected CD4⁺ T cells bearing TCR Vβ determinants by encoding a superantigen expressed by activated infected cells (JANEWAY 1991). Several recent reports have discussed attempts to indirectly reveal the presence of an HIV-associated superantigen by looking for consistent amplifications and/or deletions in the ex vivo peripheral TCR Vβ repertoire of HIV-infected individuals: a more restricted Vβ repertoire (Vβ14–Vβ20 appeared deleted) was found in HIV-infected patients with advanced disease (IMBERTI et al. 1991) and a significant increase of peripheral CD4⁺ T cells of the Vβ5.3 subfamily was reported in asymptomatic subjects (DALGLEISH et al. 1992). Perturbations in the Vβ repertoire were also found in several pairs of monozygotic twins discordant for HIV with identical MHC (SOUDEYNS et al. 1993), allowing meaningful comparisons of their Vβ repertoire since one of the major factors that influence the nature of the peripheral TCR Vβ repertoire is the MHC class II haplotype of the individual (GULWANI-AKOLKAR et al. 1991).

5.4 Vβ-Selective Anergy in T Cells from HIV-Infected Individuals

Instead of analyzing the ex vivo repertoire in peripheral T lymphocytes from HIV-infected individuals, we chose a more functional approach: since in vivo murine studies have shown that anergy of a given Vβ subset gives evidence of a previous activation of this subset by a superantigen, we searched for a selective Vβ anergy in patients' T cells. We used the bacterial superantigen streptococcal erythrogenic toxin A (ETA), known to stimulate the Vβ8 and Vβ12 subsets, to analyze the Vβ usage of peripheral T cells from asymptomatic HIV-infected subjects in response to this in vitro superantigenic activation.

Our study indicated the existence, in a large fraction of HIV-infected individuals, of a Vβ-specific anergy affecting both CD4⁺ and CD8⁺ T cells expressing the Vβ8 TCR element (GOUGEON et al. 1993c; DADAGLIO et al. 1994). We have characterized this Vβ-specific anergy and shown that it was not the consequence of a defective presentation of ETA, since in the same cultures the Vβ12⁺ population were always normally stimulated by this antigen. In fact, experiments performed with other superantigens and anti-Vβ8 antibodies indicated that this Vβ8 anergy represented an intrinsic functional defect.

Several observations are in favor of a direct involvement of HIV in the Vβ-specific anergy observed in asymptomatic HIV-infected individuals (DADAGLIO et al. 1993):

1. Anergy can be observed very early in the course of HIV infection (CDC stage 1); and comparison of clinical status of responder vs anergic patients

showed no correlation with previous viral or bacterial infections, suggesting that anergy is not induced by opportunistic pathogens.

2. A strong proliferation was induced by in vitro stimulation of normal peripheral lymphocytes with inactivated HIV; and concomitantly the selective expansion of Vβ8⁺ T cells was reproducibly detected.

Characterization of the Vβ8-specific superantigenic activity associated with HIV is currently under investigation.

5.5 Potential Roles of Superantigens in HIV Infection

Except for the acute pathogenic variant of simian immunodeficiency virus (SIV) PBj14 (Fᴜʟᴛᴢ 1991), no report has shown until now the ability of HIV to activate normal peripheral T cells. A selective expansion of superantigenic reactive T cells is known to precede anergy, thus, one can speculate that in vivo infection of CD4⁺ cells will induce viral protein expression that, in association with MHC class II molecules, will activate in a superantigenic way followed by anergy subsets bearing the cognate Vβ determinants. It is interesting to note that the putative viral superantigen involved in the Vβ8 anergy has no selective tropism for CD4⁺ T cells since both CD4⁺ and CD8⁺ T cells are found anergic in patients and both are responsive to the in vitro Vβ8-specific activation by HIV. A recent report described the dependence of HIV1 replication on a superantigen and it concerned particularly the Vβ12⁺ CD4⁺ cell subset, which replicated more efficiently HIV in vitro and which was found enriched for gp120 expressing cells in vivo (Lᴀᴜʀᴇɴᴄᴇ et al. 1992). Our study is probably concerned with another superantigen. Interestingly enough, as found by others, we could not confirm the massive deletions of a large proportion of Vβ families described (Iᴍʙᴇʀᴛɪ et al. 1991), including the Vβ8 subset.

Table 1. Potential causes of immune deficiency and CD4 cell depletion in AIDS pathogenesis

Loss of T helper function
Anergy
CD4 cross-linking by gp120
Defect in antigen presentation, cosignaling
Diminished CD4⁺ helper cell number
Imbalance Th1->Th2 cytokine profil
Loss of CD4⁺ T helper cells
Direct
HIV-induced cell lysis
Syncytia formation
Immune anti-HIV response
Indirect
Spontaneous apoptosis (chronic activation, cytokines)
Activation induced apoptosis (gp120 cross-linking)
Superantigen-induced apoptosis and anergy
Loss of impairment of the regeneration of the immune system
HIV related degradation of lymphoid organs (thymus, gut, lymph nodes, bone marrow)
Loss or functional impairment of bone marrow precursor cells

Therefore, in HIV infection it is likely that superantigens, if indeed they are present and have an impact on pathogenesis, act as potent activators of T cells contributing to virus dissemination and progressive immune failure (via anergy and apoptosis), rather than as factors directly responsible for deletion of selective subsets of T cells. The potential immunopathological mechanisms involved in AIDS pathogenesis discussed in this review are summarized in Table 1.

6 The Resistance of HIV-Infected Chimpanzees to Progression to AIDS

6.1 Naturally Infected Nonhuman Primates

Various nonhuman primates of Africa are naturally infected with lentiviruses closely related to HIV-1 and HIV-2. From sooty mangabeys (FULTZ et al. 1986), African green monkeys (OHTA et al. 1988), mandrills (TSUJIMOTO et al. 1988) and chimpanzees (PEETERS et al. 1989), lentiviruses have been isolated which do not appear to cause AIDS in their natural host but rather persist as an asymptomatic infection. Some of these primate lentivirus isolates (SIV), when transmitted to other non-African primates such as macaque species, cause a disease which resembles AIDS in humans (LETVIN et al. 1985). The mechanisms which these African primate hosts have developed to prevent progression to AIDS while maintaining a persistent lentivirus infection are important to understand for the development of new strategies for the prevention of progression to AIDS in HIV-infected individuals.

6.2 Neither Apoptosis nor Anergy in HIV1-Infected Chimpanzees

For this purpose efforts have been made in the study of chimpanzees infected with HIV-1-related strains. None of the experimentally infected chimpanzees which have been followed over 10 years in various institutes have shown any evidence of disease progression (ALTER et al. 1984). Various immunological functional studies, summarized below, did not show any functional abnormality in lymphocytes from HIV-infected chimpanzees. We were unable to demonstrate evidence of increased levels of PCD (spontaneous or activation-induced apoptosis) in lymphocytes from chimpanzees infected with HIV-1 (GOUGEON et al. 1993a) or with SIV (HEENEY et al. 1993). This was correlated with the absence of chronic activation in T lymphocytes and with the insensitivity of these lymphocytes to be primed in vitro to apoptosis by gp120. Indeed, in vitro cross-linking of CD4 molecules on T cells from healthy humans induced in these cells a priming for PCD that was revealed upon subsequent antigenic activation (BANDA et al. 1992).

When the same experiment was performed on lymphocytes from non-infected chimpanzees, no priming effect for PCD was observed (Finkel 1993). Furthermore, Th cells from healthy non-infected chimpanzees were found not to be susceptible to in vitro gp120-induced anergy, whereas in humans, coating of Th cells with gp120 induced an anergic state in response to TCR-dependent stimuli (Di Rienzo et al. 1994). Therefore, gp120 signaling via CD4 in the absence of concomitant TCR triggering has no anergic or deleting effect in chimpanzees' lymphocytes.

6.3 Other Mechanisms of Disease Resistance

Other mechanisms of disease resistance of chimpanzees to AIDS have been proposed to play an important role. For example, the resistance of monocyte/ macrophages and other APCs to HIV infection may be associated with disease resistance. Indeed, in humans, infection of APCs may directly contribute to the impairment of Th cell immunity. Early studies suggested that HIV-1 could not infect blood monocytes (Nara et al. 1989), but more recent studies indicated that HIV-1 reisolated from chimpanzees is infectious for chimpanzee macrophages in

Table 2. Resistance of HIV-infected chimpanzees to progression to AIDS: viral and immunological parameters

	HIV-1 infected chimpanzee	HIV-1 infected human
Persistent infection	Yes	Yes
Progression to AIDS	No	Yes
Virus in circulation		
Plasma antigen	+/-	++
Plasma viremia	-	++
Cell-associated viremia	++	++
Infected PBMC	$<1/10^3$	$>1/10^3$
Replication in macrophages	+/-	Yes
Tissue distribution		
Lymph nodes	+	++
Spleen	?	++
Bone marrow	-	++
In vitro T cell responses		
Anti-CD3 mAbs	Normal	Decreased
APC-dependent	Normal	Decreased
APC-independent	Normal	Decreased
CD8-dependent inhibiting activity of HIV replications in CD4+ cells	Normal	Decreased
Persistant programmed cell death		
Spontaneous	No	Yes
Activation-induced	No	Yes
gp120-dependent signaling		
gp120-induced anergy	No	Yes
Priming for apoptosis	No	Yes

PBMC, peripheral blood monocytes; mAbs, monoclonal antibodies; APC, antigen-presenting cell.

vitro (GENDELMAN et al. 1991). However, chimpanzees infected with these isolates remain asymptomatic (WATANABE et al. 1991). Another mechanism which may be involved in disease resistance is the existence in noninfected and HIV-infected chimpanzees of a CD8$^+$-dependent activity which inhibits HIV replication in infected CD4$^+$ cells (WALKER et al. 1991; CASTRO et al. 1991). The loss of this activity in humans is correlated with disease progression. Finally, although humoral immunity of chimpanzees to HIV-1 resembles that of HIV-infected patients with regard to neutralizing antibodies and the emergence of viral escape mutants (NARA et al. 1990), in vivo infection in chimpanzees appears to be related with suppression of virus and limited virus load. Therefore, HIV-1-infected chimpanzees remain resistant to disease progression, maintaining their T cell immunity comparable to that of noninfected chimpanzees. This is summarized in Table 2.

7 Conclusions

The immunopathogenic mechanisms underlying HIV infection are more complex than was thought several years ago; viral burden is substantial, particularly in lymphoid organs, inappropriate immune activation contributes to the pathogenic process, profound immune suppression finally occurs associated with the destruction of the immune environment and preventing the spontaneous regeneration of the immune system. Therefore, any strategy must consider the complexity of these pathogenic mechanisms and should not be unidirected (FAUCI 1993). Strategies aimed at blocking virus dissemination and inhibiting virus replication must be considered. Since the persistence of chronic activation may lead to immune dysfunction, as discussed above, inhibiting immune activation early in the course of HIV infection may be effective. Considering that the triggering of CD4$^+$ T cells by gp120 alone or complexed with antibodies delivers an anergic signal or primes them for subsequent activation-induced apoptosis, strategies aimed at blocking the gp120-CD4 interaction might be beneficial, for example, by administration of soluble CD4. In addition, strategies to block the apoptotic event by administration of cytokines could be envisaged. If a putative microbial superantigen contributes to AIDS pathogenesis, treatment of this microbe (antibiotic or anti-retroviral if it is coded by HIV) should be taken into account when it is identified. Finally, one should consider the recently reported progressive imbalance in Th cells corresponding to a selective defect in Th 1 responses mediated by IL-2 and interferon-γ and correlated with a predominance of Th2 responses mediated by IL-4, IL-6, and IL-10 (CLERICI and SHEARER 1993). The supposedly positive influence of Th1 type cytokines contributing to an efficient cellular T cell response, as opposed to Th2-type cytokines, shown in vitro to modulate the expression of HIV in infected cells, should also be considered in clinical trials.

References

Akbar AN, Borthwick N, Salmon M et al. (1993) The significance of low bcl-2 expression by CD45RO T cells in normal individuals and patients with acute viral infections. The role of apoptosis in T cell memory. J Exp Med 178: 427–438

Ameisen JC, Capron A (1991) Cell dystunction and depletion in AIDS: the program cell death hypothesis. Immunol Today 12: 102–105

Alter H, Eichberg JW, Masur H et al. (1984) Transmission of HTLV-III from human plasma to chimpanzees: an animal model for AIDS. Science 226: 549–552

Arends MJ, Morris RG, Wyllie AH (1990) Apoptosis. The role of the endonuclease. Am J Pathol 136: 593–608

Bagasra O, Hauptman SP, Lischner HW, Sachs M, Pomerantz RJ (1993) Detection of HIV type 1 provirus in mononuclear cells by in situ PCR. N Engl J Med 326: 1385–1391

Banda NM, Bernier J, Kurahara DK et al. (1992) Cross-linking by HIV gp120 primes T cells for activation-induced apoptosis. J Exp Med 176: 1099–1106

Barre-Sinoussi F, Cherman JC, Rey F et al. (1983) Isolation of a T-lymphotropic retrovirus from a patient at risk for acquired immune deficiency syndrome (AIDS). Science 220: 868–871

Castro BA, Walker CM, Eichberg JW, Levy JA (1991) Suppression of HIV replication by CD8[+] cells from HIV-infected and uninfected chimpanzees. Cell Immunol 132: 246–255

Chams V, Jouault T, Fenouillet E, Gluckman JC, Klatzmann D (1988) Detection of anti-CD4 autoantibodies in the sera of HIV-infected patients using recombinant soluble CD4 molecules. AIDS 2: 353–361

Choi Y, Kappler JW, Marrack P (1991) A superantigen encoding in the open reading frame of the 3' long terminal repeat of mouse mammary tumour virus. Nature 350: 203–207

Clerici M, Shearer GM (1993) A Th1 to Th2 switch is a critical step in the etiology of HIV infection. Immunol Today 14: 107–111

Crise B, Rose JK (1990) Human immunodeficiency type I glycoprotein precursor retains CD4-p56[lck] complex in the endoplasmic reticulum. J Virol 66: 2296–2301

Dadaglio G, Poccia F, Garcia S, Müller-Alouf H, Roue R, Montagnier L, Gougeon ML (1993) Vβ specific T cell clonal anergy in HIV infected individuals with possible involvement of an HIV component. In: Girard M, Valette L (eds) Retroviruses of human AIDS and related animal disease. Foundation Mérieux, Lyon, pp 17–22

Dadaglio G, Garcia S, Montagnier L, Gougeon ML (1994) Selective anergy of Vβ8[+] T cells in human immunodeficiency virus-infected individuals. J Exp Med 179: 413–424

Dalgleish A, Wilson S, Gompels M, Ludlam C, Gazzard B, Coates A, Habeshaw J (1992) T-cell receptor variable gene products and early HIV-1 infection. Lancet 339: 824

Davis MM, Bjorkman PJ (1988) T cell antigen receptor genes and T cell recognition. Nature 334: 395–402

Di Rienzo AM, Furlini, G, Olivier R, Ferris S, Heeney J, Montagnier L (1994) Different proliferative response of human and chumpanzee lymphocytes after contact with HIV-1. Eur J Immunol 24: 34–40

Embretson J, Zupancic M, Ribas JL et al. (1993) Massive covert infection of helper T lymphocytes and macrophages by HIV during the incubation period of AIDS. Nature 362: 359–362

Fauci AS (1988) The human immunodeficiency virus: infectivity and mechanisms of pathogenesis. Science 239: 617–622

Fauci AS (1993) Multifactorial nature of HIV disease: implications for therapy. Science 262: 1011–1018

Finkel T (1993) Apoptosis as a mechanism of CD4[+] T cell death in HIV infection. Conference on apoptosis in AIDS and cancer, Paris, Dec 1993

Fultz PN (1991) Replication of an acutely lethal simian immunodeficiency virus activates and induces prolifefation of lymphocytes. J Virol 65: 4902–4910

Fultz PN, McClure HM, Anderson DC et al. (1986) Isolation of T lymphotropic retrovirus from naturally infected sooty mangabeys monkeys (Cercocebus atys.). Proc Natl Acad Sci USA 83: 5286–5290

Gendelman HE, Ehrlich GD, Baca LM et al. (1991) The inability of HIV to infect chimpanzee monocytes can be overcome by serial passage in vivo. J Virol 65: 3853–3863

Gougeon ML, Montagnier L (1993) Apoptosis in AIDS. Science 260: 1269–1270

Gougeon ML, Montagnier L (1994) Apoptosis in peripheral T lymphocytes during HIV-infection: influence of superantigens and correlation with AIDS pathogenesis. In: Tomei D, Cope FO (eds) Apoptosis II: the molecular basis of apoptosis in disease. Cold Spring Harbor, New York

Gougeon ML, Olivier R, Garcia S, Guétard D, Dragic T, Dauguet C, Montagnier L (1991) Evidence for an engagement process towards apoptosis in lymphocytes of HIV infected patients. C R Acad Sci Paris 312: 529–537

Gougeon ML, Garcia S, Guétard D, Olivier R, Dauguet C, Montagnier L (1992) Apoptosis as a mechanism of cell death in peripheral lymphocytes from HIV1-infected individuals. In: Janossy G, Autran B, Miedema F (eds) Immunology of HIV infection. Karger, Basel, pp 115–126

Gougeon ML, Garcia S, Heeney J, Tschopp R, Lecoeur H, Guetard D, Rame V, Dauguet C, Montagnier L (1993a) Programmed cell death of T lymphocytes in AIDS related HIV and SIV infections. AIDS Res Hum Retroviruses 9: 553–563

Gougeon ML, Laurent-Crawford AG, Hovanessian AG, Montagnier L (1993b) Direct and indirect mechanisms mediating apoptosis during HIV infection: contribution to in vivo CD4 T cell depletion. Semin Immunol 5: 187–194

Gougeon ML, Dadaglio G, Garcia S, Müller-Alouf H, Roue R, Montagnier L (1993c) Is a dominant superantigen involved in AIDS pathogenesis. Lancet 342: 50–51

Groux H, Torpier G, Monté D, Mouton Y, Capron A, Ameisen JC (1992) Activation induced death by apoptosis from human immunodeficiency virus-infected asymtomatic individuals. J Exp Med 175: 331–340

Gulwani-Akolkar B, Posnett DN, Janson CH, Grunwald J, Wigzell H, Akolkar P, Gregersen PK, Silver J (1991) T cell receptor V-segment frequencies in peripheral blood T cells correlate with human leucocyte antigen type. J Exp Med 174: 11–39

Harper ME, Marselle LM, Gallo RC, Wong-Stall F (1986) Detection of lymphocytes expressing human T-lymphotropic virus type III in lymph nodes and peripheral blood from infected-individuals by in situ hybridization. Proc Natl Acad Sci USA 83: 772–776

Heeney J, Jonker J, Koornstra W, Dubbes R, Niphuis H, Di Rienzo AM, Gougeon ML, Montagnier L (1993) The resistance of HIV-infected chimpanzees to progression to AIDS correlates with absence of HIV-related dysfunction. J Med Primatol 22: 194–200

Ho DD, Mougdil T, Alam M (1989) Quantitation of HIV1 in the blood of infected persons. N Engl J Med 321: 1621–1625

Hsia K, Spector SA (1991) HIV DNA is present in a high percentage of CD4$^+$ lymphocytes of seropositive individuals. J Infect Dis 164: 470–475

Hugin A, Vacchio M, Morse H (1991) A virus-encoded "superantigen" in a retrovirus-induced immunodeficiency syndrome of mice. Science 252: 424

Imberti L, Sottini A. Bettinardi A, Puoti M, Primi D (1991) Selective depletion in HIV infection of T cells that bear their specific T cell receptor V beta, sequences. Science 254: 860

Janeway C (1991) Immune recognition. Mls: makes a little sense. Nature 349: 459–460

Jenkinson EJ, Kingston CA, Smith CA, Williams GT, Owen JJT (1989) Antigen-induced apoptosis in developing T cells: a mechanism for negative selection of the TCR repertoire. Eur J Immunol 19: 2175–2180

Kanagawa O, Nussrallah BA, Wiebenga ME, Murphy KM, Morse HC, Carbone FR (1992) Murine AIDS superantigen reactivity of the T cells bearing Vβ5 T cell antigen receptor. J Immunol 149: 9–15

Kappler J, Kotzin B, Herron L, Gelfand E, Bigler R , Boylston A, Carrel S, Posnett D, Choi Y, Marrack P (1989) V beta-specific stimulation of human T cells by staphylococcal toxins. Science 244: 811

Kawabe Y, Ochi A (1991) Programmed cell death and extrathymic reduction of Vβ8$^+$CD4$^+$ T cells in mice tolerant to Staphylococcus aureus enterotoxin B. Nature 349: 245–248

Kizaki H, Tadakuma T, Okada C, Muramatsu J, Ishimura C (1989) Activation of a suicide process of thymocytes through DNA fragmentation by calcium ionophores and phorbol esters. J Immunol 43: 1790–1794

Klatzmann D, Bare-Sinoussi F, Nugeyre MT et al. (1984) Selective tropism of LAV for helper-inducer T lymphocytes. Science 225: 59–62

Korman AJ, Bourgarel P, Meo T, Rieckhof GF (1992) The mouse mammary tumor virus long terminal repeat encodes a type II transmembrane glycoprotein. EMBO J 11: 1901

Lanzavecchia A, Roosnek E, Gregory T et al. (1988) T cells can present antigen such as HIV gp120 targeted to their own surface molecules. Nature 334: 530–532

Laurence J, Hodtsev AS, Posnett DN (1992) Superantigen implicated in dependence of HIV-1 replication in T cells on TCR V beta expression. Nature 358: 255

Laurent-Crawford AG, Krust B, Muller S, Rivière Y, Rey-Cuillé MA, Béchet JM, Montagnier L, Hovanessian A (1991) The cytopathic effect of HIV is associated with apoptosis. Virology 185: 829–839

Laurent-Crawford AG, Krust B, Rivière Y, Muller S, Kieny MP, Dauguet C, Hovanessian AG (1992) Membrane expression of HIV envelope glycoproteins triggers apoptosis in CD4 cells. AIDS Res Hum Retroviruses

Letvin NL, Daniel MD, Seghal PK et al. (1985) Induction of AIDS like disease in macaque monkeys with T cell tropic retrovirus STLV-III. Science 230: 71–73

Levy JA (1993) Pathogenesis of human immunodeficiency virus infection. Microb Rev 57: 183–289

MacDonald H, Baschieri S, Lees R (1991) Clonal expansion precedes anergy and death of V beta 8+ peripheral T cells responding to staphylococcal enterotoxin B in vivo. Eur J Immunol 21: 1963

Marrack P, Kappler J (1990) The staphylococcal enterotoxins and their relatives. Science 248: 1066

McConkey DJ, Hartzell P, Amador-Perez JF, Orrenius S, Jondal M (1989) Ca++ dependent killing of immature thymocytes by stimulation via the CD3/TCR complex. J Immunol 143: 1801–1806

Meyaard L, Otto SA, Jonker RR, Mijnster MJ, Keet R, Miedema F (1992) Programmed death of T cells in HIV-1 infection. Science 257: 217–219

Miedema F, Petit AJC, Terpstra FG et al. (1988) Immunological abnormalities in human immunodeficiency virus (HIV) infected asymptomatic homosexual men. HIV affects the immune system before CD4+ T helper cell depletion. J Clin Invest 82: 1908–1915

Montagnier L, Guétard D, Rame V, Olivier R, Adams M (1989) Virological and immunological factors of AIDS pathogenesis. In: Girard M, Valette L (eds) Retroviruses of human AIDS and related animal diseases. Fondation Mérieux, Lyon, pp 11–17

Nara P, Hatch W, Kessler MJ et al. (1989) The biology of HIV-1 IIIB infection in the chimpanzee: in vivo and in vitro correlations. J Med Primatol 18: 343–355

Nara P, Smit L, Dunlop N, Hatch W et al. (1990) Emergence of virus resistant to neutralization by V3-specific antibodies in experimental HIV-1 IIIB infection of chimpanzees. J Virol 64: 3779–3791

Newell MK, Haughn LJ, Maroun CH, Julius M (1990) Death of mature T cells by separate ligation of CD4 and the T cell receptor for antigen. Nature 347: 286–288

Ohta Y, Masuada T, Tsujimoto H et al. (1988) Isolation of simian immunodeficiency virus from African green monkeys and seroepidemiological survey of the virus in various non-human primates. Int J Cancer 41: 115–122

Pantaleo G, De Maria A, Koenig S et al. (1990) CD8+ T lymphocytes of patients with AIDS maintain normal broad cytoloytic function despite the loss of HIV-specific cytotoxicity. Proc Natl Acad Sci USA 87: 4818–4822

Pantaleo G, Graziosi C, Demarest JF et al. (1993) HIV infection is active and progressive in lymphoid tissue during the clinically latent stage of disease. Nature 362: 355–358

Patterson BK, Till M, Otto P et al. (1993) Detection of HIV-1 DNA and mRNA in individual cells by PCR-driven in situ hybridzation and flow cytrometry. Science 260: 976–979

Peeters M, Honore C, Huet T et al. (1989) Isolation and partial characterization of an HIV-related virus occuring naturally in chimpanzees in Gabon. AIDS 3: 625–630

Poli G, Fauci AS (1992) The effects of cytokines and pharmacologic agents on chronic HIV infection. AIDS Res Hum Retroviruses 8: 191–197

Riviere Y, Tanneau-Salvadori F, Regnault A et al. (1989) HIV-specific cytotoxic responses of seropositive individuals: distinct types of effector cells mediate killing of targets expressing gag and env proteins. J Virol 63: 2270–2277

Rosenberg ZF, Fauci AS (1990) Immunopathogenic mechanisms of HIV infection: cytokine induction of HIV expression. Immunol Today 11: 176–180

Schnittman SM, Greenhouse JJ, Psallidopoulos MC et al. (1990) Increasing viral burden in CD4+ T cells from patients with HIV infection reflects rapidly progressive immunosuppression and clinical disease. Ann Intern Med 113: 438–443

Shearer GM, Clerici M (1991) Early T-helper cell defect in HIV infection. AIDS 5: 245–255

Smith CA, Williams GT, Kingston R, Jenkinson EJ, Owen JJT (1989) Antibodies to CD3/T cell receptor complex induce death by apoptosis in immature T cells in thymic cultures. Nature 337: 181–184

Soudeyns H, Rebai N, Pantaleo GP, Ciurli C, Boghossian T, Sékali RP, Fauci AS (1993) The T cell receptor Vβ repertoire in HIV-1 infection and disease. Semin Immunol 5: 175

Terai C, Kornbluth RS, Pauza CD, Richman DD, Carson DA (1991) Apoptosis as a mechanism of cell death in cultured T lymphoblasts acutely infected with HIV-1. J Clin Invest 87: 1710–1715

Toure-Balde, Michel P, Faye A et al. HIV-2 related Programmed cell death in West Africans infected individuals (submitted to publication)

Tsujimoto H, Cooper RW, Kodama T et al. (1988) Isolation and characterization of simian immunodeficiency virus from mandrills in Africa and its relationship to other human and simian immunodeficiency viruses. J Virol 62: 4044–4050

Uehara T, Miyawaki T, Ohta K et al. (1992) Apoptotic cell death of primed CD45RO⁺ T lymphocytes in Epstein-Barr virus-induced infectious mononucleosis. Blood 80: 452–458

Walker CM, Erickson AL, Hsueh FC, Levy JA (1991) Inhibition of HIV replication in acutely infected CD4⁺ cells by CD8⁺ cells involves a noncytotoxic mechanism. J Virol 65: 5921–5927

Watanabe M, Ringler DJ, Fultz PN, Mackey JJ et al. (1991) A chimpanzee-passaged HIV isolate is cytopathic for chimpanzee cells but does not induce disease. J Virol 65: 3344–3348

Webb S, Morris C, Sprent J (1990) Extrathymic tolerance of mature T cells: clonal elimination as a consequence of autoimmunity. Cell 63: 1249–1256

Weinhold KJ, Lyerly HK, Stanley SD et al. (1989) HIV-1-gp120 mediated immune response and lymphocyte destruction in the absence of viral infection. J Immunol 142: 3091–3097

Wyllie AH, Kerr JFR, Currie AR (1984) Cell death: the significance of apoptosis. Int Rev Cytol 68: 251–267

Wyllie AH, Morris RG, Smith AL, Dunlop D (1991) Chromatin cleavage in apoptosis: association with condensed chromatin morphology and dependence on macromolecular synthesis. J Pathol 142: 67–77

Programmed Cell Death and AIDS Pathogenesis: Significance and Potential Mechanisms

J.C. Ameisen[1,2], J. Estaquier[1], T. Idziorek[1], and F. De Bels[1]

1 Introduction

Human immunodeficiency virus (HIV) infection leads, within about 10 years, to acquired immunodeficiency syndrome (AIDS), characterized by cell loss in several organs, including CD4+ T cells in the immune system, hematopoietic progenitors in the bone marrow and neurons in the brain (Everall et al. 1991; Fauci 1988, 1993; Levy 1993b). In the immune system, cell dysfunction is obseved before cell depletion is detected. These qualitative defects are characterized by a selective loss of CD4+ T cell memory function that includes, in vivo, a failure of CD4+ T cells to mediate delayed-type hypersensitivity reactions to self MHC class II-restricted recall antigens and, in vitro, a selective loss of the ability of T cells to proliferate in response to T cell receptor (TCR) stimulation by these recall antigens antibodies directed to the CD3/TCR complex, or defined polyclonal activators such as pokeweed mitogen (Clerici et al. 1989a,b; Hofmann et al. 1989; Lane et al. 1985; Miedema et al. 1988; Shearer et al. 1986). An additional and paradoxical

---

[1] Unité INSERM U 415 Pathogenèse du sida et des infections à tropisme immunitaire et nerveux, Institut Pasteur, 1 rue du Professeur Albert Calmette, 59019 Lille, France
[2] Faculté de Médecine, Université de Lille II, Place de Verdun, 59045 Lille, France

feature of cell dysfunction in HIV-infected persons is the chronic activation state of the immune system, in spite of an apparent complete lack of CD4$^+$ T helper (Th) cell function, that involves both HIV permissive (monocytes) and nonpermissive (B cells and CD8$^+$ T cells) cell populations (FAUCI 1988,1993).

The tropism of HIV for CD4$^+$ T cells and its cytopathic effect in vitro initially suggested that the pathogenesis of AIDS is solely related to direct virus-mediated cell destruction of HIV-infected cells. However, this concept has been challenged in recent years by a series of observations. First, CD4$^+$ T cell dysfunction is observed at a time when few peripheral blood CD4$^+$ T cells are infected (BRINCHMAN et al. 1991, SCHNITTMAN et al. 1989). Second, neuronal loss is observed in the brain (EVERALL et al. 1991), leading to brain atrophy and dementia; however neurons, in contrast to CD4$^+$ T cells, do not seem to be targets for HIV infection since HIV in the central nervous system is expressed in cells of the macrophage lineage (KOENIG et al. 1986; MICHAELS et al. 1988). Finally, chimpanzees, the only primate model that can be productively and chronically infected with HIV-1, do not, in contrast to HIV-1 infected humans, develop any AIDS-related disease (JOHNSON et al. 1993), even when infected with HIV-1 isolates that are cytopathic in vitro for chimpanzee CD4$^+$ T cells (WATANABE et al. 1991).

These findings suggested the possibility that indirect mechanisms may play an important role in AIDS. The pathogenesis at immune and nerve cell loss has become a major problem in AIDS research, with obvious potential therapeutic implications. The topic of this chapter is to discuss, in the context of a growing amount of experimental evidence, our earlier proposal (AMEISEN and CAPRON 1991), that cell dysfunction and cell depletion in HIV-infected persons may be related to a single mechanism, the abnormal induction in several cell populations of a physiological cell suicide process, programmed cell death (PCD), in response to activation signals that normally promote cell survival, differentiation or proliferation (AMEISEN 1992; AMEISEN and CAPRON 1991).

2 The Programmed Cell Death Hypothesis of AIDS Pathogenesis

A long prevailing concept in cell biology has been that all forms of premature cell death, in particular pathological cell death caused by infectious agents, are a passive consequence of cell injury leading to necrosis. The study of embryonic development, however, has led to the identification of a different process of premature cell death, one that occurs in the absence of disease, is physiologically regulated, and has been termed programmed cell death (PCD) (GLUCKSMAN 1951; SAUNDERS 1966) or apoptosis (KERR et al. 1972). During normal development, two apparently contradictory events take place at the same time; (1) extensive cell proliferation, differentiation, and migration and (2) massive episodes of PCD that play an essential role in form sculpturing, and in the shaping and maturation of the immune system (BLACKMAN et al. 1990) and the brain (OPPENHEIM 1991). PCD also occurs in adult tissues (COHEN 1993; DUVALL and WYLLIE 1986; RAFF 1992). Unlike necrosis, PCD is regulated by signals provided by the local environment, and its induction or prevention depends on the expression of defined genes. These have been characterized in primitive invertebrates and are beginning to be identified in mammals and humans (reviewed in: ELLIS et al. 1991; SCHWARTZ and OSBORNE 1993; WILLIAMS and SMITH 1993).

There are two major groups of pathogenic retroviruses: oncoretroviruses, that cause cancers, and lentiviruses, including HIV, that cause AIDS. It has been known for 20 years that oncoviruses cause cancer by inducing the inappropriate expression in adult cells of genes that play an essential role during normal development in cell proliferation, differentiation and migration. In a paper first submitted to *Immunology Today* in May 1990, we proposed that lentiviruses may cause cell depletion and tissue atrophy in the immune system and the brain by a converse capacity to dysregulate in adult cells the expression of genes that play an essential role during normal development in the induction of PCD (AMEISEN and CAPRON 1991). The idea presented was that most immunological and nonimmunological defects in HIV-infected persons, including CD4$^+$ T cell depletion, brain atrophy and dementia, could be related to the inappropriate induction in various cell populations of an activation-induced cell suicide process by PCD, caused by indirect interference of HIV with intracellular signaling. This hypothesis made several testable predictions, based on the assumption that both early in vitro dysfunction and late in vivo depletion of CD4$^+$ T cells are due to PCD and that modulation of cell signaling may have therapeutic implications by preventing premature cell death and restoring normal cell functions.

When a cell undergoes necrosis, the only way to prevent death is to remove the causative agent; when a cell undergoes PCD, cell death can be prevented, in most cases, only by modulation of cell signaling. The oldest illustration of the potential implication for therapeutic intervention in an active cell suicide process in disease is provided by the Greek legend of the Sirens. The song of Sirens was said to lead the sailors to death, and Ulysses used two ways to prevent it: he put wax in the ears of his sailors, so that they could not hear the song (signal transduction was cut); he asked to be attached to the mast, so that he could hear the song but could no longer react to it (this is achieved in cells that have received a death signal by blocking gene expression or protein synthesis). A third way to prevent death was used by the poet Orpheo, when, nearing the Sirens, he began to play the lyra, letting his music merge with the Siren song (the addition of an appropriate activation cosignal counteracts the effect of a death-inducing signal).

Beyond the scope of AIDS pathogenesis, our hypothesis also questioned the validity of two concepts that prevailed at that time. The first one was that PCD always represents a beneficial and physiological form of cell death, including PCD that is induced by effector cytotoxic lymphocytes and natural killer cells in target cells infected by intracellular pathogens (CLOUSTON and KERR 1985; GOLSTEIN et al. 1991). Our model implied, however, that, in the absence of any effector cells, cell suicide in response to inappropriate activation signals could lead to disease.

The second prevailing concept was that the outcome of T cell receptor (TCR) occupancy strictly depends on the developmental stage of the T cell, and that TCR stimulation could lead in mature T cells to either proliferation or clonal anergy, but not, as in immature thymocytes, to clonal deletion (BLACKMAN et al. 1990). In our model of AIDS pathogenesis, however, a cell death program could remain functional in mature CD4⁺ T cells, and be expressed in response to inappropriate T cell activation. After our paper was first submitted for publication, two in vitro experimental observations from murine models were published that indicated that TCR stimulation could indeed lead to PCD in mature T cells. One showed that antibody-mediated ligation of the CD4 molecule from resting mature CD4⁺ T cells primes them for PCD upon further stimulation of the TCR (NEWELL et al. 1990); the other indicated that TCR restimulation of a mature murine CD4⁺ T-cell clone in the absence of cosignal provided by accessory cells induces PCD (LIU and JANEWAY 1990). These findings led us to propose two potential candidate mechanisms for the induction of CD4⁺ T cell PCD in HIV-1-infected persons: (1) the ligation of the CD4 molecule by the gp120 HIV envelope, by gp120-anti-gp120 antibody immune complexes, or by cross-reactive anti-CD4 autoantibodies; (2) an inhibitory effect of HIV on the functions of accessory cells such as monocytes, leading to a lack of appropriate cosignal delivery by accessory cells to activated CD4⁺ T cells (AMEISEN and CAPRON 1991).

3 Experimental Findings and Potential Significance

Since publication of our paper, the experimental validity of our hypothesis has been suggested by reports from more than 15 laboratories, including ours, showing a relationship between HIV infection, AIDS, and PCD induction in mature T cells, thymocytes, and hematopoietic progenitor cells (BANDA et al. 1992; BISHOP et al. 1993; BONYHADI et al. 1993; CAMERON et al. 1994; COHEN et al. 1992; DEL LLANO et al. 1993; ESTAQUIER et al. 1994; GOUGEON et al. 1991,1993; GROUX et al. 1991, 1992; LAURENT-CRAWFORD et al. 1991, 1993; MARTIN et al. 1994; MEYAARD et al. 1992; MOSIER et al. 1993, 1994; OYAIZU et al. 1993; SARIN et al. 1994; SCHUITEMAKER et al. 1993; TERAI et al. 1991; TIAN et al. 1993; ZAULI et al. 1994). These include the findings that: (a) the in vitro the dysfunction of peripheral blood CD4⁺ and CD8⁺ T cells from HIV-1-infected persons is related to abnormal induction of PCD that can be prevented either by protein synthesis inhibitors, cyclosporin A (CsA) or the addition of activation cosignals (GOUGEON et al. 1991, 1993; GROUX et al. 1991, 1992; MEYAARD et al. 1992; OYAIZU et al. 1993; SARIN et al. 1994); (b) the in vitro cytopathogenic effect of HIV-1 in CD4⁺ T cells is related to PCD induction (CAMERON et al. 1994; LAURENT-CRAWFORD et al. 1991; MARTIN et al. 1994; TERAI et al. 1991); (c) the cross-linking of the CD4 molecule by anti-CD4 antibodies or by the HIV-1 envelope protein, either expressed at the surface of infected cells (COHEN et al. 1992; LAURENT-CRAWFORD et al. 1993; TIAN et al. 1993) or in the form of antibody bound immune complexes (BANDA et al. 1992; OYAIZU et al. 1993) can trigger PCD in uninfected human CD4$^+$ T cells; (d) HIV-1 infection of severe combined immunodeficiency (SCID) mice reconstituted with adult human T cells can lead to CD4$^+$ T cell depletion through in vivo induction of PCD (MOSIER et al. 1993, 1994); (e) HIV-1-mediated PCD may also impair the renewal of CD4$^+$ T cells, as indicated by experiments of HIV-1 infection of SCID mice reconstituted with human thymuses (BONYHADI et al. 1993) and by experiments in which HIV was shown to induce PCD in vitro in uninfected bone marrow hematopoietic progenitor cells (ZAULI et al. 1994); (f) finally, the relevance of these findings to AIDS pathogenesis has been further extended by the observations of abnormal levels of in vitro peripheral blood T cell PCD in primate (AMEISEN 1992; DEL LLANO et al. 1993; ESTAQUIER et al.1994; GOUGEON et al. 1993) and feline models (BISHOP et al. 1993) of pathogenic lentiviral infections that induce AIDS-related diseases, but not in HIV-1-infected chimpanzees that do not develop disease (AMEISEN 1992; ESTAQUIER et al. 1994; GOUGEON et al. 1993; SCHUITEMAKER et al. 1993).

Initial findings in our laboratory indicated that abnormal levels of T cell PCD only involved the CD4$^+$ T cell subset (GROUX et al. 1991, 1992). Subsequent work performed in our laboratory by other investigators (ESTAQUIER et al. 1994) has shown, in accordance with findings from other laboratories (GOUGEON et al. 1991, 1993; MEYAARD et al. 1992; SARIN et al. 1994), that PCD also involves CD8$^+$ T cells. The observation that both CD4$^+$ and CD8$^+$ T cells from HIV-infected persons undergo abnormal PCD in vitro whereas selective CD4$^+$ T cell depletion is an in vivo feature of progression to AIDS raises the question of the significance of these in vitro findings of PCD. One possibility is that in vitro T cell PCD does not reflect the in vivo fate of the T cells; preliminary findings suggest, however, that abnormally high levels of T cell PCD occur in vivo in lymph nodes from HIV-1-infected persons at various stages of the disease (FAUCI 1994). Another possibility is that both CD4$^+$ and CD8$^+$ T cells undergo continuous PCD in vivo in HIV-1-infected persons, but that renewal of CD4$^+$ T cells is selectively impaired. Such a possibility is supported by two recent observations. The first one suggests an intrinsic inequality in the capacity of CD4$^+$ and CD8$^+$ T cells to renew, or to expand in the periphery after initial depletion. Whole body irradiation of primates infected with the simian immunodeficiency virus (SIV) that have retained normal numbers of CD4$^+$ T cells induces an identical profound depletion of B cells, CD4$^+$ and CD8$^+$ T cells that is followed by a rapid reappearance of normal numbers of B cells and CD8$^+$ T cells in the peripheral blood, but by a prolonged state of CD4$^+$ T cell depletion (FULTZ et al. 1994). The second observation points to a possible additional effect of pathogenic lentiviruses on the survival of CD4$^+$ thymocytes; HIV-1 infection of SCID mice reconstituted with human fetal thymuses leads to a profound thymocyte depletion, related to in vivo PCD induction, that affects CD4$^+$ thymocytes more than CD8$^+$ thymocytes (BONYHADI et al. 1993). Similar thymic depletion is observed in vivo in rhesus macaques infected with a pathogenic strain of SIV but not in macaques infected with a nonpathogenic molecular clone of SIV (LACKNER et al. 1994).

We initially proposed that inappropriate T cell PCD induction may be neither pathogenic per se nor unique to HIV-1 infection, and may even have a beneficial inhibitory effect on viral production (AMEISEN 1992; AMEISEN and CAPRON 1991). Accordingly, it is the induction in HIV-1-infected persons of a prolonged and ongoing process of T cell PCD that might lead to disease by interfering with the maintenance of memory T cells and the renewal of effector T cells (AMEISEN 1992; AMEISEN and CAPRON 1991). Recent findings have indeed indicated that transient abnormal priming of T cells for PCD can be observed during several acute viral infections that lead to transient immunosuppression (AKBAR et al. 1993; RAZVI and WELSH 1993; TAMARU et al. 1993; UEHARA et al. 1992). Therefore, an essential question that has remained unresolved for most abnormal features identified so far in HIV-1-infected persons is whether T cell PCD plays a central role in AIDS pathogenesis, or is it merely a consequence of an ongoing and ineffective stimulation of the immune system in a chronic lentiviral infection.

Primate models of pathogenic and nonpathogenic chronic lentiviral infection allow one to address this question. Natural SIV infection of primates such as African green monkeys results in a stable nonpathogenic viral-host interaction that does not lead to disease (MÜLLER et al. 1993), indicating that lentiviruses do not need to destroy the immune system to persist in the infected host and to spread in a wide proportion of the infected species. Experimental infection of primate species with lentiviruses that do not infect them naturally has led to two opposite outcomes: an absence of disease in HIV-1-infected chimpanzees (JOHNSON et al. 1993) and AIDS development in SIV-infected rhesus macaques (DESROSIERS 1990). Rhesus macaques represent a very powerful model for the investigation of events involved in lentiviral pathogenesis. They can be experimentally infected either with viral strains or recombinant molecular clones of SIVmac that induce AIDS, or with viral recombinant molecular clones of SIVmac that do not lead to disease (DESROSIERS 1990; KESTLER et al. 1990). We have compared in vitro T cell PCD induction in HIV-1-infected persons and in these various primate models (ESTAQUIER et al. 1994). Abnormal levels of activation-induced PCD of the CD8$^+$ T cell-depleted peripheral blood mononuclear cells (PBMCs) (containing the CD4$^+$ T cells) were only observed in the two models leading to AIDS: HIV-1-infected humans and rhesus macaques infected with a pathogenic strain of SIVmac. In contrast, enhanced in vitro levels of activation-induced PCD in CD4$^+$ T cell-depleted PBMCs (containing the CD8$^+$ T cells) were detected after in vitro stimulation in both pathogenic and nonpathogenic models of chronic lentiviral infections and in some uninfected primate controls (ESTAQUIER et al. 1994).

These findings suggest that the priming of CD4$^+$ and CD8$^+$ T cells for PCD that occurs in HIV-1-infected humans may represent two different processes, with distinct significance. The one involving CD4$^+$ T cells is closely related to AIDS pathogenesis; the other, involving CD8$^+$ T cells, could be an indirect consequence of immune stimulation that may occur during both pathogenic and nonpathogenic lentiviral infections and in other circumstances. In some HIV-1-infected persons characterized as long-term nonprogressors or long-term survivors (LEVY 1993a), who are infected for more than 6 years and have retained normal CD4$^+$ T cell counts and CD4$^+$ T cell functions, a similar pattern of T cell PCD restricted to the CD8$^+$ T cells has been observed (Estaquier et al., manuscript in preparation).

4 Candidate Mechanisms
for Programmed CD4⁺ T Cell Death in AIDS

The primate models of chronic lentiviral infection indicate that pathogenesis does not solely result from the ability of a lentivirus to infect CD4⁺ T cells, a property shared by all these viruses. Rather, pathogenesis may involve an additional interplay between defined retroviral gene sequences, as indicated by the differing outcomes of macaque infection with various SIVmac molecular clones, and defined genes of the host species, as shown by the opposing outcomes of HIV-1 infection in humans and chimpanzees.

4.1 HIV Envelope-Mediated CD4 Signaling

The first mechanism that we proposed, inappropriate signaling through the binding of the HIV-1 envelope protein to the CD4 molecule (AMEISEN and CAPRON 1991), has now been explored and represents a candidate for PCD induction in CD4⁺ T cells (BANDA et al. 1992; COHEN et al. 1992; LAURENT-CRAWFORD et al. 1993; OYAIZU et al. 1993; TIAN et al. 1993). Death of uninfected CD4⁺ T cells consecutive to an interaction between the HIV envelope protein expressed by infected cells and the CD4 molecule expressed by uninfected CD4⁺ T cells was shown to be due to PCD induction (COHEN et al. 1992; LAURENT-CRAWFORD et al. 1993; TIAN et al. 1993), a process that can be prevented either by selective inhibitors of T cell activation (COHEN et al. 1992; TIAN et al. 1993), or by CD4 antibodies that do not inhibit binding of HIV envelope to the CD4 molecule but may act by modifying CD4-mediated signal transduction (LAURENT-CRAWFORD et al. 1993). Cross-linking of the CD4 molecule by anti-CD4 antibodies or by gp120 plus anti-gp120 antibodies was also shown either to prime purified normal human CD4⁺ T cells for PCD, in response to subsequent TCR stimulation (BANDA et al. 1992), or in the absence of TCR stimulation, provided that accessory cells are present in the culture (OYAIZU et al. 1993).

An extreme interpretation of these findings is that HIV-mediated cytopathic effect requires, in all cases, an interaction between "effector" infected cells, expressing the HIV envelope, and neighboring "target" CD4⁺ T cells. Limiting dilution experiments will be required in order to assess whether HIV infection can induce cell death at the level of a single infected cell. An alternate interpretation is that some level of cytopathic effect is induced by all lentiviruses in vivo, regardless of whether they are pathogenic or not, but that cytopathogenicity is not sufficient to induce cell depletion and disease in the absence of additional indirect mechanisms of envelope-CD4-mediated PCD induction.

Together, these findings suggest that the HIV envelope protein and the immune response to it may participate in CD4⁺ T cell PCD in HIV-infected persons. Although the HIV envelope protein is a tempting candidate, such a possibility has to be considered in the broader context of the primate models of pathogenic and nonpathogenic lentiviral infection mentioned above. In these models, the capacity of the viral envelope to bind the CD4 molecule is a feature that ensures CD4+ T cell infection and is shared by all lentiviruses. Therefore, one is forced to postulate that subtle differences in the CD4 molecules (in humans and chimpanzees) or the lentiviral envelopes (in pathogenic and nonpathogenic molecular clones of SIVmac) are sufficient to account for radical differences in the capacity of envelope–CD4 interaction to induce PCD. Such a possibility awaits to be addressed experimentally in an in vitro system using the various envelopes from these lentiviruses and the CD4+ T cells from different primate species.

4.2 HIV-Mediated Accessory Cell Dysfunction

The second mechanism that we originally proposed as a candidate for the priming of CD4+ T cells for PCD in HIV-1-infected persons was related to a general property of lentiviruses, their tropism for accessory cells such as monocytes/macrophages (AMEISEN and CAPRON 1991). If lentiviral infection of monocytes/macrophages induces a defect in their accessory cell function, this could lead to T cell PCD through abnormal delivery of the activation cosignals that are required for appropriate T cell activation (AMEISEN and CAPRON 1991). The two-signal model of lymphocyte activation, a paradigm in cellular immunology for more than 20 years, implies that mature T cell proliferation requires both TCR stimulation by antigen and appropriate cosignaling provided by antigen-presenting accessory cells (BRETSCHER and COHN 1970; JANEWAY 1992; JENKINS 1992). Mature TCR stimulation in the absence of appropriate accessory cell cosignaling leads to a state of T cell desensitization that has been termed anergy (JENKINS 1992). During the last 4 years, it was found that TCR stimulation in the presence of inappropriate cosignaling can also induce programmed T cell death (GROUX et al. 1993; LENARDO 1991; LIU and JANEWAY 1990; SAMBHARA and MILLER 1991; WANG et al. 1993). These findings have led to a blurring of the frontiers between anergy and PCD induction and to a progressive reassessment of the outcome of T cell activation in conditions that do not lead to T cell proliferation. They suggest that a death program is functional in mature T cells and that its expression may depend both on the degree of T cell activation and on the nature of the environmental cosignals provided to the T cell by the accessory cell. Such a death program might have a physiological role in the prevention of autoimmunity and in the ending of a normal immune response to foreign antigens. In the latter case, PCD may occur in low affinity or bystander-activated T cells and in terminally differentiated effector T cells, sparing memory T cells.

An extreme view of T cell survival regulation is that any activation of the T cell will lead to PCD induction, unless environmental cosignals adapted to the activation state of the T cell are provided that will prevent PCD and allow differentiation and proliferation to proceed. In such a view, the two-signal model of T cell activation would only represent a particularly well studied example of the general control exerted on cell survival by the environment (RAFF 1992). T cell anergy may represent an intermediate case of cell survival, in which the initial

TCR activating signal is not strong enough to induce PCD and therefore does not require additional cosignaling in order to allow T cell survival. The possibility, however, that anergy represents a state of priming for PCD that will lead to death upon further restimulation is suggested by recent findings that anergic B cells have a reduced lifespan in vivo (FULCHER and BASTEN 1994).

Provided that pathogenic lentiviruses have the capacity to modify accessory cell function, they may lead to T cell PCD by "default", simply by altering the balance of activation signals required to prevent PCD induction in activated T cells (AMEISEN and CAPRON 1991). Such an inbalance could be achieved in two opposite ways: (1) a reduction in the availability of the accessory cell cosignals required to prevent T cell death in response to a given activation signal; or (2) an increase in the intensity or duration of the initial T cell activation that will render inoperative the normal amount of preventive accessory cell cosignals that are present in the T cell environment.

Recent findings suggest that HIV-1 infection of accessory cells may indeed play a role in T cell PCD induction: In chimpanzees infected with various syncytia-inducing (SI) or nonsyncytia-inducing (NSI) strains of HIV-1, the lack of in vivo pathogenesis of HIV-1 and the lack of priming of peripheral blood T cells for in vitro PCD induction in this model is not related to a lack of cytopathic properties of these viruses (since some of the SI strains are cytopathic in vitro for chimpanzee CD4$^+$ T cells) (SCHUITEMAKER et al. 1993), but may involve a lack in the capacity of all these HIV strains to infect chimpanzee monocytes; however, the validity of this observation has been questioned (MANNHALTER et al. 1994) and remains to be confirmed. In addition, in SCID mice reconstituted with adult human T cells and monocytes, infection with NSI monocytotropic molecular clones of HIV-1, which are noncytopathic in vitro for CD4$^+$ T cells, leads to a more rapid and profound in vivo depletion of CD4$^+$ T cells than does infection with SI clones of HIV-1 that are highly cytopathic in vitro for CD4$^+$ T cells, but are poorly tropic for monocytes (Mosier et al. 1993). This phenomenon seems related to in vivo PCD induction in CD4$^+$ T cells (MOSIER et al. 1994). Finally, recent preliminary findings indicate that T helper (Th) 1 and Th2 cytokines secreted by accessory cells may participate in the regulation of PCD in T cells from HIV-1-infected persons: the (Th) 2 cytokine interleukin (IL)-10, by inducing T cell PCD; and the Th1 cytokine IL-12, by preventing T cell PCD (CLERICI et al. 1994; ESTAQUIER and AMEISEN 1994; AMEISEN et al. 1994).

4.3 Accessory Cell Dysfunction, T Helper (Th)1/Th2 Cytokines, and Programmed T Cell Death

G. Shearer and M. Clerici have recently proposed that in HIV-infected persons, the functional defects of cell-mediated immunity that involve the CD4+ Th1 cell population may be related to a progressive shift of CD4$^+$ T cells from Th1 to Th2, characterized by a loss of IL-2 and interferon-γ (IFN-γ) production, concomitant with increases in IL-4 and IL-10 secretion (CLERICI et al. 1993; CLERICI and SHEARER

1993). It was subsequently suggested that such a process may involve accessory cell dysfunction in HIV-infected persons (MEYAARD et al. 1993). In this context, we have investigated the Th1/Th2 cytokine secretion profile in HIV-infected persons and the possible role these cytokines may play in T cell PCD. Our results indicate that stimuli that induce PCD in T cells from HIV-infected persons induce in vitro levels of IL-2 and IFN-γ secretion that are similar in HIV-infected persons and healthy controls, when measured 24 h after in vitro activation. No significant IL-4 or IL-5 secretion was detected in most HIV-infected persons, up to 4 days after in vitro stimulation, and IL-10 secretion was similar in activated T cells from HIV-infected persons and controls (ESTAQUIER and AMEISEN 1994). Our findings of a lack of Th2 cell expansion are consistent with preliminary cytokine messenger RNA analysis in vivo in the lymph nodes of HIV-infected persons (FAUCI 1994). Therefore, we favor the interpretation that the progressive loss of sustained in vitro and in vivo Th1 cell response that characterize AIDS progression is not related to an absence of Th1 CD4+ cells (nor to a down-regulation of Th1 cells by an expanding Th2 cell population) but to the fact that the stimulation of Th1 cells from HIV-infected persons induces their rapid death by apoptosis.

The addition of antibodies against the type 2 cytokines IL-10 or IL-4 or the addition of the type 1 cytokine IL-12 has been reported to restore the defective in vitro proliferative response of T cells from HIV-infected persons to stimuli (CLERICI et al. 1993). Recent results from our laboratory and G. Shearer's indicate that the addition of antibodies to IL-10 or IL-4 or the addition of IL-12 have a preventive effect on T cell PCD induction in response to in vitro stimulation in HIV-infected persons.

Together, our results imply that T cells from HIV-infected persons may have an abnormal susceptibility to type 2 cytokines. The preventive effect of IL-12 on T cell PCD suggests the possibility that this abnormal susceptibility may be related to the reported defect of IL-2 secretion by accessory cells from HIV-infected persons (CHEHIMI et al. 1994).

Type 1 and type 2 cytokines exert a major role in the regulation of Th1/Th2 balance by inducing the respective expansion of Th1 or Th2 CD4+ cell population and by down-regulating the converse Th cell population (MOSMANN and COFFMAN 1989). In this context, our results have two important implications. The first one is that cytokine-mediated T cell PCD may play a general role in the regulation of Th1/Th2 responses. In other words type 1 or type 2 cytokines may contribute to Th1 or Th2 CD4+ cell expansion by inducing PCD in the converse Th CD4+ cell population. The second implication directly relates to AIDS pathogenesis. If accessory cell dysfunction represents, as we initially proposed (AMEISEN 1992; AMEISEN and CAPRON 1991), a major pathological feature in HIV-infected persons, it is tempting to speculate that accessory cells may secrete the cytokines or express the cell surface molecules that normally play an essential role in Th1 PCD induction during an efficient process of Th1 to Th2 switch, but may not secrete the cytokines that normally allow the concomitant expansion of the Th2 CD4+ cells. Accordingly, a progressive loss of Th1 CD4+ cells would occur in HIV-infected persons in the absence of a compensatory expansion of Th2 CD4+ cells.

Such an abortive Th1/Th2 switch could at least partly account for both the progressive CD4⁺ T cell dysfunction and depletion that lead to AIDS. An obvious implication would be that therapeutic strategies designed to prevent Th1 CD4⁺ cell PCD and to allow Th1 CD4⁺ cell expansion may have an important role in the treatment of AIDS. But this concept also has a paradoxical implication. In contrast to several infectious or parasitic diseases, in which an imbalance between Th1/Th2 cells has been shown to directly lead to disease (SCOTT and KAUFMANN 1991), it is important to remember that AIDS is primarily a direct consequence of CD4⁺ T cell depletion. If an ineffective abortive Th1/Th2 switch is involved in AIDS pathogenesis, one may not exclude the provocative possibility that therapeutic strategies designed to allow an efficient Th2 CD4⁺ cell switch and to induce Th2 CD4⁺ cell expansion may also have beneficial effects during the course of CD4⁺ T cell depletion and AIDS (ESTAQUIER and AMEISEN 1994; AMEISEN et al. 1994).

4.4 Accessory Cells and Programmed Cell Death in Immunological and Nonimmunological Organs

Several other mechanisms could be involved in the priming or induction of T cell PCD in HIV-infected persons. Candidates include putative superantigens that may be encoded by HIV (JANEWAY 1991) or the binding of self-molecules expressed on the surface of activated T cells, such as Fas/APO-1 (KRAMMER et al. 1994), that may result either from the generation of autoantibodies or from an abnormal expression of Fas or the Fas ligand on the surface of lymphocytes or accessory cells.

Whatever mechanism may be involved, it is important to consider that the T cells that are recirculating in the peripheral blood represent, at any given time, less than 2% of the total lymphocyte pool in the body that is essentially present in the lymphoid organs (WESTERMANN and PABST 1990). It is possible that T cell PCD mainly occurs in the lymphoid organs, a place in which both the immune response develops and most of the viral burden is located (FAUCI 1993). Therefore, the presence of T cells primed for PCD in the peripheral blood of HIV-infected persons that does not appear to increase with progression to disease (MEYAARD et al. 1994) could only represent a very indirect consequence of two major additive events that may play an essential role in AIDS pathogenesis and occur outside the peripheral blood: mature T cell deletion following PCD induction in the lymph nodes and impairement of T cell renewal by PCD induction in progenitor cells in both the thymus and the bone marrow.

The progressive depletion of accessory cells, such as the follicular dendritic cells in the lymph nodes and the epithelial cells in the thymus, could lead at a late stage of the disease to irreversible T cell PCD due to a complete absence of appropriate accessory cell cosignal delivery. Mechanisms involved in accessory cell death in HIV-infected persons remain unknown. It is possible that activated accessory cells require signals from activated T cells in order to survive. If this were true, HIV-mediated interference with intercellular signaling could play a role in the progressive collapse of lymphoid organs that occur at the late stage of the disease (Fauci 1993). An essential question that remains to be addressed is whether therapeutic modulation of cell signaling may have any effect in preventing cell death in these organs.

Finally, it remains to be investigated whether cell loss and tissue atrophy that occur in non-immunological organs from HIV-infected persons are also related to inappropriate induction of PCD. The ultrastructural observation of abnormal levels of epithelial cell apoptosis in rectal crypts of AIDS patients (Kotler et al. 1986) supports the possibility that abnormal PCD induction in AIDS involves a wide range of cell populations.

In the immune system, both CD4$^+$ T cells and accessory cells are targets for HIV infection. HIV-infected accessory cells include macrophages and dendritic cells in the lymph nodes, macrophages in the bone marrow, and epithelial cells in the thymus (Fauci 1993; Levy 1993b). Therefore, it is difficult to assess whether deletion of CD4$^+$ T cells, thymocytes and hematopoietic progenitor cells is due to direct HIV-mediated cytopathogenic effect or to indirect mechanisms triggered by accessory cell infection.

In the brain, however, neuronal loss (Everall et al. 1991) is observed, although neurons, in contrast to CD4$^+$ T cells, do not seem to be targets for HIV-1 infection. HIV-1 in the central nervous system is expressed primarily in cells of the macrophage lineage (Koenig et al. 1986; Michaels et al. 1988), and HIV-infected macrophages have been shown to be able to induce neuronal cell death (Genis et al. 1992). Similar to T cells, neurons normally depend on signals provided by other cell populations in order to prevent PCD induction (Raff et al. 1993). Therefore, the abnormal induction of PCD in CD4$^+$ T cells, hematopoietic progenitors and neurons through inappropriate delivery of accessory cell survival signal may represent a unifying mechanism by which HIV- or SIV-infected macrophages could be involved in the pathogenesis of both immunological and non-immunological defects leading to AIDS. If this were true, the tropism of a given lentivirus for accessory cells and the nature of the functional changes that this virus induces in the infected accessory cells would represent the critical features that distinguish lentiviral infections leading to AIDS from nonpathogenic lentiviral infections.

5 Programmed Cell Death Dysregulation and Disease

Further studies will be required to assess whether PCD dysregulation is central to AIDS pathogenesis, to identify the viral and host genes involved in PCD dysregulation, and to explore to what extent therapeutic strategies aimed at the in vivo modulation of PCD may contribute to the prevention of AIDS.

Abnormal expression of genes involved in the physiological regulation of PCD can result either in premature cell death or in extended cell survival and oncogenesis (Raff 1992; Schwartz and Osborne 1993; Williams and Smith 1993).

Therefore, therapeutic strategies aimed at preventing abnormal T cell PCD in HIV-infected persons might not be devoid of deleterious effects, including inducing an increase in viral production, breaking of self-tolerance, dysregulation of the immune response, or development of tumors. Animal models of AIDS-related diseases will be required to investigate the possible consequences of in vivo treatment designed to prevent PCD. A hopeful note may be inferred from findings indicating that the PCD suppressor gene *bcl*-2 does not prevent all forms of PCD in the T cell lineage and does not appear to favor T cell oncogenesis (KORSMEYER 1992). It is possible therefore that prevention of abnormal HIV-mediated T cell PCD induction may be achieved by treatments that will not lead to T cell immortalization.

Another important question is whether tumors, such as Kaposi's sarcoma and B-cell lymphoma, that are frequent in HIV-infected individuals are the sole consequence of the progressive immunodeficiency that occurs in these patients or are they related to PCD dysregulation. AIDS Kaposi's sarcoma cells could provide a relevant model, since they appear to depend on growth factors released by other cells in order to become transformed (NAIR et al. 1992). It is possible that AIDS represents a range of diseases in which retroviral-mediated interference with cell signaling leads, at the same time but in different cell populations, to PCD induction and cell loss and to PCD prevention and cell immortalization (AMEISEN 1992). An example for such a concept may be provided by the murine mammary tumor virus, an oncoretrovirus that induces both PCD in a cell population (the T cell) and cell immortalization and cancer in another one (the mammary gland epithelial cell) (HELD et al. 1994).

Recent evidence supports the general concept that the persistence and pathogenesis of several viruses may be related to their capacity to subvert PCD regulation in various cell populations (WILLIAMS and SMITH 1993). For 20 years, oncoretroviruses have represented very powerful probes to identify how the inappropriate expression of host genes leads to oncogenesis, a mechanism that also occurs in the absence of any viral infection. Inappropriate induction of PCD similar to that involved in AIDS may represent a paradigm for the pathogenesis of other diseases leading, in the absence of any viral infection, to cell dysfunction, cell loss and tissue atrophy. If this were true, in vivo modulation of cell signaling may have wide ranging implications for therapy.

References

Akbar AN, Borthwick N, Salmon M, Gombert W, Bofill M, Shamsadeen N, Pilling D, Pett S, Grundy JE, Janossy G (1993) The significance of low bcl-2 expression by CD45RO+ T cells in normal individuals and patients with acute viral infections. The role of apoptosis in T cell memory. J Exp Med 178: 427–438

Ameisen JC (1992) Programmed cell death and AIDS: from hypothesis to experiment. Immunol Today 13: 388–391

Ameisen JC, Capron A (1991) Cell dysfunction and depletion in AIDS: the programmed cell death hypothesis. Immunol Today 12: 102–105

Ameisen JC, Estaquier J, Idziorek T (1994) From AIDS to parasite infection: pathogen-mediated subversion of programmed cell death as a mechanism for immune dysregulation. Immunol Rev 142: 9–51

Banda NK, Bernier J, Kurahara DK, Kurrle R, Haigwood N, Sekaly RP, Finkel TH (1992) Crosslinking CD4 by HIV gp120 primes T cells for activation-induced apoptosis. J Exp Med 176: 1099–1106

Bishop SA, Gruffydd-Jones TJ, Harbour DA, Stokes CR (1993) PCD (apoptosis) as a mechanism of cell death in PBMC from cats infected with feline immunodeficiency virus (FIV). Clin Exp Immunol 93: 65–71

Blackman M, Kappler J, Marrack P (1990) The role of TCR in positive and negative selection of developing T cells. Science 248: 1335–1341

Bonyhadi ML, Rabin L, Salimi S, Brown DA, Kosek J, McCune JM, Kaneshima H (1993) HIV induces thymus depletion in vivo. Nature 363: 728–732

Bretscher P, Cohn M (1970) A theory of self-nonself discrimination. Science 169: 1042–1049

Brinchman JE, Albert J, Vartdal F (1991) Few infected CD4+ T cells but a high proportion of replication-competent provirus copies in asymptomatic HIV-1 infection. J Virol 65: 2019–2023

Cameron PU, Pope M, Gezelter S, Steinman RM (1994) Infection and apoptotic cell death of CD4 T cells during an immune response to HIV-1-pulsed dendritic cells. AIDS Res Hum Retroviruses 10: 61–71

Chehimi J, Starr SE, Frank I, D'Andrea A, Ma X, MacGregor RR, Sennelier J, Trinchieri G (1994) Impaired IL-12 production in HIV-infected patients. J Exp Med 179: 1361–1366

Clerici M, Shearer G (1993) A TH1/TH2 switch is a critical step in the etiology of HIV infection. Immunol Today 14: 107–111

Clerici M, Stocks NI, Zajac RA, Boswell RN, Bernstein DC, Mann DL, Shearer GM, Berzofsky JA (1989a) Interleukin-2 production used to detect antigenic peptide recognition by T-helper lymphocytes from asymptomatic HIV-seropositive individuals. Nature 339: 383–385

Clerici M, Stocks NI, Zajac RA, Neal Boswell R, Lucey DR, Via CS, Shearer GM (1989b) Detection of three distinct patterns of T helper cell dysfunction in asymptomatic, human immunodeficiency virus-seropositive patients. J Clin Invest 84: 1892–1899

Clerici M, Lucey P, Berzofsky J, Pinto L, Wynn T, Blatt S, Dolan M, Hendrix C, Wolf S, Shearer G (1993) Restoration of HIV specific cell-mediated immune response by IL-12 in vitro. Science 262: 1721–1724

Clerici M, Sarin A, Henkart P, Shearer GM (1994) Cell mediated immunity in HIV infection. Biotechnology 9: 53 (Special issue: Proceedings of "Biotech AIDS'94", Florence)

Clouston WM, Kerr JFR (1985) Apoptosis, lymphocytotoxicity and the containment of viral infections. Med Hypothesis 18: 399–404

Cohen DI, Tani Y, Tian H, Boone E, Samelson L, Lane HC (1992) Participation of tyrosine phosphorylation in the cytopathic effect of HIV-1. Science 256: 542–545

Cohen JJ (1993) Apoptosis. Immunol Today 14: 126–130

Del Llano AM, Amieiro-Puig JP, Kraiselburd EN, Kessler MJ, Malaga CA, Lavergne JA (1993) The combined assessment of cellular apoptosis, mitochondrial function and proliferative response to pokeweed mitogen has prognostic value in SIV infection. J Med Primatol 22: 194–200

Desrosiers RC (1990) The simian immunodeficiency viruses. Annu Rev Immunol 8: 557–578

Duvall E, Wyllie AH (1986) Death and the cell. Immunol Today 7: 115–119

Ellis RE, Yuan J, Horvitz HR (1991) Mechanisms and functions of cell death. Annu Rev Cell Biol 7: 663–698

Estaquier J, Ameisen JC (1994) Programmed cell death (apoptosis) and AIDS: is Th1 dysfunction and deletion related to an abortive Th1/Th2 switch process? In: Romagnani S, Abbas K (eds) Challenges in modern medicine. Third international conference on cytokines: basic principles and practical applications. Serono Symposia Publications, Rome, pp195–201

Estaquier J, Idziorek T, De Bels F, Barré-Sinoussi F, Hurtrel B, Aubertin AM, Venet A, Mehtali M, Muchmore E, Michel P, Mouton Y, Girard M, Ameisen JC (1994) PCD and AIDS: the significance of T-cell apoptosis in pathogenic and non pathogenic primate lentiviral infections. Proc Natl Acad Sci USA 91: 9431–9435

Everall IP, Luthert PJ, Lantos PL (1991) Neuronal loss in the frontal cortex in HIV-infection. Lancet 337: 1119–1121

Fauci AS (1988) The human immunodeficiency virus: infectivity and mechanisms of pathogenesis. Science 239: 617–620

Fauci AS (1993) Multifactorial nature of HIV disease: implications for therapy. Science 262: 1011–1018

Fauci AS (1994) Multifactorial and multiphasic components of the immunopathogenic mechanisms of HIV disease. In: Girard M, Valette L (eds) Retroviruses of human AIDS and related animal diseases. VIIIth Colloque des Cent Gardes, 1993. Fondation Mérieux, Lyon, pp 81–85

Fulcher DA, Basten A (1994) Reduced life span of anergic self-reactive B cells in a double-transgenic model. J Exp Med 179: 125–134

Fultz PN, Schwiebert RS, Su LY, Salter MM (1994) Total lymphoid irradiation as a novel therapeutic approach for treatment of HIV-induced disease. In: Girard M, Valette L (eds) Retroviruses of human AIDS and related animal diseases. VIIIth Colloque des Cent Gardes, 1993. Fondation Mérieux, Lyon, pp 245–249

Genis P, Jett M, Bernton EW, Boyle T, Gelbard HA, Dzenko K, Keane RW, Resnick L, Mizrachi Y, Volsky DJ, Epstein LG, Gendelman HE (1992) Cytokines and arachidonic metabolites produced during HIV-infected macrophage-astroglia interactions: implications for the neuropathogenesis of HIV disease. J Exp Med 176: 1703–1718

Glucksman A (1951) Cell deaths in normal vertebrate ontogeny. Biol Rev 26: 59–86

Golstein P, Ojcius DM, Young JDE (1991) Cell death mechanism and the immune system. Immunol Rev 121: 29–65

Gougeon M-L, Olivier R, Garcia S, Guetard D, Dragic T, Dauguet C, Montagnier L (1991) Mise en évidence d'un processus d'engagement vers la mort cellulaire par apoptose dans les lymphocytes de patients infectés par le VIH. C R Acad Sci (Paris) 312: 529–537

Gougeon M-L, Garcia S, Heeney J, Tschopp R, Lecoeur H, Guétard D, Rame V, Dauguet C, Montagnier L (1993) PCD in AIDS-related HIV and SIV infections. AIDS Res Hum Retroviruses 9: 553–563

Groux H, Monté D, Bourez JM, Capron A, Ameisen JC (1991) L'activation des lymphocytes TCD4+ de sujets asymptomatiques infectés par le VIH entraîne le déclenchement d'un programme de mort lymphocytaire par apoptose. C R Acad Sci (Paris) 312: 599–606

Groux H, Torpier G, Monté D, Mouton Y, Capron A, Ameisen JC (1992) Activation-induced death by apoptosis in CD4+ T Cells from HIV-infected asymptomatic individuals. J Exp Med 175: 331–340

Groux H, Monté D, Plouvier B, Capron A, Ameisen JC (1993) CD3-mediated apoptosis of human medullary thymocytes and activated peripheral T cells: respective roles of IL-1, IL-2, IFNγ and accessory cells. Eur J Immunol 1623–1629

Held W, Acha-Orbea H, McDonald HR, Waanders GA (1994) Superantigens and retroviral infection: insights from MMTV. Immunol Today 15: 184–190

Hofmann B, Jakobsen KD, Odum N, Dickmeiss E, Platz P, Ryder LP, Pedersen C, Mathiesen L, Bygbjerg I, Faber V, Svejgaard A (1989) HIV-induced immunodeficiency relatively preserved PHA as opposed to decreased PWM responses via CD2/PHA pathway. J Immunol 142: 1874–1880

Janeway CA Jr (1991) MLS: makes a little sense. Nature 349: 459–461

Janeway CA Jr (1992) The immune system evolved to discriminate infectious nonself from non infectious self. Immunol Today 13: 11–16

Jenkins M (1992) The role of cell division in the induction of clonal anergy. Immunol Today 13: 69–73

Johnson BK, Stone GA, Godec MS, Asher DM, Gajdusek DC, Gibbs CJ Jr (1993) Long-term observations of HIV-infected chimpanzees. AIDS Res Hum Retroviruses 9: 375–378

Kerr JFR, Wyllie AH, Currie AR (1972) Apoptosis: a basic biological phenomenon with wide-ranging implications in tissue kinetics. Br J Cancer 26: 239–257

Kestler H, Kodama T, Ringler D, Marthas M, Pedersen N, Lackner A, Regier D, Sehgal P, Daniel M, King N, Desrosier R (1990) Induction of AIDS in rhesus monkeys by molecularly cloned SIV. Science 248: 1109–1112

Koenig S, Gendelman HE, Orenstein JM, Dal Canto MC, Pezeshkpour GH, Yungbluth M, Janotta F, Aksamit A, Martin MA, Fauci AS (1986) Detection of AIDS virus in macrophages in brain tissue from AIDS patients with encephalopathy. Science 233: 1089–1093

Korsmeyer S (1992) Bcl-2: a repressor of lymphocyte death. Immunol Today 13: 285–288

Kotler DP, Weavor SC, Torzakio JA (1990) Ultrastructural features of epithelial cell degeneration in rectal crypts of patients with AIDS. Am J Surg Pathol 10: 531–538

Krammer PH, Behrmann I, Daniel P, Dhein J, Debatin KM (1994) Regulation of apoptosis in the immune system. Curr Opin Immunol 6: 279–289

Lackner AA, Vogel P, Hoogenboom E, Luge JD, Marthas M (1994) Pathogenic (SIVmac-239) and nonpathogenic (SIVmac-1A11) molecular clones of SIV have distinct tissue distributions that vary with length of infection. In: Girard M, Valette L (eds) Retroviruses of human AIDS and related animal diseases. VIIIth Colloque des Cent Gardes, 1993. Fondation Mérieux, Lyon, pp 27–34

Lane HC, Depper JM, Greene WC, Whalen G, Waldmann TA, Fauci AS (1985) Qualitative analysis of immune function in patients with the acquired immunodeficiency syndrome. N Engl J Med 313: 79–84

Laurent-Crawford AG, Krust B, Muller S, Rivière Y, Rey-Cuillé MA, Béchet JM, Montagnier L, Hovanessian AG (1991) The cytopathic effect of HIV is associated with apoptosis. Virology 185: 829–839

Laurent-Crawford AG, Krust B, Rivière Y, Desgranges C, Muller S, Kieny MP, Dauguet C, Hovanessian AG (1993) Membrane expression of HIV envelope glycoproteins triggers apoptosis in CD4 cells. AIDS Res Hum Retroviruses 9: 761–773

Lenardo MJ (1991) IL-2 programs mouse $\alpha\beta$ T-lymphocytes for apoptosis. Nature 353: 858–861

Levy JA (1993a) HIV pathogenesis and long-term survival. AIDS 7: 1401–1410

Levy JA (1993b) Pathogenesis of human immunodeficiency virus infection. Microbiol Rev 57: 183–289

Liu Y, Janeway CA (1990) INFγ plays a crucial role in induced cell death of effector T cell: a possible third mechanism of self tolerance. J Exp Med 172: 1735–1741

Mannhalter JW, Husch B, Küpcü Z, Eibl MM (1994) Capacity of HIV-1 to infect chimpanzee monocyte in vitro. Infect Dis 169: 1407–1409

Martin SJ, Matear P, Vyakarnam A (1994) HIV-1 infection of human CD4 T cells in vitro. Differential induction of apoptosis in these cells. J Immunol 152: 330–342

Meyaard L, Otto SA, Jonker RR, Mijnster MJ, Keet RPM, Miedema F (1992) PCD of T cells in HIV-1 infection. Science 257: 217–219

Meyaard L, Otto SA, Schuitemaker H, Miedema F (1993) T-cell dysfunction in HIV infection: anergy due to defective antigen-presenting cell function? Immunol Today 14: 161–164

Meyaard L, Otto SA, Keet IPM, Roos MTL, Miedema F (1994) PCD of T cells in HIV infection. No correlation with progression to disease. J Clin Invest 93: 982–988

Michaels J, Sharer LR, Epstein LG (1988) HIV-1 infection of the nervous system: a review. Immunodefic Rev 1: 71–104

Miedema F, Petit AJC, Terpestra FG, Eeftinck Schattenkerk JKM, Dewolf F, Al BJM, Roos M, Lange JMA, Danner SA, Goudsmit J, Schellekens PTA (1988) Immunological abnormalities in human immunodeficiency virus (HIV)-infected asymptomatic homosexual men. J Clin Invest 82: 1908–1914

Mosier DE, Gulizia RJ, MacIsaac PD, Torbett BE, Levy JA (1993) Rapid loss of CD4+ T cells in human-PBL-SCID mice by noncytopathic HIV isolates. Science 260: 689–692

Mosier D, Gulizia R, Rochford R, Tenner-Racz K, Racz P (1994) Local patterns of CD4 T-cell depletion in lymph nodes of human PBL-SCID mice infected with macrophage-tropic HIV strains. J Cell Biochem [Suppl] 18B (abstract J257)

Mosmann T, Coffman R (1989) TH1 and TH2 cells: different patterns of lymphokine secretion lead to different functional properties. Annu Rev Immunol 7: 145–173

Müller MC, Saksena NK, Nerrienet E, Chappey C, Herve VMA, Durand JP, Legal-Campodonico P, Lang MC, Digoutte JP, Georges AG, Georges-Courbot MC, Sonigo P, Barre-Sinoussi F (1993) Simian immunodeficiency viruses from Central and Western Africa: evidence for a new species-specific lentivirus in tantalus monkeys. J Virol 67: 1227–1235

Nair BC, De Vico AL, Nakamura S, Copelans TD, Chen Y, Patel A, O'Neil T, Oroszlan S, Gallo RC, Sarngadharan MG (1992) Identification of a major growth factor for AIDS-Kaposi's sarcoma cells as oncostatin M. Science 255: 1430–1432

Newell MK, Haughn LJ, Maroun CR, Julius MH (1990) Death of mature T cells by separate ligation of CD4 and the TCR for antigen. Nature 347: 286–289

Oppenheim RW (1991) Cell death during development of the nervous system. Annu Rev Neurosci 14: 453–501

Oyaizu N, McCloskey TW, Coronesi M, Chirmule N, Pahwa S (1993) Accelerated apoptosis in PBMCs from HIV-1 infected patients and in CD4 cross-linked PBMCs from normal individuals. Blood 82: 3392–3400

Raff M (1992) Social controls on cell survival and cell death. Nature 356: 397–400

Raff MC, Barres BA, Burne JF, Coles HS, Ishizaki Y, Jacobson MD (1993) PCD and the control of cell survival: lessons from the nervous system. Science 262: 695–700

Razvi ES, Welsh RM (1993) PCD of T lymphocytes during acute viral infection: a mechanism for virus-induced immunodeficiency. J Virol 67: 5754–5765

Sambhara S, Miller R (1991) PCD of T cells signaled by the TCR and the α3 domain of class-I MHC. Science 252: 1424–1427

Sarin A, Clerici M, Blatt SP, Hendrix CW, Shearer GM, Henkart PA (1994) Inhibition of activation-induced programmed cell death and restoration of defective immune responses of HIV+ donars by cysteine protease inhibitors. J Immunol 153: 862–872

Saunders JWJ (1966) Death in the embryonic systems. Science 154: 604–612

Schnittman SM, Psallidopoulos MC, Lane HC, Thompson L, Baseler M, Massari F, Fox CH, Salzman NP, Fauci AS (1989) The reservoir for HIV-1 in human peripheral blood is a cell that maintains expression of CD4. Science 245: 305–308

Schuitemaker H, Meyaard L, Kootstra NA, Otto SA, Dubbes R, Tersmette M, Heeney JL, Miedema F (1993) Lack of T-cell dysfunction and PCD correlates with inability of HIV-1 to infect chimpanzee monocytes. J Infect Dis 168: 1140–1147

Schwartz LM, Osborne BA (1993) Programmed cell death, apoptosis, and killer genes. Immunol Today 14: 582–590

Scott P, Kaufmann SHE (1991) The role of T-cell subsets and cytokines in the regulation of infection. Immunol Today 12: 346–348

Shearer GM, Bernstein DC, Tung KSK, Via CS, Redfield R, Salahuddin SZ, Gallo RC (1986) A model for the selective loss of major histocompatibility complex self-restricted T cell immune responses during the development of acquired immunodeficiency syndrome (AIDS). J Immunol 137: 2514–2521

Tamaru Y, Miyawaki T, Iwai T, Nibu R, Yachie A, Koizumi S, Taniguchi N (1993) Absence of bcl-2 expression by activated CD45RO⁺ T lymphocytes in acute infectious mononucleosis supporting their susceptibility to PCD. Blood 82: 521–527

Terai C, Kornbluth RS, Pauza CD, Richman DD, Carson DA (1991) Apoptosis as a mechanism of cell death in cultured T lymphoblasts acutely infected with HIV-1. J Clin Invest 87: 1710–1715

Tian H, Kolesnitchenko V, Donoghue E, Shaw G, Lane C, Cohen D (1993) HIV envelope-directed CD4 T-cell degeneration represents a novel cell death program. In: The first national conference on human retroviruses and related infections. American Society for Microbiology, Washington DC, abstract 275

Uehara T, Miyawaki T, Ohta K, Tamaru Y, Yokoi T, Nakamura S, Taniguchi N (1992) Apoptotic cell death of primed CD45RO⁺ T lymphocytes in Eptein-Barr virus-induced infection mononucleosis. Blood 80: 452–458

Wang R, Murphy K, Loh D, Weaver C, Russell J (1993) Differential activation of antigen-stimulated suicide and cytokine production pathways in CD4 T cells is regulated by the antigen-presenting cell. J Immunol 150: 3832–3842

Watanabe M, Ringler DJ, Fultz PN, MacKey JJ, Boyson JE, Levine CG, Letvin NL (1991) A Chimpanzee-passaged HIV isolate is cytopathic for chimpanzee cells but does not induce disease. J Virol 65: 3344–3348

Westermann J, Pabst R (1990) Lymphocyte subsets in the blood: a diagnostic window on the lymphoid system? Immunol Today 11: 406–410

Williams GT, Smith CT (1993) Molecular regulation of apoptosis: genetic controls on cell death. Cell 74: 777–779

Zauli G, Vitale M, Re MC, Furlini G, Zamai L, Falcieri E, Gibellini D, Visani G, Davis BR, Capitani S, La Placa M (1994) In vitro exposure to HIV-1 induces apoptotic cell death of the factor-dependent TF-1 hematopoietic cell line. Blood 83: 167–175

Programmed Death of T Cells in HIV Infection: Result of Immune Activation?

L. Meyaard and F. Miedema

1 T Cell Function and Apoptosis in HIV Infection

Programmed cell death (PCD) of T cells has been proposed to explain CD4+ T cell loss and impaired T cell function in HIV infection. Several laboratories have demonstrated that T cells from HIV-infected individuals indeed die upon culture due to apoptosis. Here we discuss the data and give our view on the different hypotheses proposed regarding the mechanism of this type of cell death.

Infection with HIV is characterized by an asymptomatic phase of variable length and a decline in CD4+ T cells, eventually leading to AIDS. Importantly, even before the number of CD4+ T cells starts to decline, functional abnormalities of T cells can be demonstrated in asymptomatic individuals. Both CD4+ and CD8+ T cell function such as interleukin-2 (IL-2) production and proliferation after stimulation with recall antigens and CD3 antibodies is affected (Clerici et al. 1989; Miedema et al. 1988; Schellekens et al. 1990). The mechanism by which HIV is capable of affecting immune function at a stage of infection in which the number of infected cells is low (Schnittman et al. 1989) remains to be elucidated. In recent years, PCD of (CD4+) T cells has been investigated as a cause of T cell dysfunction and CD4+ T cell depletion in HIV-infected individuals.

In 1991, Ameisen and Capron first proposed that in HIV infection interaction of soluble gp120 with CD4, previously shown to lead to impaired lymphocyte

Department of Clinical Viro-Immunology, Central Laboratory of the Netherlands Red Cross Blood Transfusion Service and Laboratory of Experimental and Clinical Immunology of the University of Amsterdam, Plesmanlaan 125, 1066 CX Amsterdam, The Netherlands function (Cefai et al. 1990), would prime CD4+ T cells for PCD. This hypothesis was supported by results obtained with mature murine lymphocytes which die from PCD after stimulation via T cell receptor (TCR)/CD3 when CD4 was previously ligated by CD4 antibodies (Newell et al. 1990).

Indeed, peripheral blood mononuclear cells (PBMCs) from HIV-infected individuals die due to apoptosis in vitro (Groux et al. 1992; Meyaard et al. 1992; Gougeon et al. 1993; Oyaizu et al. 1993; Pandolfi et al. 1993; Meyaard et al. 1994; Lewis et al. 1994). When cultured overnight, PBMCs from HIV-infected persons display the typical electron microscopic morphology characteristic of PCD. Cells have extensive peripheral chromatin condensation, dilation of the endoplasmatic reticulum and preservation of mitochondrial structures, all features of PCD (Wyllie et al. 1980). Low molecular weight DNA fractions isolated from lysed cells and subjected to gel electrophoresis exhibit the DNA cleavage pattern specific for apoptosis. The fragmentation can be prevented by Zn^{2+}, which inhibits endonuclease activity (Duke et al. 1983). PCD can be enhanced by activation in vitro with TCR/CD3 monoclonal antibodies (Mabs), lectins, superantigens or ionomycin (Meyaard et al. 1992; Groux et al. 1992; Gougeon et al. 1993; Oyaizu et al. 1993). PCD occurs in both CD4+ and CD8+ T cells and phenotypical analysis suggests that CD8+ cells die at higher percentages (Gougeon et al. 1993; Meyaard et al. 1992, 1994).

In primary HIV infection the increased percentage of T cells dying due to apoptosis after overnight culture is high (up to 60%) and parallels increased numbers of CD8+ cells. Since they form the largest fraction of T cells, numerically the majority of cells dying during primary infection are activated CD8+CD45RO+ cells. However, all CD8+ T cell subsets contain cells dying due to PCD and there is no evidence for preferential death in one specific subset of cells (Meyaard et al. 1994; Lewis et al. 1994). In the asymptomatic phase of HIV infection there is a variable but, compared to HIV-negative controls, consistently increased percentage of cells dying due to PCD (Gougeon et al. 1993; Meyaard et al. 1992, 1994). In our hands, PCD does not correlate with CD4+ T cell numbers in asymptomatic individuals, nor with T cell function as measured by proliferation in response to CD3 mAb, arguing against dramatic changes in the extent of PCD with progression to disease. Longitudinal analysis of four individuals throughout infection also demonstrates a variation but not a consistent increase or decrease in the number of cells in apoptosis over time. There is no correlation of the numbers of cells dying due to PCD after in vitro culture with virus load or presence of syncytia-inducing (SI) and nonsyncytia-inducing (NSI) HIV variants (Meyaard et al. 1994).

2 Mechanisms of T Cell Apoptosis in HIV Infection

In HIV-infected chimpanzees, which do not develop clinical symptoms, the proportion of T cells dying due to PCD does not exceed that in noninfected animals (Gougeon et al. 1993; Schuitemaker et al. 1993). This could imply either a function for PCD in HIV pathogenesis in humans, or that PCD is a reflection of immunopathogenic events. Several hypothesis on the cause of increased PCD of T cells in human HIV infection and the contribution to AIDS pathogenesis have been proposed, including direct virus infection of cells, CD4 ligation by gp120 and excessive immune activation. These are discussed below.

2.1 Direct Viral Infection

In vitro infection of T cells and T cell lines with HIV results in cell death associated with apoptosis (TERAI et al. 1991; LAURENT-CRAWFORD et al. 1991; MARTIN et al. 1994) and pulsing of dendritic cells with HIV results in infection and apoptosis of cocultured CD4[+] T cells (CAMERON et al. 1994). The capacity of HIV to induce apoptosis in vitro is related to the cell line and virus strain used and is, at least in part, associated with the efficiency of virus replication in these cells (MARTIN et al. 1994)

Direct virus-induced cell death, however, can be excluded as the main cause of PCD of peripheral T cells in asymptomatic HIV infection. Not only is the frequency of infected cells during asymptomatic infection too low to explain the cell death observed, but there also seems to be no clear-cut relation between elevated virus load during both acute and asymptomatic infection and increases in PCD (MEYAARD et al. 1994). However, in later stages of infection, with a high viral burden in T cells in lymph nodes, direct infection of cells leading to apoptosis might contribute to CD4[+] T cell depletion. Moreover, HIV infection of thymocytes might lead to increased apoptotic death in the thymus, thereby affecting regeneration of the peripheral T cell compartment (BONYHADI et al. 1993).

2.2 Ligation of CD4 by gp120

The initial hypothesis on PCD in HIV infection was that interaction of soluble HIV envelope protein gp120 with CD4 could prime T cells for PCD (AMEISEN and CAPRON 1991). Mature murine lymphocytes die from PCD after stimulation via TCR/CD3 when CD4 was previously ligated by CD4 antibodies (NEWELL et al. 1990). Furthermore, addition of gp120 in vitro impairs T cell function (OYAIZU et al. 1990; MANCA et al. 1990; CEFAI et al. 1990). Indeed, in human cells, cross-linking of CD4 mAb or bound gp120 on human CD4[+] T cells followed by signaling through the TCR results in apoptosis in vitro (BANDA et al. 1992; OYAIZU et al. 1993). Expression of gp160 in a CD4[+] T cell line causes down-regulation of CD4 and single cell killing due to apoptosis (LU et al. 1994) and in vitro exposure to HIV without infection of a CD34[+] hematopoietic progenitor cell line induces apoptotic cell death (ZAULI et al. 1994).

These data all point to a role for gp120 in inducing T cell deficiency and apoptosis, and gp120-CD4 ligation might be a mechanism for apoptosis of CD4[+]T cells in vivo. The relative contribution to the PCD observed in HIV infection, however, is hard to assess.

2.3 Immune Activation

A CD4-dependent mechanism for PCD is not likely to be the only explanation. First, both CD4+ and CD8+ cells, with a preference for CD8+ cells, die due to apoptosis (MEYAARD et al. 1992; GOUGEON et al. 1993). Second, during primary HIV infection, the number of cells dying exceeds by far the percentage of CD4+ cells present at that time (MEYAARD et al. 1994). CD8+ T cells from HIV-infected individuals have increased expression of activation markers such as CD38, HLA-DR and CD57, suggestive of continuous immune activation (STITES et al. 1989, SALAZAR-GONZALEZ et al. 1985). CD8+ cells expressing activation markers have severely decreased proliferative responses and clonogenic potential (PANTALEO et al. 1990) and are reported to die in culture (PRINCE and JENSEN 1991). Since the percentage of cells dying due to PCD in primary HIV infection parallels the CD8+ T cell expansion, it is tempting to speculate that PCD in HIV infection reflects turnover of activated immune cells, although PCD is not confined to a specific subset expressing activation markers (MEYAARD et al. 1994; LEWIS et al. 1994).

PCD as a result of massive immune activation following acute virus infection is not specific for HIV infection since it was also demonstrated for cytomegalovirus (CMV) infection in humans (VAN DEN BERG et al. 1994) and acute lymphocytic choriomeningitis virus (LCMV) infection in mice (RAZVI and WELSH 1993), correlating with hyporesponsiveness as a result of hyperactivation of T cells in vivo. Similar findings have been reported for Epstein-Barr virus (EBV) infection in humans (MOSS et al. 1985; UEHARA et al. 1992), in which both CD4+ and CD8+ cells die upon culture.

Dying cells were confined to the CD45RO+ T cell population and cell death could be prevented by culture in the presence of cytokines such as IL-2 (BISHOP et al. 1985; UEHARA et al. 1992). Also, in that condition, PCD was suggested to affect the population of activated T cells which expands during the acute phase of the infection.

We propose that PCD in acute HIV infection is a reflection of immune activation leading to a high turnover of cells, as is observed in acute virus infections in general. Large numbers of apoptotic cells in the early stage of infection are followed by moderately increased numbers of cells dying during the asymptomatic phase, as is also observed in the asymptomatic phase of feline immunodeficiency virus infection in cats (BISHOP et al. 1993). In asymptomatic HIV infection, PCD reflects a continuous activation leading to priming for death and deletion of responding T cells.

3 Turnover of Activated T Cells by Apoptotic Cell Death

The mechanism by which T cells in HIV infection are driven towards apoptosis might reflect a general phenomenon of termination of the immune response upon activation. WESSELBORG et al. (1993) described that, while freshly isolated T cells from healthy individuals are resistant to PCD, the susceptibility of these cells to induced death increases upon activation and culture. In agreement with the observations in HIV infection, in these experiments no correlation between susceptibility to death and expression of a specific activation marker could be demonstrated.

Several cascades of events can be envisaged by which the immune system will set stop at an initiated immune response. As suggested by findings in murine LCMV infection (RAZVI and WELSH 1993) and experimental autoimmune encephalomyelitis (CRITCHFIELD et al. 1994), T cell death might be a physiological response to IL-2 stimulation after massive immune activation or high antigen doses. The apoptosis-related Fas/APO-1 antigen is known to be preferentially expressed on previously activated or memory T cells (MIYAWAKI et al. 1992). Interaction of Fas with its ligand might play a role in the elimination of excessive immune cells. CD45RO$^+$ cells in acute EBV infection, known to undergo apoptosis, have increased expression of Fas (UEHARA et al. 1992) and of a new activation antigen, presumably with similar function (UEHARA et al. 1993).

The proto-oncogene bcl-2 has been identified as a controller of PCD in a variety of cell types (KORSMEYER 1992). It was proposed that the regulation of bcl-2 expression within the CD45RO$^+$ T cell population regulates cell death and survival and is a mechanism for the removal of unwanted T cells after resolution of viral disease (AKBAR et al. 1993b). After repeated stimulation, primed T cells lose bcl-2 expression, gain Fas expression and become more susceptible to death (SALMON et al. 1994). In acute human EBV infection, associated with cell death of CD4$^+$ and CD8$^+$ T cells in culture, CD45RO$^+$ T cells have decreased levels of bcl-2 expression compared to cells from uninfected controls. These cells were demonstrated to undergo apoptosis in vitro (TAMARU et al. 1993). In addition, a significant correlation between cells with low bcl-2 expression and apoptosis in culture was observed in other acute viral infections (AKBAR et al. 1993a). The mechanism of induction of apoptosis-related antigens or repression of survival genes in activated T cells remains to be elucidated.

Importantly, we and others observed that, in HIV infection, T cells die irrespective of the expression of activation markers (MEYAARD et al. 1994; LEWIS et al. 1994). Massive immune activation could lead to exhaustion of growth and survival factors and subsequently result in PCD. Our finding that growth factors in vitro could not prevent cells of HIV-infected individuals from dying does not exclude such a mechanism but may indicate that these cells are already irreversibly primed for PCD in vivo (MEYAARD et al. 1994). Other groups, however, reported rescue of cells of HIV-infected individuals from apoptosis by combinations of growth factors (GROUX et al. 1992; PANDOLFI et al. 1993; GOUGEON et al. 1993). Furthermore, oxidative stress has been proposed as a mediator of apoptosis (BUTTKE and SANDSTROM 1994). Activated CD4$^+$ and CD8$^+$ T cells from HIV-infected individuals have a glutathione deficiency (DRÖGE et al. 1992; STAAL et al. 1992) and therefore might be less capable of withstanding oxidative stress and thereby death due to PCD.

Antigen presenting cell (APC) function, regulating either proliferation and cytokine production or cell death of the responding T cell, was proposed as a mechanism to shape a given immune response (WANG et al. 1993). Increased prostaglandin E2 production by HIV-infected human macrophages induces apoptosis in cocultured noninfected lymphocytes (MASTINO et al. 1993). We have previously argued that APC dysfunction as a result of HIV infection may cause T cell dysfunction (MEYAARD et al. 1993), which is also based on observations in HIV-infected chimpanzees. Chimpanzees can become persistently infected with HIV without development of clinical symptoms. T cells of HIV-infected chimpanzees have a normal response to stimulation in vitro and the proportion of T cells dying due to PCD in infected animals does not exceed that in noninfected animals (SCHUITEMAKER et al. 1993; GOUGEON et al. 1993). Interestingly, HIV does not infect chimpanzee monocytes and only T cell tropic variants are isolated from infected animals (SCHUITEMAKER et al. 1993; GENDELMAN et al. 1991). The absence of infected APCs in chimpanzees might be the explanation for the fact that enhanced PCD of T cells does not occur (SCHUITEMAKER et al. 1993).

4 Concluding Remarks

As in other viral infections, PCD in acute HIV infection might be a reflection of immune activation by a so far unknown mechanism, leading to decreased expression of survival genes and increased expression of apoptosis genes and turnover of immune cells. The increased numbers of cells dying during the asymptomatic phase might be the result of continuous activation and priming for death to maintain T cell homeostasis.

Although PCD in early asymptomatic infection merely reflects the activated immune system rather than being pathogenic in itself, virus-induced apoptosis might contribute to CD4$^+$ cell depletion: first, by infection of thymocytes, thus affecting renewal of the T cell compartment; and second, when the viral burden increases, by HIV-induced apoptosis of infected CD4$^+$ T cells.

Acknowledgement.   The authors thank Dr Hanneke Schuitemaker for critical reading of the manuscript.

References

Akbar AN, Borthwick NJ, Salmon M, Gombert W, Bofill M, Shamsadeen N, Pilling D, Pett S, Grundy JE, Janossy G (1993a) The significance of low bcl-2 expression by CD45RO T cells in normal individuals and patients with acute viral infections. The role of apoptosis in T cell memory. J Exp Med 178: 427–438

Akbar AN, Salmon M, Savill J, Janossy G (1993b) A possible role for bcl-2 in regulating T cell memory—a 'balancing act' between cell death and survival. Immunol Today 14: 526–532

Ameisen JC, Capron A (1991) Cell dysfunction and depletion in AIDS: the programmed cell death hypothesis. Immunol Today 12: 102–105

Banda NK, Bernier J, Kurahara DK, Kurrle R, Haigwood N, Sekaly R-P, Helman Finkel T (1992) Crosslinking CD4 by Human Immunodeficiency Virus gp120 primes T cells for activation-induced apoptosis. J Exp Med 176: 1099–1106

Bishop CJ, Moss DJ, Ryan JM, Burrows SR (1985) T lymphocytes in infectious mononucleosis. II. Response in vitro to interleukin-2 and establishment of T cell lines. Clin Exp Immunol 60: 70–77

Bishop SA, Gruffydd-Jones TJ, Harbour DA, Stokes CR (1993) Programmed cell death (apoptosis) as a mechanism of cell death in peripheral blood mononuclear cells from cats infected with feline immunodeficiency virus (FIV). Clin Exp Immunol 93: 65–71

Bonyhadi ML, Rabin L, Salimi S, Brown DA, Kosek J, McCune JM, Kaneshima H (1993) HIV induces thymus depletion in vivo. Nature 363: 728–732

Buttke TM, Sandstrom PA (1994) Oxidative stress as a mediator of apoptosis. Immunol Today 15: 7–10

Cameron PU, Pope M, Gezelter S, Steinman RM (1994) Infection and apoptotic cell death of CD4+ T cells during an immune response to HIV-1 pulsed dendritic cells. AIDS Res Hum Retroviruses 10: 61–71

Cefai D, Debre P, Kaczorek M, Idziorek T, Autran B, Bismuth G (1990) Human immunodeficiency virus-1 glycoproteins gp120 and gp160 specifically inhibit the CD3/T cell-antigen receptor phosphoinositide transduction pathway. J Clin Invest 86: 2117–2124

Clerici M, Stocks N, Zajac RA, Boswell RN, Lucey DR, Via CS, Shearer GM (1989) Detection of three different patterns of T helper cell dysfunction in asymptomatic, human immundeficiency virus-seropositive patients. J Clin Invest 84: 1892–1899

Critchfield JM, Racke MK, Zúñiga-Pflücker JC, Cannella B, Raine CS, Goverman J, Lenardo MJ (1994) T cell deletion in high antigen dose therapy of autoimmune encephalomyelitis. Science 263: 1139–1143

Dröge W, Eck HP, Mihm S (1992) HIV-induced cysteine deficiency and T cell dysfunction—a rationale for treatment with N-acetylcysteine. Immunol Today 13: 211–214

Duke RC, Chervenak R, Cohen JJ (1983) Endogenous endonuclease-induced DNA fragmentation: An early event in cell-mediated cytolysis. Proc Natl Acad Sci USA 80: 6361–6365

Gendelman HE, Ehrlich GD, Baca LM, Conley S, Ribas J, Kalter DC, Meltzer MS, Poiesz BJ, Nara P (1991) The inability of human immunodeficiency virus to infect chimpanzee monocytes can be overcome by serial viral passage in vivo. J Virol 65: 3853–3863

Gougeon M, Garcia S, Heeney J, Tschopp R, Lecoeur H, Guetard D, Rame V, Dauguet R, Montagnier L (1993) Programmed cell death in AIDS-related HIV and SIV infections. AIDS Res Hum Retroviruses 9: 553–563

Groux H, Torpier G, Monté D, Mouton Y, Capron A, Ameisen JC (1992) Activation-induced death by apoptosis in CD4+ T cells from Human Immunodeficiency Virus-infected asymptomatic individuals. J Exp Med 175 331–340

Korsmeyer SJ (1992) Bcl-2: a repressor of lymphocyte death. Immunol Today 13: 285–288

Laurent-Crawford AG, Krust B, Muller S, Rivière Y, Rey-Cuillé MA, Béchet J-M, Montagnier L, Hovanessian AG (1991) The cytopathic effect of HIV is associated with apoptosis. Virology 185: 829–839

Lewis DE, Ng Tang DS, Adu-Oppong A, Schober W, Rodgers JR (1994) Anergy and apoptosis in CD8+ T cells from HIV-infected individuals. J Immunol 153: 412–420

Lu Y-Y, Koga Y, Tanaka K, Sasaki M, Kimura G, Nomoto K (1994) Apoptosis induced in CD4+ cells expressing gp160 of human immunodeficiency virus type 1. J Virol 68: 390–399

Manca F, Habeshaw JA, Dalgleish AG (1990) HIV envelope glycoprotein, antigen specific T-cell responses, and soluble CD4. Lancet 335: 811–815

Martin SJ, Matear PM, Vyakarnam A (1994) HIV-1 infection of human CD4+ T cells in vitro. Differential induction of apoptosis in these cells. J Immunol 152: 330–342

Mastino A, Grelli S, Placentini M, Oliverio S, Favalli C, Perno CF, Garaci E (1993) Correlation between induction of lymphocyte apoptosis and prostaglandin E2 production by macrophages infected with HIV. Cell Immunol 152: 120–130

Meyaard L, Otto SA, Jonker RR, Mijnster MJ, Keet RPM, Miedema F (1992) Programmed death of T cells in HIV-1 infection. Science 257: 217–219

Meyaard L, Schuitemaker H, Miedema F (1993) T-cell dysfunction in HIV infection: Anergy due to defective antigen presenting cell function? Immunol Today 14: 161–164

Meyaard L, Otto SA, Keet IPM, Roos MThL, Miedema F (1994) Programmed death of T cells in HIV-1 infection: no correlation with progression to disease. J Clin Invest 93: 982–988

Miedema F, Petit AJC, Terpstra FG, Schattenkerk JKME, De Wolf F, Al BJM, Roos MThL, Lange JMA, Danner SA, Goudsmit J, Schellekens PTA (1988) Immunological abnormalities in human immunodeficiency virus (HIV)-infected asymptomatic homosexual men. HIV affects the immune system before CD4+ T helper cell depletion occurs. J Clin Invest 82: 1908–1914

Miyawaki T, Uehara T, Nibu R, Tsuji T, Yachie A, Yonehara S, Taniguchi N (1992) Differential expression of apoptosis-related Fas antigen on lymphocyte subpopulations in human peripheral blood. J Immunol 149: 3753–3758

Moss DJ, Bishop CJ, Burrows SR, Ryan JM (1985) T lymphocytes in infectious mononucleosis. I. T cell death in vitro. Clin Exp Immunol 60: 61–69

Newell MK Haughn LJ, Maroun CR, Julius MH (1990) Death of mature T cells by separate ligation of CD4 and the T-cell receptor for antigen. Nature 347: 286–288

Oyaizu N, Chirmule N, Kalyanaraman VS, Hall WW, Good RA, Pahwa S (1990) Human immunodeficiency virus type 1 envelope glycoprotein gp120 produces immune defects in CD4⁺ T lymphocytes by inhibiting interleukin 2 mRNA. Proc Natl Acad Sci USA 87: 2379–2383

Oyaizu N, McCloskey TW, Coronesi M, Chirmule N, Kalyanaraman VS, Pahwa S (1993) Accelerated apoptosis in peripheral blood mononuclear cells (PBMCs) from human immunodeficiency virus type-1 infected patients and in CD4 cross-linked PBMCs from normal individuals. Blood 82: 3392–3400

Pandolfi F, Oliva A, Sacco G, Polidori V, Liberatore D, Mezzaroma I, Kurnick JT, Aiuti F (1993) Fibroblast-derived factors preserve viability in vitro of mononuclear cells isolated from subjects with HIV-1 infection. AIDS 7: 323–329

Pantaleo G, Koenig S, Baseler M, Clifford Lane H, Fauci AS (1990) Defective clonogenic potential of CD8⁺ T lymphocytes in patients with AIDS. Expansion in vivo of a nonclonogenic CD3⁺CD8⁺DR⁺CD25-T cell population. J Immunol 144: 1696–1704

Prince HE, Jensen ER (1991) HIV-related alterations in CD8 cell subsets defined by in vitro survival characteristics. Cell Immunol 134: 276–286

Razvi ES, Welsh RM (1993) Programmed cell death of T lymphocytes during acute viral infection: a mechanism for virus-induced immune deficiency. J Virol 67: 5754–5765

Salazar-Gonzalez JF, Moody DJ, Giorgi JV, Martinez-Maza O, Mitsuyasu RT, Fahey JL (1985) Reduced ecto-5'-nucleotidase activity and enhanced OKT 10 and HLA-DR expression on CD8 lymphocytes in the aquired immune deficiency syndrome: evidence of CD8 cell immaturity. J Immunol 135: 1778–1785

Salmon M, Pilling D, Borthwick NJ, Viner N, Janossy G, Bacon PA, Akbar AN (1994) The progressive differentiation of primed T cells is associated with an increasing susceptibility to apoptosis. Eur J Immunol 24: 892–899

Schellekens PThA, Roos MThL, De Wolf F, Lange JMA, Miedema F (1990) Low T-cell responsiveness to activation via CD3/TCR is a prognostic marker for AIDS in HIV-1 infected men. J Clin Immunol 10: 121–127

Schnittman SM, Psallidopoulos MC, Lane HC, Thompson L, Baseler M, Massari F, Fox CH, Salzman NP, Fauci AS (1989) The reservoir for HIV-1 in human peripheral blood is a T cell that maintains expression of CD4. Science 245: 305–308

Schuitemaker H, Meyaard L, Kootstra NA, Otto SA, Dubbes R, Tersmette M, Heeney JL, Miedema F (1993) Lack of T-cell dysfunction and programmed cell death in human immunodeficiency type-1 infected chimpanzees correlates with absence of monocytotropic variants. J Infect Dis 168: 1140–1147

Staal FJT, Ela SW, Roederer M, Anderson MT, Herzenberg LA (1992) Gluthatione deficiency and human immunodeficiency virus infection. Lancet 339: 909–912

Stites DP, Moss AR, Bacchetti P, Osmond D, McHugh TM, Wang YJ, Hebert S, Colfer B (1989) Lymphocyte subset analysis to predict progression to AIDS in a cohort of homosexual men in San Francisco. Clin Immunol Immunopathol 52: 96–103

Tamaru Y, Miyawaki T, Iwai K, Tsuji T, Nibu R, Yachie A, Koizumi S, Taniguchi N (1993) Absence of bcl-2 expression by activated CD45RO⁺ T lymphocytes in acute infectious mononucleosis supporting their susceptibility to programmed cell death. Blood 82: 521–527

Terai C, Kornbluth RS, Pauza CD, Richman DD, Carson DA (1991) Apoptosis as a mechanism of cell death in cultured T lymphoblasts acutely infected with HIV-1. J Clin Invest 87: 1710–1715

Uehara T, Miyawaki T, Ohta K, Tamaru Y, Yokoi T, Nakamura S, Taniguchi N (1992) Apoptotic cell death of primed CD45RO⁺ T lymphocytes in Epstein-Barr virus induced infectious mononucleosis. Blood 80: 452–458

Uehara T, Miyawaki T, Natsuume-Sakai S, Nibu R, Hasui M, Yachie A, Shimizu S, Taniguchi N (1993) A novel activation antigen identified by monoclonal IMN3.1 antibody and expressed preferentially on human T cells susceptible to apoptotic cell death. J Immunol 150: 3243–3253

Van den Berg AP, Meyaard L, De Leij LHFM, Otto SA, Mesander G, Van Son WJ, Klompmaker IJ, Miedema F, The TH (1995) Decreased T cell proliferative capacity and increased rate of programmed cell death of lymphocytes in human cytomegalovirus infection (submitted for publication)

Wang R, Murphy KM, Loh DY, Weaver C, Russell JH (1993) Differential activation of antigen-stimulated suicide and cytokine production pathways in CD4⁺ T cells is regulated by the antigen presenting cell. J Immunol 150: 3832–3842

Wesselborg S, Janssen O, Kabelitz D (1993) Induction of activation-driven death (apoptosis) in activated but not resting peripheral blood T cells. J Immunol 150: 4338–4345

Wyllie AH, Kerr JFK, Currie AR (1980) Cell death: The significance of apoptosis. Int Rev Cytol 68: 251–306

Zauli G, Vitale M, Re MC, Furlini G, Zamai L, Falcieri E, Gibellini D, Visani G, Davis BR, Capitani S, La Placa M (1994) In vitro exposure to human immunodeficiency virus type 1 induces apoptotic cell death of the factor-dependent TF-1 hematopoietic cell line. Blood 83: 167–175

Maintenance of the T Lymphocyte Pool by Inhibition of Apoptosis: A Novel Strategy of Immunostimulation?

G. Kroemer, N. Zamzami, P. Marchetti, and M. Castedo

1 Introduction

Throughout development T lymphocytes are constantly confronted with a series of vital options. First, T cells can decide to either ignore an antigen or to become activated upon antigenic stimulation. Second, T cell activation may have rather disparate consequences, T lymphocytes may become productively activated to proliferate, differentiate or exert an effector function. Alternatively, they can become activated in an abortive or aberrant fashion, leading to anergy or apoptosis. The option that the T cell will choose among these possibilities is not only dictated by the conformation and concentration of the antigenic peptide/MHC complex, but depends also on a series of further circumstances: the particular context of cosignals perceived via receptors interacting with the antigen-presenting cell (APC), bystander cells (Ding and Shevach 1994), cell matrix proteins and soluble factors (cytokines and hormones), and the particular subpopulation to which the responding cells belong and their differentiation stage and (pre-) activation state. In this sense, T cells function as semiotic entities; they integrate

CNRS-UPR420, 19 rue Guy Môquet, B.P.8, 94801 Villejuif, France signals from a complex universe as a function of their previous experiences to act in a quantitatively and qualitatively graded, rather than all-or-nothing, fashion.

T lymphocytes are notoriously prone to undergo apoptosis, especially during intrathymic differentiation, before acquiring the phenotype of mature peripheral T cells, and later after antigen priming and acquisition of a memory phenotype (SALMON et al. 1994). Similar to other cell populations, e.g., enterocytes, it may be postulated that in T cells each mitosis is compensated for by the programmed cell death (PCD) of two other T cells to maintain the homeostasis of T cell populations. Apoptosis—in the morphological sense of the term—is probably only a default pathway of PCD in vivo. According to current understanding, cells that are programmed to die in vivo are efficiently recognized, engulfed and rapidly degraded by adjacent phagocytes before major structural changes occur and before intracellular macromolecules are broken down by catabolic enzymes. The biochemical events that occur in cells that are prone to death, before apoptosis and associated oligonucleosomal DNA fragmentation, are poorly understood. Nonetheless, it is clear that changes in the propensity of lymphocytes to undergo PCD may have major functional consequences. Thus, resistance to apoptosis, as it is induced, for example, by a null mutation in the Fas antigen (WATANABE-FUKUNAGA et al. 1992; RUSSEL et al. 1993; BOSSU et al. 1993; MÖRÖY et al. 1993) or a structural mutation in the *bcl*-2 proto-oncogene (GARCHON et al. 1994; LEIJON et al. 1994), may be involved in the pathogenesis of autoimmune diseases. In contrast, an enhanced apoptotic decay of peripheral T lymphocytes accounts for the numeric and functional deficiency of T lymphocytes caused by HIV infection in humans (GOUGEON et al. 1991; GROUX et al. 1992; MEYAARD et al. 1992). In this context, drugs that either induce or inhibit apoptosis induction in T lymphocytes may be of the utmost clinical importance. The purpose of this chapter is to discuss the possibilities of interfering with apoptosis induction in vivo.

2 Pathological Conditions Coupled to Enhanced Apoptotic Turnover of T Lymphocytes

An abnormally high apoptotic turnover of lymphoid cells causing a severe immunodeficiency may be related to two different groups of causes: (1) Genetic manipulations of mice may have this effect. Thus, genetic knock-out of the anti-apoptotic *bcl*-2 proto-oncogene causes disappearance of the lymphoid system after birth (VEIS et al. 1993; NAKAYAMA et al. 1993). Similarly, expression of the homeobox fusion gene E2A-PBX1 under control of the Ig heavy chain enhancer causes a reduction of thymocytes and bone marrow B lineage progenitors that is due to increased cell death (DEDERA et al. 1993). (2) A number of viruses cause a transient or chronic immunodeficiency mediated by enhanced lymphocyte apoptosis. This concerns mice infected with murine lymphocytic choriomeningitis virus (LCMV) (RAZVI and WELSH 1993) or vaccinia virus (GONZALO et al. 1994c) and humans manifesting one of the diseases caused by Epstein-Barr virus (EBV), namely, infectious mononucleosis (UEHARA et al. 1992). Mice infected with the *tst*1 mutant of Moloney murine leukemia virus manifest a progressive apoptosis-mediated lymphocyte depletion (SAHA et al. 1994). Patients infected with HIV also demonstrate an enhanced apoptotic decay of lymphocytes cultured in vitro (see chapters by Gougeon, Meyaard, and Ameisen, this volume). Enhanced DNA fragmentation is observed both in nonstimulated and mitogen-stimulated T lymphocytes from HIV-infected donors. An abnormal tendency to undergo apoptosis may be expected to negatively affect immune function in a dual fashion: (1) by reducing the total number of T cells, thus causing a *numeric* immunodeficiency, and (2) by converting specific immune responses that should lead to immune activation into tolerizing ones, thus causing a *functional* immunodeficiency.

A further pathological condition in which apoptosis could play a role is septic shock, a critical clinical condition that is caused by bacterial endo- and exotoxins, and by cytokines induced by these (GLAUSER et al. 1991; COHEN and GLAUSER 1991). The apoptosis-inducing substances implicated in septic shock include lipopolysaccharide (endotoxin), which induces PCD of monocytes (MANGAN and WAHL 1991) and endothelial cells (ABELLO et al. 1994); superantigenic exotoxins, which induce PCD by activating T cells via the T cell receptor (TCR) (KOTZIN et al. 1993), nonsuperantigenic exotoxins, such as staphylococcal α-toxin, which induce apoptosis via membrane permeabilization (JONAS et al. 1994); tumor necrosis factor-α (TNF-α) (HERNÁNDEZ-CASELLES and STUTMAN 1993) and interferon-γ (IFN-γ) (LIU and JANEWAY 1990), both of which may cause apoptosis of T cells. Substances that cause T cell apoptosis such as D-galactosamine (GONZALO et al. 1993b) augment the toxicity of lipopolysaccharide and bacterial superantigen in vivo.

3 The Rationale of Blocking Lymphocyte Apoptosis by Pharmacological Interventions

Experimental inhibition of apoptosis has a twofold impact: (1) manipulation of PCD is of theoretical interest and may contribute to the elucidation of pathways involved in the induction and/or execution of apoptosis in an obligatory fashion. (2) Drugs that selectively inhibit lymphocyte death could be clinically employed as immunostimulators. Inhibition of apoptosis should enhance the probability that a lymphocyte that is interacting with a target antigen, e.g., virus-encoded peptides presented by infected cells or mutant antigens presented by tumor cells, will mount a productive immune response instead of being deleted. In addition, anti-apoptotic medication may counteract the depletion of T cells during certain acute viral infections or more importantly, in HIV-infected individuals (see above). Inhibition of T cell apoptosis thus should prevent and/or attenuate both the numeric and the functional immunodeficiency observed in AIDS.

Nevertheless, pharmacological inhibition of lymphocyte apoptosis may be a two-edged sword. Theoretically, interference with cell death might cause the development of autoimmune lesions when the physiological deletion of lymphocytes and their precursors is prevented. In addition, suppression of cell death may entail an increase in the overall lymphocyte number, thus leading to an unwarranted lymphoproliferation with a consecutive lymphadenopathy and splenomegaly. Examples how genetically conferred inhibition of apoptosis causes such autoimmune or lymphoproliferative phenomena are well known. Thus, the possibility that external agents that inhibit apoptosis might cause similar phenomena should not be neglected.

4 Different Levels of Apoptosis Inhibition

The inhibition of apoptosis can be achieved on rather different levels (KROEMER and MARTÍNEZ-A 1994; KROEMER 1995): (a) inhibition of the apoptosis-inducing stimulus before or during interactions with specific receptors, (b) reprogramming of the cell by providing costimuli, (c) inhibition of signal transduction, (d) interventions in the cell cycle, and (e) inhibition of catabolic pathways involved in carrying out the death program.

4.1 Interception of Apoptosis-Inducing Stimuli

The induction of T cell apoptosis requires an interaction between the apoptosis-inducing substance and specific receptors. This applies to antigen driven deletion and to the action of certain endogenous inducers of apoptosis, namely, contact-dependent stimuli targeted to potential apoptosis-triggering receptors (CD2, Apo-1/Fas/CD95), as well as glucocorticoids and lymphokines. A trivial possibility to inhibit apoptosis consists of neutralizing the relevant pro-apoptotic stimulus or impeding its interaction with the relevant receptor. Thus, neutralization of super-antigens with antibody or elimination of superantigen presentation by specialized APCs such as B lymphocytes (GOLLOB and PALMER 1993) will impede super-antigen-induced T cell deletion, and blockade of glucocorticoid receptors will inhibit glucocorticoid-induced thymocyte apoptosis (SCHWARTZMAN and CIDLOWSKI 1991). In a less trivial fashion, blockade of glucocorticoid receptors with RU-38486 also prevents superantigen-induced deletion in vivo (GONZALO et al. 1993b), a finding that unravels cooperative interactions between two PCD pathways, that triggered via the TCR and that involving glucocorticoids.

4.2 Costimuli

The efficiency of apoptosis induction depends to a large extent on the context of the signal received by a T cell. Thus, certain additional stimuli can prevent T cells from undergoing apoptosis. As an example interleukin (IL)-2 and IL-4 prevent the glucocorticoid-induced apoptosis of Th1 and Th2 cells, respectively (ZUBIAGA et al. 1992). Retinol derivatives, which are cofactors of T cell stimulation (Garbe et al. 1992), inhibit the activation-induced death of T cell hybridomas (Iwata et al. 1992; Yang et al. 1993). In a similar fashion, addition of APCs can impede apoptosis induction via the TCR (Kabelitz and Wesselborg 1992).

4.3 Signal Transduction

Pro-apoptotic stimuli trigger cascades of intracellular second messenger systems whose blockade can abolish the apoptosis-inducing effect. Thus, elevation of cAMP (Lee et al. 1993), chelation of intracellular calcium (Caron-Leslie and Cidlowski 1991) or cyclosporin A (Shi et al. 1989) inhibit the activation-induced death of T cell hybridomas. These manipulation will either desensitize the TCR (cAMP), grossly interfere with T cell activation (calcium chelation) or inhibit most consequences of TCR ligation (cyclosporin A). Given the variety of different signal transduction pathways that may be triggered upon occupancy of a particular receptor type, blockade of a single second messenger system may abolish the induction of apoptosis without interfering with further biological effects triggered by a given extracellular stimulus. Thus, interference with G protein-mediated signal transduction by means of pertussis toxin can abolish the induction of apoptosis by T cell activation without interfering with other consequences of TCR ligation such as lymphokine production or cellular proliferation (Gonzalo et al. 1994b; Ramírez et al. 1994). This apoptosis-inhibiting effect critically depends on the ADP-ribosylating activity carried out by the S1 subunit of pertussis toxin.

4.4 Cell Cycle Arrest

The possibility to induce apoptosis is likely to critically depend on a G_0/G_1 cell cycle transition and/or on regulatory molecules involved in this transition. This explains the ability to interfere with apoptosis induction by inhibiting the action of p34^{cdc2} (Shi et al. 1994) and the effect of oligodeoxynucleotides that inhibit expression of c-myc (Shi et al. 1992) or other immediate activation genes (c-fos, c-jun) (Colotta et al. 1992). It remains to be determined in response to which apoptosis-inducing stimuli is G_0/G_1 transition a prerequisite for apoptosis.

4.5 Inhibition of Catabolic Processes

Inhibition of the different catabolic processes can also inhibit and/or delay PCD in vivo. Antioxidants such as N-acetylcysteine (Buttke and Sandstrom 1994), inhibitors of proteases (Bruno et al. 1992; Weaver et al. 1993) and inhibitors of endonucleases participating in DNA fragmentation (zinc, aurintricarboxylic acid, etc.) have been reported to inhibit or to delay PCD in vitro.

Table 1. Effect of different pharmacological treatments on the deletion of peripheral T cells in vivo

Inhibitor of PCD	Agent causing T lymphocyte depletion or anergy				Reference
	Dex	SEB-induced early phase (12–24 h) of deletion of thymocytes and splenocytes	SEB-triggered late phase (4–10 days) of deletion of splenocytes	SEB-induced anergy of splenocytes	
RU-38486	Competes with DEX for GC receptor occupancy and neutralizes biological effects of DEX	Impedes early deletion when administered simultaneously with SEB	No effect on deletion when administered from day 4	No effect	GONZALO et al. 1993b
All-trans retinol	No effect	Postpones early deletion by 12 h in spleen, lymph nodes and thymus	Partial inhibition of deletion when administered from day 4	No effect	GONZALO et al. 1994a
Linomide	Impedes PCD of peripheral T cells (not or to a lower degree in thymocytes)	Partially inhibits deletion of peripheral T cells when administered 3 days before SEB	Partially inhibits deletion of peripheral T cells when administered before SEB injection	No effect	GONZALO et al. 1994c
Aurintricarboxylic acid	Not tested	Inhibits deletion during 24 h	Not tested	Not tested	MOGIL et al. 1994
Pertussis toxin	No effect	Inhibits deletion of $CD4^+V\beta8^+$ (not $CD8^+$) spleen cells and $V\beta8^+$ thymocytes	Partially inhibits deletion of $CD4^+$ and $CD8^+$ spleen and lymph node cells when administered together with SEB; no effect when injected on day 4	Inhibits anergy induction and abolishes established anergy	GONZALO et al. 1994b
Closporin A	No effect	No effect	No effect or slight enhancement of deletion, when SEB is administered repeatedly	Impedes induction of anergy	GONZALO et al. 1992, VANIER and PRUD'HOMME 1992
Interleukin-2	No effect	No effect	No effect	Partial reversal of anergy	Unpublished
Cycloheximide	Not determined	Not tested	Slight delay (0.5 days)	Impedes induction of anergy at doses that suppress protein synthesis in vivo during 12 h after SEB injection	YUH et al. 1993

PCD, programmed cell death; DEX, dexamethasone; SEB, staphylococcal enterotoxin B; GC, glucocorticoid.

5 Inhibition of T Cell Apoptosis In Vivo

In vitro inhibition of PCD has furnished important information on the cellular and molecular mechanisms of apoptosis. However, most of the procedures employed in vivo, especially those concerning postreceptive events of the PCD-inducing cascade, cannot be applied to in vivo systems. Thus, gross inhibition of signal transduction, interference with cell cycle progression, inhibition of the action of oxygen radicals, proteases and nucleases will affect all cell types and thus will provoke important side effects.

Table 1 summarizes the in vivo effects of pharmacological agents on dexamethasone-induced PCD of T cells or on peripheral clonal deletion of Vβ8⁺ T cells induced by staphylococcal enterotoxin B (SEB) in vivo. SEB induces apoptosis of peripheral T cells in two phases. A first phase of deletion (lowest level of Vβ8⁺ T cells 12–18 h after injection of SEB) is followed by a period of proliferation (24–72 h postinjection) and a second phase of deletion (>3 days after injection) (KAWABE and OCHI 1991; MACDONALD et al. 1991; WAHL et al. 1993; GONZALO et al. 1993b, 1994a). The spectrum of activity of different substances that inhibit apoptosis induction reveals that these three types of PCD obey different principles. Thus, the glucocorticoid receptor-blocking agent RU-38486 only inhibits the early, not the late, phase of SEB driven deletion. In contrast, retinol and pertussis toxin, which inhibit both phases of SEB-induced apoptosis, have no effect on the glucocorticoid-mediated depletion of peripheral T cells in vivo (Table 1). Linomide affects all three types of apoptosis in the peripheral T cell compartment but fails to exert a PCD inhibitory function on thymocytes. Other substances that have been reported to inhibit T cell apoptosis or deletion fail to affect PCD induction by SEB in vivo: cyclosporin A, IL-2 and cycloheximide. These three agents, however, specifically interfere with the induction of T cell anergy by SEB (Table 1).

As to the mode of action by which these substances act (Table 2), it appears that there are several levels of apoptosis inhibition, RU-38486 neutralizes the effect of endogenous glucocorticoids; *all*-trans-retinol is likely to act as a costimulator, pertussis toxin interferes with signal transduction at the level of GTP binding proteins, and both aurintricarboxylic acid and *N*-acetylcysteine inhibit the effector phase of apoptosis. In contrast, as will be discussed in the following section, the mode of action of linomide remains largely elusive.

Table 2. Putative mode of action of anti-apoptotic drugs acting on the staphylococcal enterotoxin B-driven deletion of thymocytes or peripheral T cells in vivo

Substance	Level of action			
	Neutralization of stimulus	Costimulus (functional)	Signal transduction	Catabolic metabolism
RU-38486	+	–	–	–
All-trans retinol	–	+	–	–
Pertussis toxin	–	–	+	–
Aurintricarboxylic acid	–	–	–/?	+
N-acetylcysteine	–	–	–	+

For details and references consult main text.

6 Linomide—A Novel Inhibitor of T Cell Apoptosis

Linomide (N-phenylmethyl-1,2 dihydro-4-hydroxyl-1-methyl-2-oxoquinoline-3-car-boxamide) is a quinoline 3-carboxamide that inhibits apoptosis of peripheral T cells induced by two rather different stimuli: superantigen and glucocorticoids (Table 2). It thus appears that linomide blocks an early event of the PCD-inducing cascade that is common to disparate modes of deletion. Accordingly, in a model of dexamethasone (DEX)-induced splenic T cell death, it interferes with the zinc-resistant DNA fragmentation into high molecular weight fragments (>50 kbp) and abolishes early apoptotic changes in cellular morphology (ZAMZAMI et al. 1995), suggesting that it does inhibit the first steps of PCD.

Linomide interferes with an alteration of cellular electrophysiology that can be detected in cells committed to PCD. Before T cells demonstrate the apoptosis-associated alterations of chromosomal DNA, they exhibit a reduction of their mitochondrial potential, as determined by means of 3,3'-dihexyloxacabocyanine iodide ($DiOC_6(3)$) (PETIT et al. 1990). Following injection of dexamethasone or SEB, splenic T cells isolated ex vivo do not exhibit DNA fragmentation, but demonstrate already a reduction in $DiOC_6(3)$ incorporation (ZAMZAMI et al. 1995). Only after a short period of in vitro culture at 37 °C splenic T cells ($\geq$60 min) do exhibit DNA fragmentation and a loss in chromosomal material (KAWABE and OCHI 1991). Linomide prevents the loss in mitochondrial potential and thus inhibits the earliest PCD-associated event that can be detected (Table 3). In consequence, linomide acts differently from zinc and aurintricarboxylic acid, which both reduce $DiOC_6(3)$ uptake and inhibit a rather late step of the apoptotic cascade, namely,

Table 3. Effects of different agents on different phases of glucocorticoid-induced T cell apoptosis

Substance	Signs of PCD		
	Loss of mitochondrial potential among viable cells	DNA fragmentation	Loss of viability
Linomide	Restoration of normal potentials in dexamethasone or SEB-treated splenic T cells	Inhibition of all types of DNA fragmentation	Inhibition
Zinc	Reduces potential by itself	Inhibition of oligonucleasomal DNA fragmentation, no effect on generation of large (>50 kbp) fragments	Delay
Aurintricarboxylic acid	Reduces potential by itself	Inhibition of oligonucleosomal but not large DNA fragmentation	Delay
N-acetyl cysteine	Partial inhibition of loss in potential in thymocytes exposed to glucocorticoids	Inhibition	Inhibition

PCD, programmed cell death; SEB, staphylococcal enterotoxin B.
Mitochondrial potentials were measured by means of the dye DIOC6(3) (PETIT et al. 1990) Data from ZAMZAMI et al. (1995).

the activation of endonucleases. This may explain why linomide has a higher potential of preserving T cell viability than substances interfering with DNA fragmentation only.

7 Biological Consequences of Apoptosis Inhibition

As outlined in the introduction to this paper, T cells are constantly confronted with a vital decision: survival or suicide? Accordingly, reducing the probability of apoptosis may be expected to augment the chance that T cells will become productively activated and not deleted during an immune reaction. In this context it is not surprising that linomide has a series of immunostimulatory effects on cellular immune reactions. It enhances transplantation rejections, delayed type hypersensitivity, aggravates certain experimentally induced or spontaneously developing signs of autoimmunity, and enhances anti-cancer immunity in vivo. Furthermore, linomide inhibits the acute toxicity of bacterial exo- and endotoxins in a model of septic shock (Table 4).

Three other substances that inhibit T ciell deletion in vivo can abolish immune tolerance. Cyclosporin A, which has the capacity of inhibiting self-peptide-specific intrathymic deletion in vivo (URDAHL et al. 1994), may induce autoimmune

Table 4. Immunostimulatory effects of apoptosis inhibiting drugs

Substance	Immunostimulatory effects	Reference
Retinol	Abolishes neonatal allotransplantation tolerance when administered together with first graft	MALKOVSKY et al. 1985
Cyclosporin A	Syngeneic graft vs host reaction	JONES et al. 1989
Pertussis toxin	Adjuvant effect in oral immunization	WILSON et al. 1993
	Adjuvant for the induction of experimental autoimmune reactions	BROEKHUYSE et al. 1992
	Acceleration of spontaneous autoimmune disease	GOVERMAN et al. 1993
Linomide	Enhancement of natural killer cell activity	KALLAND et al. 1985
	Enhancement of mitogn responses of T and B cells ex vivo	LARSSON et al. 1987, ILBÄCK et al. 1989
	Aggravation of collagen type II-induced arthritis	KLEINAU et al. 1989
	Enhancement of sialadenitis in MRL/lpr mice and decrease of other autoimmune phenomena	JONSSON et al. 1988, TARKOWSKI et al. 1986
	Augmentation of delayed type hypersensitivity reactions	STÅLHANDSKE and KALLAND 1986
	Enhanced anti-cancer immunity	KALLAND 1986
	Acceleration of cardiac allograft rejections	WANDERS et al. 1989
	Prevention of vaccinia-virus-induced T cell lymphopenia	GONZALO et al. 1994c
	Inhibition of septic shock-like acute death induced by bacterial endo- and exotoxins	GONZALO et al. 1993a

symptoms after withdrawal of the drug. Retinol abolishes neonatal allotransplantation tolerance. Finally, pertussis toxin has a strong adjuvant effect and enhances the development of both experimental and spontaneous autoimmune diseases (Table 4). These data suggest a correlation between apoptosis-inhibiting effects and immunostimulatory and/or pro-autoimmune effects.

A particular clinical condition in which the inhibition of T cell apoptosis may be useful is viral infection, Accordingly, linomide prevents the lymphopenia caused by infection with vaccinia virus and reduces the degree of endonucleolysis observed in purified CD4$^+$ and CD8$^+$ T cells from virus-infected Balb/c mice (Gonzalo et al. 1994c). This clearly illustrates that apoptosis inhibition can counteract virus-induced immunosuppression in vivo. Attempts to prevent lymphocyte apoptosis in HIV-infected persons have been performed in vitro. Cyclosporin A, anti-CD28 antibodies (Groux et al. 1992), fibroblast-derived cytokines (Pandolfi et al. 1993), and antioxidants such as N-acetylcysteine (Buttke and Sandstrom 1994), catalase, vitamin E or 2-mercaptoethanol (Sandstrom et al. 1993) have been shown to inhibit lymphocyte death from HIV carriers in vitro. Clinical trials will soon unravel whether PCD-inhibiting substances can be successfully employed in HIV carriers to inhibit lymphocyte PCD and to maintain the lymphocyte pool in vivo and whether this will help prevent the deterioration of immune function.

Acknowledgements. We thank Drs. Terie Kalland, Gunnar Hedlund (Lund, Sweden), and Anne Senik (Villejuif, France) for constant support. This work was partially supported by grants from the Association pour la Recherche sur le Cancer (ARC), Association Nationale pour la Recherche sur le SIDA (ANRS), Fondation pour la Recherche Médicale (FRM), Institut National de la Santé et de la Recherche Médicale (INSERM), Leo Foundation, and Sidaction (to G.K.). N.Z. receives a fellowship from Roussel-Uclaf.

References

Abello PA, Fidler SA, Bulkley GB, Buchman TG, Solomkin JS, Maie RV (1994) Antioxidants modulate induction of programmed endothelial cell death (apoptosis) by endotoxin. Arch Surg 129: 134–141

Bossu P, Singer GG, Andres P, Ettinger R, Marshak-Rothstein A, Abbas AK (1993) Mature CD4$^+$ T lymphocytes from MRL/lpr mice are resistant to receptor-mediated tolerance and apoptosis. J Immunol 151: 7233–7239

Broekhuyse RM, Kuhlmann ED, Winkens HJ (1992) Experimental autoimmune posterior uveitis accompanied by epitheloid cell accumulations (EAPU). A new type of experimental ocular disease induced by immunization with PEP-65, a pigment epithelial polypeptide preparation. Exp Eye Res 55: 819–829

Bruno S, Del-Bino G, Lassota P, Giaretti W, Darzynkiewicz Z (1992) Inhibitors of proteases prevent endonucleolysis accompanying apoptotic death of HL-60 leukemia cells and normal thymocytes. Leukemia 6: 1113–1120

Buttke TM, Sandstrom PA (1994) Oxidative stress as a mediator of apoptosis. Immunol Today 15: 7–10

Caron-Leslie LA, Cidlowski JA (1991) Similar actions of glucocorticoids and calcium on the regulation of apoptosis in S49 cell. Mol Endocrinol 5: 1169–1179

Cohen J, Glauser MP (1991) Septic shock: treatment. Lancet ii: 736–739

Colotta F, Polentarutti N, Sironi M, Mantovani A (1992) Expression and involvement of c-fos and c-jun protooncogenes in programmed cell death induced by growth factor deprivation in lymphoid cell lines. J Biol Chem 267: 18278–18283

Dedera DA, Waller EK, LeBrun DP, Sen-Majumdar A, Stevens ME, Barsh GS, Cleary ML (1993) Chimeric homeobox gene E2A-PBX1 induces proliferation, apoptosis, and malignant lymphomas in transgenic mice. Cell 74: 833–843

Ding L, Shevach EM (1994) Activation of CD4⁺ T cells by delivery of the B7 costimulatory signal on bystander antigen-presenting cells (trans-costimulation). Eur J Immunol 24: 859–866

Garbe A, Buck J, Hämmerling U (1992) Retinoids are important cofactors in T cell activation. J Exp Med 176: 109–117

Garchon H-J, Luan J-J, Eloy L, Bédossa P, Bach J-F (1994) Genetic analysis of immune dysfunction in non-obese diabetic (NOD) mice: mapping of a susceptibility locus to the Bcl-2 gene correlates with increased resistance of NOD T cells to apoptosis induction. Eur J Immunol 24: 380–384

Glauser MP, Zanetti G, Baumgartner JD, Cohen J (1991) Septic shock: pathogenesis. Lancet ii: 732–739

Gollob KJ, Palmer E (1993) Aberrant induction of T cell tolerance in B cell suppressed mice. J Immunol 150: 3705–3712

Gonzalo JA, Moreno de Alborán I, Alés-Martínez JE, Martínez-AC, Kroemer G (1992) Expansion and clonal deletion of peripheral T cells induced by bacterial superantigen is independent of the interleukin 2 pathway. Eur J Immunol 22: 1007–1011

Gonzalo JA, González-García A, Hedlund G, Kalland T, Martínez-AC, Kroemer G (1993a) Linomide, a novel immunomodulator that prevents death in four models of septic shock. Eur J Immunol 23: 2372–2374

Gonzalo JA, González-García A, Martínez-AC, Kroemer G (1993b) Glucocorticoid-mediated control of the clonal deletion and activation of peripheral T cells in vivo. J Exp Med 177: 1239–1246

Gonzalo JA, Baixeras E, González-García A, George-Chandy A, van Rooijen N, Martínez-AC, Kroemer G (1994a) Differential in-vivo-effects of a superantigen and an antibody targeted to the same T cell receptor: activation-induced cell death versus passive macrophage-dependent deletion. J Immunol 152: 1597–1608

Gonzalo JA, González-García A, Baixeras E, Zamzami N, Tarazona T, Rino Rappuoli R, Martínez-AC, Kroemer G (1994b) Pertussis toxin interferes with superantigen-induced deletion of peripheral T cells without affecting T cell activation in vivo. Inhibition of deletion and associated programmed cell death depends on ADP-ribosyltransferase activity. J Immunol 152: 4291–4299

Gonzalo JA, González-García A, Hedlund G, Kalland T, Martínez-AC, Kroemer G (1994c) Linomide inhibits programmed cell death of peripheral T cells in vivo. Eur J Immunol 24: 48–52

Gougeon ML, Olivier R, Garcia S, Guetard D, Dragic T, Dauguet C, Montagnier L (1991) Mise en évidence d'un processus d'engagement vers la mort cellulaire par apoptose dans les lymphocytes de patients infectés par le VIH. C R Acad Sci [III] 312: 529–535

Goverman J, Woods A, Larson L, Weiner LP, Hood L, Zaller DM (1993) Transgenic mice that express a myelin basic protein-specific T cell receptor develop spontaneous autoimmunity. Cell 72: 551–560

Groux H, Torpier G, Monté D, Mouton Y, Capron A, Ameisen JC (1992) Activation-induced death by apoptotis in CD4⁺ T cells from human immunodeficiency virus-infected asymptomatic individuals. J Exp Med 175: 331–340

Hernández-Caselles T, Stutman O (1993) Immune functions of tumor necrosis factor. I. Tumor necrosis factor induces apoptosis of mouse thymocytes and can also stimulate or inhibit IL-6-induced proliferation depending on the concentration of mitogenic costimulation. J Immunol 151: 3999–4012

Ilbäck NG, Fohlman J, Friman G (1989) Effects of the immunomodulator LS 2616 on lymphocytes subpopulations in murine coxsackievirus B3 myocarditis. J Immunol 142: 3225–3228

Iwata M, Mukai M, Nakai Y, Iseki R (1992) Retinoic acids inhibit activation-induced apoptosis in T cell hybridomas and thymocytes. J Immunol 149: 3202–3208

Jonas D, Walev I, Borger T, Liebetrau M, Palner M, Bhakdi S (1994) Novel path to apoptosis—small transmembrane pores created by staphylococcal alpha-toxin in T lymphocytes evoke internucleosomal DNA degradation. Infect Immun 62: 1304–1312

Jones RJ, Vogelsang GB, Hess AD, Farmer ER, Mann RB, Geller RB, Pientedosi S, Santos GW (1989) Induction of graft-versus-host-disease following autologous bone marrow transplantation. Lancet 1: 754

Jonsson R, Tarkowski A, Bäckmann K (1988) Effects of immunomodulating treatment on autoimmune sialadenitis in MRL/Mp-lpr/lpr mice. Agents Actions 25: 368–374

Kabelitz D, Wesselborg S (1992) Life and death of a superantigen-reactive human CD4⁺ T cell clone: staphylococcal enterotoxins induce death by apoptosis but simultaneously trigger a proliferative response in the presence of HLA-DR⁺ antigen-presenting cell. Int Immunol 4: 1381–1388

Kalland T (1986) Effects of the immunomodulator LS 2616 on growth and metastasis of the murine melanoma. Cancer Res 46: 3018–3022

Kalland T, Gunnar A, Stalhandske T (1985) Augmentation of mouse natural killer cell activity by LS2616, a new immunomodulator. J Immunol 134: 3956–3961

Kawabe Y, Ochi A (1991) Programmed cell death and extrathymic reduction of Vβ8⁺ CD4⁺ T cells in mice tolerant to Staphylococcus aureus enterotoxin B. Nature 349: 245–248

Kleinau S, Larsson P, Björk J, Holmdahl R, Klareskog L (1989) Linomide, a new immunomodulatory drug, shows different effects on homologous versus heterologous collagen-induced arthritis in rat. Clin Exp Immunol 78: 138–142

Kotzin BL, Leung DYM, Kappler J, Marrack P (1993) Superantigens and their potential role in human disease. Adv Immunol 54: 99–166

Kroemer G (1994) The pharmacology of T lymphocyte apoptosis. Adv Immunol (in press)

Kroemer G, Martínez-AC (1994) Pharmacological inhibition of programmed lymphocyte cell death. Immunol Today 15: 235–242

Larsson E-L, Joki A, Stålhandske T (1987) Mechanism of action of the new immunomodulator LS2616 on T cell responses. Int J Immunopharmacol 9: 425–431

Lee M-R, Liou M-L, Liou M-L, Yang Y-F, Lai M-Z (1993) cAMP analogs prevent activation-induced apoptosis of T cell hybridomas. J Immunol 151: 5208–5217

Leijon K, Hammarström B, Holmberg D (1994) Non-obese diabetic (NOD) mice display enhanced immune responses and prolonged survival of lymphoid cells. Int Immunol 6: 339–345

Liu Y, Janeway CAJ (1990) Interferon g plays a critical role in induced cell death of effector T cell: a possible third mechanism of self-tolerance. J Exp Med 172: 1735–1739

MacDonald HR, Baschieri S, Lees RK (1991) Clonal expansion precedes anergy and death of Vβ8+ peripheral T cells responding to staphylococcal enterotoxin B in vivo. Eur J Immunol 21: 1963–1966

Malkovsky M, Medawar PB, Thatcher DR, Toy J, Hunt R, Rayfield LS, Doré C (1985) Acquired immunological tolerance of foreign cells is impaired by recombinant interleukin 2 or vitamin A acetate. Proc Natl Acad Sci USA 82: 536–538

Mangan DF, Wahl SM (1991) Differential regulation of human monocyte programmed cell death (apoptosis) by chemotactic factors and pro-inflammatory cytokines. J Immunol 147: 3408–3412

Meyaard L, Otto SA, Jonker RR, Mijnster MJ, Keet RPM, Miedema F (1992) Programmed cell death in HIV-1 infection. Science 257: 217–219

Mogil RJ, Shi Y, Bissonnette RP, Bromley P, Yamaguchi I, Green DR (1994) Role of DNA fragmentation in T cell activation-induced apoptosis in vitro and in vivo. J Immunol 152: 1674–1683

Möröy T, Grzeschiczek A, Petzold S, Hartmann K-U (1993) Expression of an Pim-1 transgene accelerates lymphoproliferation and inhibits apoptosis in lpr/lpr mice. Proc Natl Acad Sci USA 90: 10734–10738

Nakayama K-I, Nakayama K, Negishi I, Kuida K, Shinkai Y, Louie MC, Fields LE, Lucas PJ, Stewart V, Alt FW, Loh DY (1993) Disappearance of the lymphoid system in Bcl-2 homozygous mutant chimeric mice. Science 261: 1584–1588

Pandolfi F, Oliva A, Sacco G, Polidori V, Liberatore D, Mezzaroma I, Giovannetti A, Kurnick JT, Aiuti F (1993) Fibroblast-derived factors preserve viability in vitro of mononuclear cells isolated from subjects with HIV-1 infection. AIDS 7: 323–329

Petit PX, O'Connor JE, Grunwald D, Brown SC (1990) Analysis of the membrane potential of rat- and mouse-liver mitochondria by flow cytometry and possible applications. Eur J Biochem 389–397

Ramírez R, Carracedo J, Zamzami N, Castedo M, Kroemer G (1994) Pertussis toxin inhibits activation-induced cell death of human thymocytes, pre-B leukemia cells and monocytes. J Exp Med 180:1147–1152

Razvi ES, Welsh RM (1993) Programmed cell death of T lymphocytes during acute viral infection: a mechanism for virus-induced immune deficiency. J Virol 67: 5754–5765

Russel JH, Rush B, Weaver C, Wang R (1993) Mature T cells of autoimmune lpr/lpr mice have a defect in antigen-stimulated suicide. Proc Natl Acad Sci USA 90: 4409–4413

Saha K, Yuen PH, Wong PKY (1994) Murine retrovirus-induced depletion of T cells is mediated through activation-induced death by apoptosis. J Virol 68: 2735–2740

Salmon M, Pilling D, Borthwick NJ, Viner N, Janossy G, Bacon PA, Akbar AN (1994) The progressive differentiation of primed T cells is associated with an increasing susceptibility to apoptosis. Eur J Immunol 24: 892–899

Sandstrom PA, Roberts B, Folks TM, Buttke TM (1993) HIV gene expression enhances T cell susceptibility to hydrogen peroxide-induced apoptosis. AIDS Res Hum Retroviruses 9: 1107–1113

Schwartzman RA, Cidlowski JA (1991) Internucleosomal deoxyribonucleic acid cleavage activity in apoptotic thymocytes: detection and endocrine regulation. Endocrinology 128: 1190–1197

Shi L, Nishioka WK, Th'ng J, Bradbury EM, Litchfield DW, Greenberg AH (1994) Premature p34^{cdc2} activation required for apoptosis. Science 263: 1143–1145

Shi Y, Sahai BM, Green DR (1989) Cyclosporin A inhibits activation-induced cell death in T cell hybridomas and thymocytes. Nature 339: 625–627

Shi Y, Glynn JM, Guibert LJ, Cotter TG, Bissonnette RP, Green DR (1992) Role for c-myc in activation-induced apoptotic cell death in T cell hybridomas. Science 157: 212–214

Stålhandske T, Kalland T (1986) Effects of the novel immunomodulator LS2616 on the delayed type hypersensitivity reaction to Bordetella pertussis in the rat. Immunopharmacology 11: 87–92

Tarkowski A, Gunnarsson K, Nilsson LÅ, Lindholm L, Stålhandske T (1986) Successfull treatment of autoimmunity in MRL/lpr mice with LS 2616, a new immunomodulator. Arthritis Rheum 29: 1405–1409

Uehara T, Miyawaki T, Ohta K, Tamaru Y, Yokoi T, Nakamura S, Taniguchi N (1992) Apoptotic cell death of primed CD45RO+ T lymphocytes in Epstein-Barr virus-induced infectious mononucleosis. Blood 80: 452–458

Urdahl KB, Pardoll DM, Jenkins MK (1994) Cyclosporin A inhibits positive selection and delays negative selection in alpha beta TCR transgenic mice. J Immunol 152: 2853–2859

Vanier LE, Prud'homme GJ (1992) Cyclosporin A markedly enhances superantigen-induced peripheral T cell deletion and inhibits anergy induction. J Exp Med 176: 37–46

Veis DJ, Sorenson CM, Shutter JR, Korsmeyer SJ (1993) Bcl-2-deficient mice demonstrate fulminant lymphoid apoptosis, polycystic kidneys and hypopigmented hair. Cell 75: 229–240

Wahl S, Miethke T, Heeg K, Wagner H (1993) Clonal deletion as direct consequence of an in vivo T cell response to bacterial superantigen. Eur J Immunol 23: 1197–1200

Wanders A, Larsson E, Gerdin B, Tufveson G (1989) Abolition of the effect of cyclosporin A on rat cardiac allograft rejection by the new immunomodulator LS 2616 (Linomide). Transplantation 47: 216–217

Watanabe-Fukunaga R, Brannan CI, Copeland NG, Jenkins NA, Nagata S (1992) Lymphoproliferation disorder in mice explained by defects in Fas antigen that mediates apoptosis. Nature 356: 314–317

Weaver VM, Lach B, Walker PR, Sikorvska M (1993) Role of proteolysis in apoptosis-involvement of serine proteases in internucleosomal DNA fragmentation in immature thymocytes. Biochem Cell Biol 71: 488–500

Wilson AD, Robinson A, Irons L, Stokes CR (1993) Adjuvant action of cholera toxin and pertussis toxin in the induction of IgA antibody response to orally administered antigen. Vaccine 11: 113–118

Yang Y, Vacchio MS, Ashwell JD (1993) 9-cis retinoic acid inhibits activation-driven T cell apoptosis: implications for retinoid X receptor involvement in thymocyte development. Proc Natl Acad Sci USA 90: 6170–6174

Yuh K, Siminovitch KA, Ochi A (1993) T cell anergy is programmed early after exposure to bacterial superantigen in vivo. Int Immunol 5: 1375–1382

Zamzami N, Marchetti P, Castedo M, Zanin C, Vayssière J-L, Petix PX, Kroemer G (1995) Reduction in mitochondrial potential constitutes on early irreversible step of programmed lymphocyte death in vivo. J Exp Med (in press)

Zubiaga AM, Munoz E, Huber BT (1992) IL-4 and IL-2 selectively rescue Th cell subsets from glucocorticoid-induced apoptosis. J Immunol 149: 107–112

Subject Index

Current Topics in Microbiology and Immunology

Volumes published since 1989 (and still available)

Vol. 157: **Swanstrom, Roland; Vogt, Peter K. (Ed.):** Retroviruses. Strategies of Replication. 1990. 40 figs. XII, 260 pp. ISBN 3-540-51895-9

Vol. 158: **Muzyczka, Nicholas (Ed.):** Viral Expression Vectors. 1992. 20 figs. IX, 176 pp. ISBN 3-540-52431-2

Vol. 159: **Gray, David; Sprent, Jonathan (Ed.):** Immunological Memory. 1990. 38 figs. XII, 156 pp. ISBN 3-540-51921-1

Vol. 160: **Oldstone, Michael B. A.; Koprowski, Hilary (Eds.):** Retrovirus Infections of the Nervous System. 1990. 16 figs. XII, 176 pp. ISBN 3-540-51939-4

Vol. 161: **Racaniello, Vincent R. (Ed.):** Picornaviruses. 1990. 12 figs. X, 194 pp. ISBN 3-540-52429-0

Vol. 162: **Roy, Polly; Gorman, Barry M. (Eds.):** Bluetongue Viruses. 1990. 37 figs. X, 200 pp. ISBN 3-540-51922-X

Vol. 163: **Turner, Peter C.; Moyer, Richard W. (Eds.):** Poxviruses. 1990. 23 figs. X, 210 pp. ISBN 3-540-52430-4

Vol. 164: **Bækkeskov, Steinnun; Hansen, Bruno (Eds.):** Human Diabetes. 1990. 9 figs. X, 198 pp. ISBN 3-540-52652-8

Vol. 165: **Bothwell, Mark (Ed.):** Neuronal Growth Factors. 1991. 14 figs. IX, 173 pp. ISBN 3-540-52654-4

Vol. 166: **Potter, Michael; Melchers, Fritz (Eds.):** Mechanisms in B-Cell Neoplasia. 1990. 143 figs. XIX, 380 pp. ISBN 3-540-52886-5

Vol. 167: **Kaufmann, Stefan H. E. (Ed.):** Heat Shock Proteins and Immune Response. 1991. 18 figs. IX, 214 pp. ISBN 3-540-52857-1

Vol. 168: **Mason, William S.; Seeger, Christoph (Eds.):** Hepadnaviruses. Molecular Biology and Pathogenesis. 1991. 21 figs. X, 206 pp. ISBN 3-540-53060-6

Vol. 169: **Kolakofsky, Daniel (Ed.):** Bunyaviridae. 1991. 34 figs. X, 256 pp. ISBN 3-540-53061-4

Vol. 170: **Compans, Richard W. (Ed.):** Protein Traffic in Eukaryotic Cells. Selected Reviews. 1991. 14 figs. X, 186 pp. ISBN 3-540-53631-0

Vol. 171: **Kung, Hsing-Jien; Vogt, Peter K. (Eds.):** Retroviral Insertion and Oncogene Activation. 1991. 18 figs. X, 179 pp. ISBN 3-540-53857-7

Vol. 172: **Chesebro, Bruce W. (Ed.):** Transmissible Spongiform Encephalopathies. 1991. 48 figs. X, 288 pp. ISBN 3-540-53883-6

Vol. 173: **Pfeffer, Klaus; Heeg, Klaus; Wagner, Hermann; Riethmüller, Gert (Eds.):** Function and Specificity of Á / ‰ TCells. 1991. 41 figs. XII, 296 pp. ISBN 3-540-53781-3

Vol. 174: **Fleischer, Bernhard; Sjögren, Hans Olov (Eds.):** Superantigens. 1991. 13 figs. IX, 137 pp. ISBN 3-540-54205-1

Vol. 175: **Aktories, Klaus (Ed.):** ADP-Ribosylating Toxins. 1992. 23 figs. IX, 148 pp. ISBN 3-540-54598-0

Vol. 176: **Holland, John J. (Ed.):** Genetic Diversity of RNA Viruses. 1992. 34 figs. IX, 226 pp. ISBN 3-540-54652-9

Vol. 177: **Müller-Sieburg, Christa; Torok-Storb, Beverly; Visser, Jan; Storb, Rainer (Eds.):** Hematopoietic Stem Cells. 1992. 18 figs. XIII, 143 pp. ISBN 3-540-54531-X

Vol. 178: **Parker, Charles J. (Ed.):** Membrane Defenses Against Attack by Complement and Perforins. 1992. 26 figs. VIII, 188 pp. ISBN 3-540-54653-7

Vol. 179: **Rouse, Barry T. (Ed.):** Herpes Simplex Virus. 1992. 9 figs. X, 180 pp. ISBN 3-540-55066-6

Vol. 180: **Sansonetti, P. J. (Ed.):** Pathogenesis of Shigellosis. 1992. 15 figs. X, 143 pp. ISBN 3-540-55058-5

Vol. 181: **Russell, Stephen W.; Gordon, Siamon (Eds.):** Macrophage Biology and Activation. 1992. 42 figs. IX, 299 pp. ISBN 3-540-55293-6

Vol. 182: **Potter, Michael; Melchers, Fritz (Eds.):** Mechanisms in B-Cell Neoplasia. 1992. 188 figs. XX, 499 pp. ISBN 3-540-55658-3

Vol. 183: **Dimmock, Nigel J.:** Neutralization of Animal Viruses. 1993. 10 figs. VII, 149 pp. ISBN 3-540-56030-0

Vol. 184: **Dunon, Dominique; Mackay, Charles R.; Imhof, Beat A. (Eds.):** Adhesion in Leukocyte Homing and Differentiation. 1993. 37 figs. IX, 260 pp. ISBN 3-540-56756-9

Vol. 185: **Ramig, Robert F. (Ed.):** Rotaviruses. 1994. 37 figs. X, 380 pp. ISBN 3-540-56761-5

Vol. 186: **zur Hausen, Harald (Ed.):** Human Pathogenic Papillomaviruses. 1994. 37 figs. XIII, 274 pp. ISBN 3-540-57193-0

Vol. 187: **Rupprecht, Charles E.; Dietzschold, Bernhard; Koprowski, Hilary (Eds.):** Lyssaviruses. 1994. 50 figs. IX, 352 pp. ISBN 3-540-57194-9

Vol. 188: **Letvin, Norman L.; Desrosiers, Ronald C. (Eds.):** Simian Immunodeficiency Virus. 1994. 37 figs. X, 240 pp. ISBN 3-540-57274-0

Vol. 189: **Oldstone, Michael B. A. (Ed.):** Cytotoxic T-Lymphocytes in Human Viral and Malaria Infections. 1994. 37 figs. IX, 210 pp. ISBN 3-540-57259-7

Vol. 190: **Koprowski, Hilary; Lipkin, W. Ian (Eds.):** Borna Disease. 1995. 33 figs. IX, 134 pp. ISBN 3-540-57388-7

Vol. 191: **ter Meulen, Volker; Billeter, Martin A. (Eds.):** Measles Virus. 1995. 23 figs. IX, 196 pp. ISBN 3-540-57389-5

Vol. 192: **Dangl, Jeffrey L. (Ed.):** Bacterial Pathogenesis of Plants and Animals. 1994. 41 figs. IX, 343 pp. ISBN 3-540-57391-7

Vol. 193: **Chen, Irvin S. Y.; Koprowski, Hilary; Srinivasan, Alagarsamy; Vogt, Peter K. (Eds.):** Transacting Functions of Human Retroviruses. 1995. 49 figs. Approx. IX, 240 pp. ISBN 3-540-57901-X

Vol. 194: **Potter, Michael; Melchers, Fritz (Eds.):** Mechanisms in B-cell Neoplasia. 1995. Approx. 152 figs. XXV, 458 pp. ISBN 3-540-58447-1

Vol. 195: **Montecucco, Cesare (Ed.):** Clostridial Neurotoxins. 1995. Approx. 28 figs. Approx. 260 pp. ISBN 3-540-58452-8

Vol. 196: **Koprowski, Hilary; Maeda Hiroshi (Eds.):** The Role of Nitric Oxide in Physiology and Pathophysiology. 1995. 21 figs. IX, 90 pp. ISBN 3-540-58214-2

Vol. 197: **Meyer, Peter (Ed.):** Gene Silencing in Higher Plants and Related Phenomena in Other Eukaryotes. 1995. 17 figs. IX, 232 pp. ISBN 3-540-58236-3

Vol. 198: **Griffiths, Gillian M.; Tschopp, Jürg (Eds.):** Pathways for Cytolysis. 1995. 45 figs. IX, 224 pp. ISBN 3-540-58725-X

Vol. 199/I: **Doerfler, Walter; Böhm, Petra (Eds.):** The Molecular Repertoire of Adenoviruses I. 1995. Approx. 51 figs. Approx. XIII, 296 pp. ISBN 3-540-58828-0

Vol. 199/II: **Doerfler, Walter; Böhm, Petra (Eds.):** The Molecular Repertoire of Adenoviruses II. 1995. Approx. 36 figs. Approx. XIII, 296 pp. ISBN 3-540-58829-9

Vol. 199/III: **Doerfler, Walter; Böhm, Petra (Eds.):** The Molecular Repertoire of Adenoviruses III. 1995. Approx. 50 figs. Approx. XIII, 324 pp. ISBN 3-540-58987-2